Physics: Advanced Concepts and Applications

Physics: Advanced Concepts and Applications

Edited by
Daniel White

Willford Press
www.willfordpress.com

Published by Willford Press,
118-35 Queens Blvd., Suite 400,
Forest Hills, NY 11375, USA

ISBN: 978-1-68285-618-5

Cataloging-in-Publication Data

Physics : advanced concepts and applications / edited by Daniel White.
p. cm.
Includes bibliographical references and index.
ISBN 978-1-68285-618-5
1. Physics. I. White, Daniel.
QC21.3 .P49 2019
530--dc23

For information on all Willford Press publications
visit our website at www.willfordpress.com

WILLFORD PRESS

Printed in China.

Contents

Preface

This book has been an outcome of determined endeavour from a group of educationists in the field. The primary objective was to involve a broad spectrum of professionals from diverse cultural background involved in the field for developing new researches. The book not only targets students but also scholars pursuing higher research for further enhancement of the theoretical and practical applications of the subject.

Physics is the study of matter, its motion and behavior in space and time. The fundamental principles of energy and force are central to the development of physics as a discipline. Various products developed using the applied principles of physics have transformed modern society, like television, computers, domestic appliances, etc. Electromagnetism, classical mechanics, thermodynamics, statistical mechanics, quantum mechanics and special relativity are the core aspects of physics. Research in physics broadly explores phenomena in condensed matter physics, nuclear physics, particle physics, astrophysics, etc. This book provides significant information of this discipline to help develop a good understanding of physics. It brings forth some of the most innovative concepts and elucidates the unexplored aspects of this field. The topics covered in this book offer the readers new insights in the area of physics.

It was an honour to edit such a profound book and also a challenging task to compile and examine all the relevant data for accuracy and originality. I wish to acknowledge the efforts of the contributors for submitting such brilliant and diverse chapters in the field and for endlessly working for the completion of the book. Last, but not the least; I thank my family for being a constant source of support in all my research endeavours.

Editor

1

First investigation on plasma impurities of the IR-T1 tokamak

R. Alipour, M. Ghoranneviss,[a] and A. Salar Elahi
Plasma Physics Research Center, Science and Research Branch, Islamic Azad University, Tehran, Iran

One of the best ways to get information about plasma and interactions within it, is the study on visible light emitted from plasma. Detection of the emission spectrum from plasma leads to identify the spectral lines of ions and impurities inside the plasma and measurement of all related parameters to ion flux such as velocity and temperature. The first measurement of impurities into the IR-T1 tokamak plasma is done by spectroscopic technique. The CCD-based spectrometer device with the wavelength range of 400-550 nm is used to detection the spectrum emitted from the plasma. The impurities into the plasma in the intended wavelength range are identified. Furthermore, the Doppler broadening of the most intense peaks of impurities and their temperatures are measured. The methods in this paper can be used on all fusion devices.

I. INTRODUCTION

Tokamak is one of the several types of magnetic confinement devices being developed to contain the hot plasma needed for producing controlled thermonuclear fusion power. Based on various purposes, tokamaks can be divided to two groups, large and small. The main purposes of large tokamaks such as TFTR (Zweben et al. 1989), DIII-D (Burrell et al. 2004), JET (Von Hellermann et al. 1990), JT-60U (Kumagai et al. 2000) and Tore Supra (Hennequin et al. 2006), is to study the behavior of large plasmas, which have parameters approaching those required for a thermonuclear plasma, with a view to predicting reliably the characteristics of future demonstration reactors (Orlinkij et al. 1988). Small tokamaks such as CASTOR (Weinzettl et al. 2003), STOR-M (Onchi et al. 2013), COMPASS (Naydenkova et al. 2009) and ISTTOK (Gomes et al. 2003) have close relationship with large tokamaks. Because of high cost and difficulty in testing conditions, most considered experiments before doing on large tokamaks do first on small tokamaks. Small tokamaks have proved their ability in response to testing new ideas and have pioneered in several important results, which have already been used on large tokamaks. IR-T1 tokamak is a small tokamak that used in this experiment. The main specifications of this tokamak are given in "Experimental setup" section. Many diagnostics such as Langmuir and rake probes, limiter biasing system (Ghoranneviss et al. 2013; Alipour et al. 2017; Meshkani et al. 2017), soft and hard x-ray detectors (Roudaki et al. 2015; Hasanvand et al. 2016) and Mirnov coils (Ghanbari et al. 2011; Ghasemloo et al. 2013) are launched on IR-T1 tokamak. Study of plasma parameters under different conditions is the main purpose of experiments on small tokamaks. One of the best ways to get information about plasma and interactions within it, is the study on visible light emitted from plasma. Detection of the emission spectrum from plasma leads to identify the spectral lines of ions and impurities inside the plasma and measurement of all related parameters to ion flux such as velocity and temperature. The main purpose of using the spectroscopy system on the IR-T1 tokamak is to study on the visible radiation emitted

[a]Corresponding author: Ghorannevis@hotmail.com

from the plasma, monitoring hydrogen recycling and impurities flux in the plasma. From many years ago, the spectrophotometer with a photomultiplier tube (PMT) was installed on the IR-T1 tokamak (Hojabri et al. 2005). PMT-based spectroscopy systems have high accuracy and resolution to collecting and showing the emission spectrum but inability to collect the emission spectrum in a wide range of wavelengths is an important weakness of this type of spectrometer. Although, more PMT can be used, but it is not economical. Spectrometer with charge-coupled device (CCD) detectors is another type of spectrometer devices that have extensive dynamic range and commonly used to collect all emitted spectrums between 200-1100 nm. The CCD detector collects the charge, dissipated as photons strike the photoactive surface. At the end of integration time, the remaining charge is transferred to a buffer and then this signal is being transferred to the AD converter (analog to digital converter).

To investigation on the spectrum emitted from the plasma in specified wavelength range, identification spectral lines and impurities and measuring their temperature, a CCD-based spectrometer device is bought and installed on IR-T1 tokamak. Full detail of this system is explained in the next section. The contents of this paper are the results of using this spectrometer for the first time.

II. EXPERIMENTAL SETUP

A. IR-T1 tokamak

IR-T1 tokamak is a small and large aspect ratio tokamak with a circular cross section that is located at PPRC-SRBIAU, Iran (see figure 1). It works under the hydrogen atmosphere. The tokamak chamber is made of stainless steel, which is surrounded by leaden walls in all directions. It is included of two poloidal limiters with radial thickness of 2.5 cm, which are made of stainless steel, too. The main parameters of this tokamak are listed in Table I.

B. Spectroscopy system

A portable spectrometer device which is launched on the IR-T1 tokamak (for the first time) for spectral lines measurements, is AvaSpec-ULS3648-R8-USB2-RM made by AVANTES company.

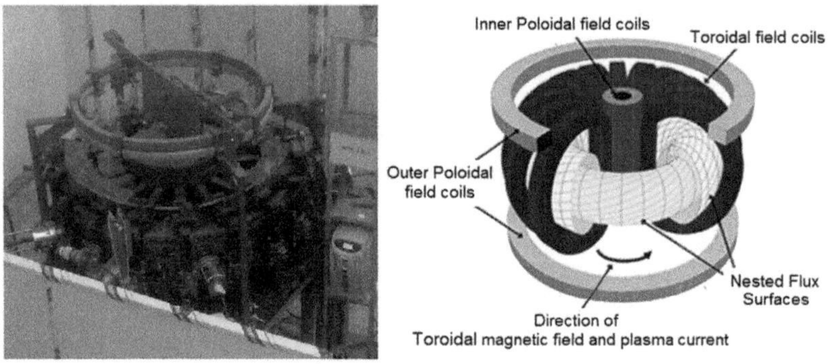

FIG. 1. IR-T1 tokamak setup.

TABLE I. Main parameters of IR-T1 tokamak.

Parameters	Value
Major radius	45 cm
Minor radius	12.5 cm
Toroidal magnetic field	< 1.0 T
Plasma current	< 40 kA
Discharge time	< 35 ms
Electron density	$0.7 - 1.5 \times 10^{13}$ cm^{-3}
Electron temperature	150-180 eV
Edge electron temperature	10-20 eV

TABLE II. The main technical parameters of AvaSpec-ULS3648-R8-USB2-RM.

Optical Bench	ULS Symmetrical Czerny-Turner, 75 mm focal length
Wavelength range	400 - 550 nm
Detector	CCD line array, 3648 pixels
Signal/Noise	350:1
Integration time	10 μs - 10 minutes
Grating	1800 lines / mm
Resolution	0.04 nm
Interface	USB 2.0 high-speed, 480 Mbps RS-232, 115.200 bps
Slit size	5 μm
Sensitivity	160000 counts / μW per ms integration time or 5 photons / count at 600 nm
Options	DCL-UV/VIS-200 (Quartz Detector Collection Lens)

The main technical data and grating efficiency curve for this spectrometer are shown in Table II and figure 2, respectively.

The visible light emitted from the plasma is transferred to the spectrometer by 2 m optical fiber with a diameter of 1000 μm. To connect the optical fiber vertically to the visible valve and

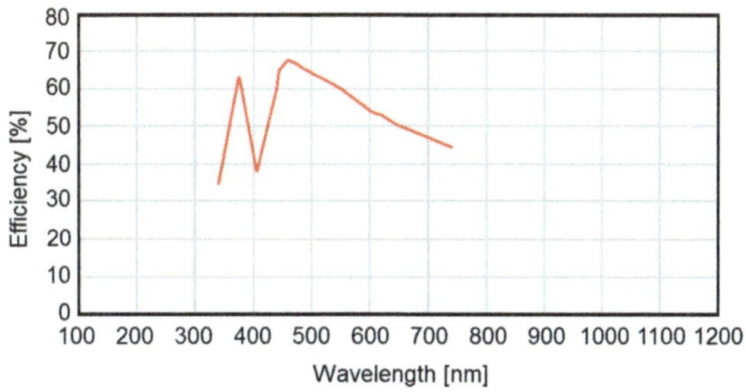

FIG. 2. The grating efficiency curve for 1800 lines/mm gratings.

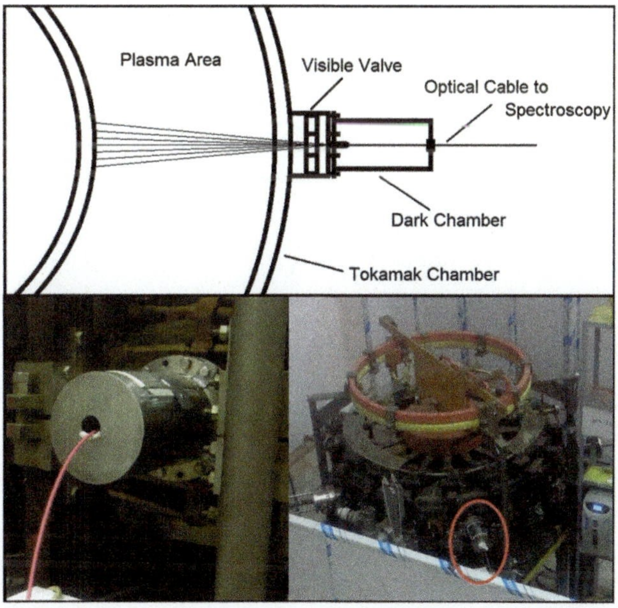

FIG. 3. Schematic of optical fiber connection to the visible valve of IR-T1 Tokamak.

to prevent from entering the ambient light into the optical fiber, a new chamber was made (see figure 3).

III. RESULTS AND DISCUSSIONS

A. IR-T1 plasma parameters

Some of the plasma parameters such as plasma current (I_p), loop voltage (V_{loop}) and plasma resistivity (R_p) during a typical discharge are shown in figure 4. The plasma resistivity (R_p) can be calculated using following equation:

$$R_p = \frac{r^2}{2R}\frac{V_{Loop}}{I_p} \tag{1}$$

Where r and R parameters are minor and major radius, respectively. The maximum plasma current reached in this experiment was about 16 kA.

B. Impurities measurement

The magnetic fields of the coils around the tokamak chamber and electromagnetic waves caused by plasma, such as x-ray radiation, can influence on electronic parts of the spectrometer device. These factors cause the spectroscopy device does not work properly and show the correct spectrum. So, the spectrometer is placed behind the leaden wall. Furthermore, a leaden chamber was made to place the device inside it. The effect of leaden chamber and leaden walls on the detected spectrum can be clearly seen in Figure 5.

As mentioned before, the IR-T1 tokamak work under Hydrogen atmosphere. So, the hydrogen peaks such as H_{beta} (486.14 nm), H_{gamma} (434.05 nm) and H_{delta} (410.18 nm) in the detected spectrum is the main characteristics of this plasma. It can be seen in figure 5 that there is almost no peak of H_{beta} in the detected spectrum before shielding. This is because of the effect of magnetic fields and electromagnetic waves on electronic parts of the spectrometer which prevents the device work properly. This problem is solved after shielding by placing the device behind the leaden wall and inside the leaden chamber. Figure 6 shows the measured spectrum in 5 wavelength ranges between 400-550 nm, with the scale of intensity to wavelength. The identified ions and their transitions are shown in Table III.

The wavelengths that are marked by an asterisk (*) have overlapped with other peaks. The data in figure 6 and Table III are based on information of the National Institute of Standards and Technology (NIST - atomic spectra database lines). The results showed that the low rate of impurities such as N,

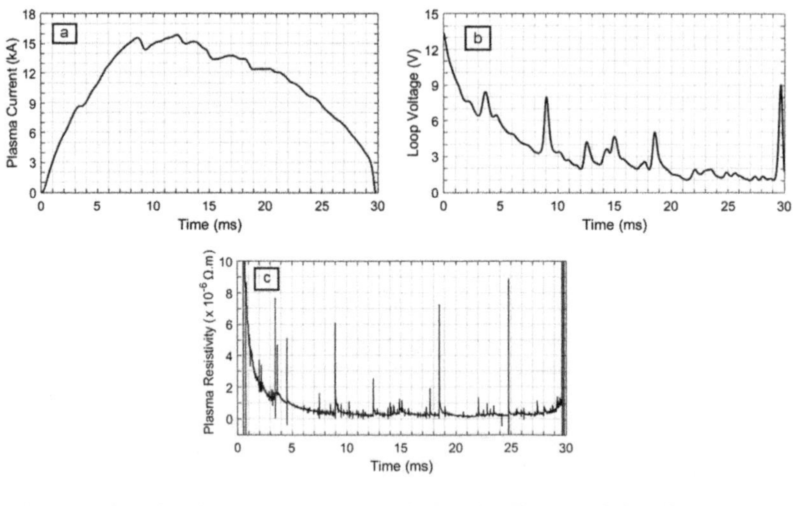

FIG. 4. Time evolution of a) plasma current, b) loop voltage and c) Plasma resistivity for a typical discharge.

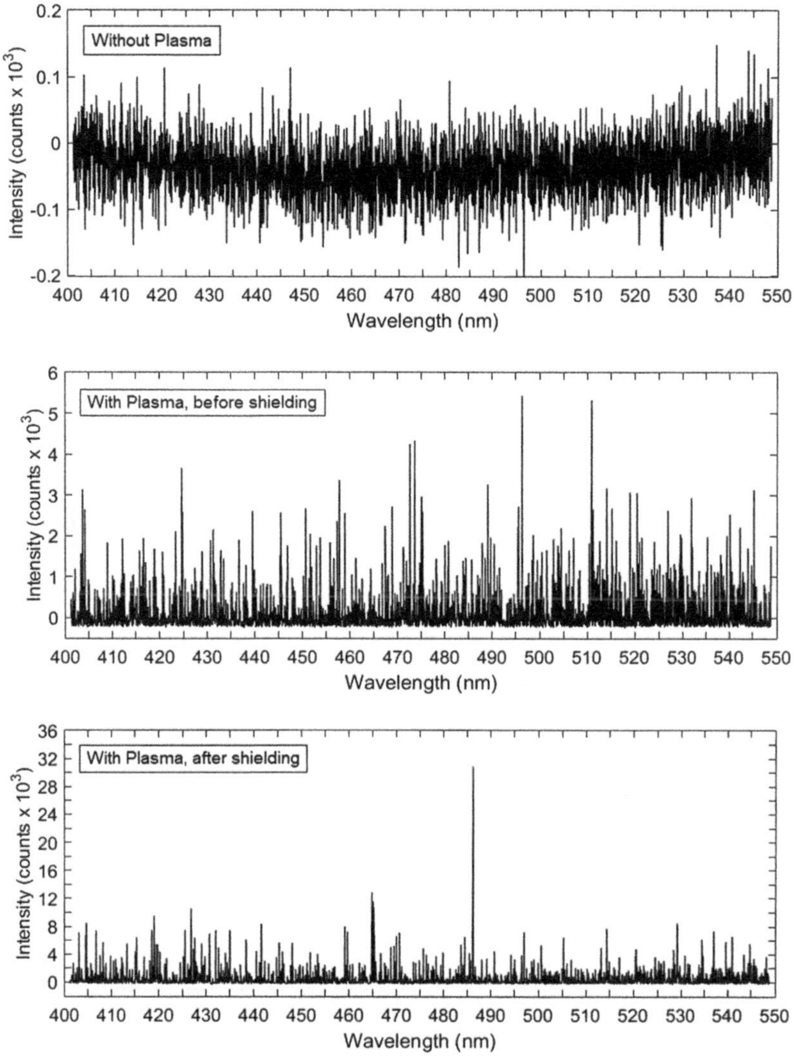

FIG. 5. The spectrum of the IR-T1 tokamak plasma measured by AVA-spectrometer between 400-550 nm, with and without plasma.

O, Cr, C, Mo, Ar, Ni, Ti, Mn, W and Fe are existed inside the plasma. The main reasons of existence any of these impurities are in Table IV.

IV. BROADENING MEASUREMENT

To investigation on the broadening phenomena of the spectrum emitted from IR-T1 tokamak plasma, the spectral lines of N III (409.731), O V (412.390), Cr I (425.434), C II (426.700), Mo I (427.726), O II (441.490), Ar II (443.018), N II (444.703), Ni I (446.264), Ti I (453.323), C III (464.742), Mn I (476.237), O IV (479.825), W I (484.382) and Fe I (537.149), which are the most intense peaks of impurities into the IR-T1 tokamak plasma, were chosen.

Total measured broadening is equal to:

$$\Delta\lambda^2_{Measured} = \Delta\lambda^2_{Natural} + \Delta\lambda^2_{Rresolution} + \Delta\lambda^2_{Doppler} + \Delta\lambda^2_{Zeeman} + \Delta\lambda^2_{Stark} \tag{2}$$

Where

$$\Delta\lambda^2_{Resolution} = \Delta\lambda^2_{Instrumental} + \Delta\lambda^2_{Diffraction} \tag{3}$$

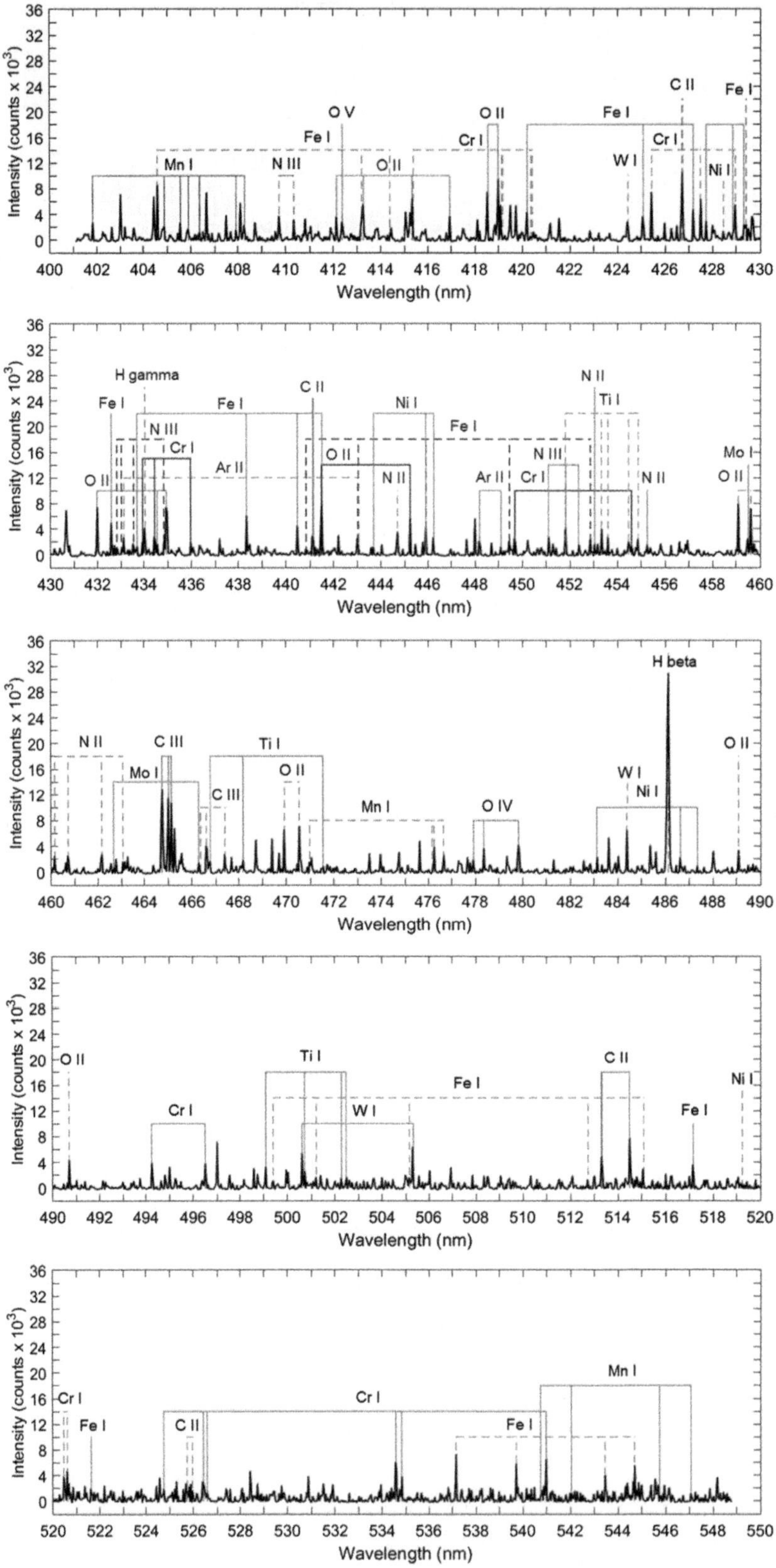

FIG. 6. The measured spectrum of a shot after shielding by AVA-spectrometer between 400-550 nm, during the whole time of discharge.

TABLE III. Identified spectrum lines observed in the IR-T1 tokamak plasma, according to the Fig. 6.

Wavelength (nm) with peak intensities ($\times 10^3$)	Ions (Multiplet Number) and Transitions
404.136, 405.554 (2.7), 406.352 (1.2), 407.027, 401.810 (2.8), 403.572, 404.875 (2.1), 405.893 (1.7), 407.924 (1.5), 408.362, 408.294 (2.3), 407.942	Mn I (5) a ^{6}D - z ^{6}D^0
404.581 (8.5), 406.359, 407.174, 396.926, 400.524, 414.387 (0.5), 413.206 (5.5)*	Fe I (43) a ^{3}F - y ^{3}F^0
409.731 (3.7), 410.337 (3.1)	N III (1) 3 ^{2}S - 3 ^{2}P0
410.18 (0.4)	H delta
412.39 (2.8)	O V (4) 3s ^{3}P^0 - 3p ^{3}D
412.652,415.381 (6.5)*, 416.362, 419.127 (0.6), 420.359 (0.4), 415.306, 416.362, 419.175 (0.2), 420.419 (0.4), 412.730, 415.381, 416.316, 419.068	Cr I (35) a ^{5}G - y ^{5}G^0
416.923 (3.6), 414.074, 412.148 (3.6), 415.654, 412.934, 415.330 (6.5)*, 413.280 (5.5)*	O II (19) 3p ^{4}P^0 - 3d ^{4}P
418.978 (9.5), 418.545 (7.5)	O II (36) 3p^1 ^{2}F^0 - 3d^1 ^{2}G
424.437 (2.8), 468.053, 484.382 (6.5), 505.330 (6.4), 500.616 (5.3), 522.468, 551.471	W I (1) d^4 s^2 ^{5}D - d^4 sp ^{7}D^0
425.434 (7.5), 427.480 (6.4), 428.972 (5.5)	Cr I (1) 3d^5 4s a ^{7}S - 3d^5 4p z ^{7}P^0
426.727 (10.6)*, 426.702 (10.6)*	C II (6) 2s^2 3d ^{2}D - 2s^2 4f ^{2}F^0
427.176 (4.8), 430.790, 432.576 (4.9), 420.203 (4.3), 425.079 (3.6), 414.767	Fe I (42) 3d^7 4s a ^{3}F - 3d^7 4p z ^{3}G^0
427.724 (3), 428.865 (1.1), 429.322 (2.4)	Mo I (7) a ^{5}D - z ^{5}F^0
434.05 (4.2)	H gamma
434.811, 442.601, 443.018 (2.6), 426.653, 433.125 (1.8), 437.974, 417.839, 428.290	Ar II (7) 4s ^{4}P - 4p ^{4}D^0
434.836 (2), 433.553 (1.3), 432.815 (1), 432.393, 435.366, 433.952, 433.044, 433.014 (0.9), 432.393, 432.137	N III (10) 3p ^{4}D - 3d ^{4}F^0
434.942 (7.4), 433.686, 432.577, 436.689, 434.556 (1.7), 431.963 (7.4), 431.713	O II (2) 3s ^{4}P - 3p ^{4}P^0
435.177, 434.450 (2.6), 433.945 (1.9), 433.971, 438.497, 437.127, 435.963 (2), 435.105, 441.225, 437.325	Cr I (22) 3d^4 4s^2 a ^{5}D - 3d^4 4s 4p z ^{5}F^0
438.354 (6.1), 440.475 (4.5), 441.512 (1.1), 429.412 (2.4), 433.704 (1.3), 436.790, 422.976, 429.146	Fe I (41) 3d^7 4s a ^{3}F - 3d^7 4p z ^{5}G^0
440.154, 445.903 (4.3)*, 447, 048, 446.246 (2.6), 443.698 (1.1), 428.468 (1.4), 432.560, 435.958, 438.454, 416.134, 422.169, 428.519	Ni I (86) z ^{5}D^0 - e ^{5}F
441.152 (2.9)*, 441.120 (2.9)*	C II (39) 3d ^{2}D^0 - 4f ^{2}F
441.490 (8.3), 441.697, 445.237 (5.6)	O II (5) 3s ^{2}P - 3p ^{2}S^0
444.703 (3.5)	N II (15) 3p ^{1}P - 3d ^{1}D^0
448.183 (2.1), 437.076, 449.099 (1.2), 436.207	Ar II (39) 3d ^{2}D - 4p1 ^{2}D^0
449.686 (2.5), 454.595 (1.3), 458.005	Cr I (10) 3d^5 4s a ^{5}S - 3d^4 4s 4p y ^{5}P^0
451.489, 451.092 (2.8), 453.457, 452.360 (1.8), 451.818 (4.8)*, 454.734, 453.084	N III (3) 3s ^{4}P^0 - 3p ^{4}D
452.861 (2.5), 449.456 (2.1), 448.225, 445.912 (4.3)*, 444.234, 440.771, 440.841 (0.9), 443.061 (1.1)	Fe I (68) 3d^7 4s a ^{5}P - 3d^6 4s 4p x ^{5}D^0
453.04 (1.7)	N II (59) 3d ^{1}F^0 - 4f ^{1}G
453.323 (4), 453.478, 453.557, 453.592, 453.605 (2.7), 455.548, 455.245, 454.876 (2.4), 454.468 (2), 451.273, 451.802 (4.8)*, 452.279, 452.730	Ti I (42) a ^{5}F - y ^{5}F^0
455.253 (1.8)	N II (58) 3d ^{1}F^0 - 4f ^{3}G
459.097 (8), 459.617 (7.2)	O II (15) 2s^2 2p^2 3s ^{2}D - 2s^2 2p^2 3p ^{2}F^0
462.646 (0.8), 466.276 (1.5), 466.193, 452.434, 457.650, 459.516 (2.5)	Mo I (6) a ^{5}D - y ^{5}P^0
463.053 (1.7), 461.386, 464.308, 462.139 (2.7), 460.147 (2.5), 460.715 (2.6)	N II (5) 3s ^{3}P^0 - 3p ^{3}S
464.740 (12.9), 465.016 (11.5), 465.135 (10.8)	C III (1) 1s^2 2s 3s ^{3}S - 1s^2 2s 3p ^{3}P^0
466.590 (4.1), 467.391 (2.8), 466.353 (0.4)	C III (5) 1s^2 2p 3s ^{3}P^0 - 1s^2 2p 3p ^{3}P
468.190 (1.9), 466.758 (1.7), 465.646, 471.529 (0.8), 469.367	Ti I (6) a ^{3}F - z ^{3}G^0
470.535 (7.1), 469.921 (6.6), 474.171	O II (25) 3p ^{2}D^0 - 3d ^{2}F
476.237 (3.9), 476.643 (2.7), 476.585, 476.152 (1.4), 470.971 (1.6), 472.747, 473.910, 467.168, 470.115	Mn I (21) a ^{4}D - z ^{4}F^0
479.825 (4.2), 478.343 (3.7), 477.257, 481.307, 479.422, 477.909 (1.9), 482.393, 480.077	O IV (9) 3p ^{4}P - 3d ^{4}D^0
486.14 (30.9)	H beta
492.460, 490.688 (4.4), 489.093 (3.4)	O II (28) 3p ^{4}S^0 - 3d ^{4}P
494.249 (3.9), 496.492 (3.9)	Cr I (9) a ^{5}S - y ^{7}P^0
498.173, 499.106 (3.3), 499.950, 500.720 (2.6), 501.427, 501.616, 502.002, 502.287 (1.5), 502.484 (1.6), 504.540, 504.357, 504.064	Ti I (38) a ^{5}F - y ^{5}G^0

TABLE III. *Continued.*

Wavelength (nm) with peak intensities (× 10^3)	Ions (Multiplet Number) and Transitions
501.207 (1.7), 505.163 (1.6), 508.334, 510.745, 512.372, 493.969, 499.413 (1.1), 504.107, 507.974, 512.736 (1.3), 514.293, 515.084 (3), 515.191	Fe I (16) a ^{5}F - z ^{5}F^0
501.759, 499.823, 501.246, 495.320, 491.203, 486.626 (2.3), 483.118 (2.4), 487.343 (0.9), 485.738, 515.799, 519.252 (0.7), 509.687, 501.004	Ni I (111) z ^{5}F^0 - e ^{5}F
514.516 (7.7), 513.921, 513.726, 515.108, 514.349, 513.329 (5)*, 513.296 (5)*	C II (16) 3s ^{4}P^0 - 3p ^{4}P
517.159 (3.6), 519.494, 521.627 (1.9), 504.175, 510.764, 533.280	Fe I (36) a ^{3}F - z ^{3}F^0
520.843, 520.603 (4.7), 520.451 (3.8)	Cr I (7) a ^{5}S - z ^{5}P^0
525.736 (2.7), 525.962 (1.7), 524.943, 525.355	C II (30) 3d ^{4}F^0 - 4p ^{4}D
526.954, 532.804, 537.149 (7.3), 540.577, 543.452 (4), 539.713 (5.8), 542.969, 544.692 (5.7), 545.561, 550.146, 550.678, 549.751	Fe I (15) a ^{5}F - z ^{5}D^0
534.106, 542.036 (1.3), 548.139, 540.742 (0.8), 547.063 (0.9), 551.677, 545.747 (0.5), 550.586, 553.775	Mn I (4) a ^{6}D - y ^{6}P^0
540.979 (6.6), 534.580 (6.1), 529.668, 534.831 (3.8), 529.826, 526.415 (3.1), 530.074, 526.572 (1.4), 524.756 (1.9)	Cr I (18) a ^{5}D - z ^{5}P^0

The instrumental broadening is caused by dispersion and defects of the instruments while the diffraction broadening cause by the diffraction limit of grating. In this study, the effect of Zeeman and Stark broadenings are ignored for ease in calculation and measuring the ions temperature of plasma. So,

$$\Delta\lambda^2_{Measured} = \Delta\lambda^2_{Instrumental} + \Delta\lambda^2_{Diffraction} + \Delta\lambda^2_{Doppler} \quad (4)$$

A. Instrumental broadening measurement

Instrumental broadening could be measured from following equation:

$$\Delta\lambda^2_{Instrumental} = R_d \times W \quad (5)$$

Where R_d and W are the reciprocal linear dispersion and the input slit width, respectively. The reciprocal linear dispersion equation is:

$$R_d = \frac{\Delta\lambda}{\Delta l} = \frac{d \times \cos(\beta)}{f \times m} \quad (6)$$

Where f is the focal length, β is the diffraction angle, m is the diffraction order and d is the grating space which can be calculated as:

TABLE IV. The main reasons of existence some impurities inside the IR-T1 tokamak plasma.

Impurities	The main reasons
O	1. From the diffusion pump. 2. From the air (the oxygen in the air can penetrate through the pores during the changing or replacing probes).
Ar	• The Ar is the cleaning factor of tokamak.
Fe, Ni, Cr, Mn	1. From the diagnostic probes which are made of stainless steel. 2. From the two stainless steel physical limiters inside the tokamak. 3. From the inner wall of the tokamak which is made of stainless steel.
W	1. From the filament inside the tokamak. 2. From the diagnostic probes which are made of tungsten.
Mo	• From the diagnostic probes which are made of molybdenum.
C	• From the diagnostic probes.
N	1. From the air (the nitrogen in the air can penetrate through the pores during the changing or replacing probes). 2. By inserting nitrogen-containing compounds inside the tokamak.
Ti	• By inserting titanium-containing compounds inside the tokamak.

$$d = \frac{1\,mm}{1800\,grooves} = 5.556 \times 10^{-4}\,mm = 555.6\,nm$$

With assuming of small diffraction angle ($\cos(\beta) \approx 1$), when $f = 75\ mm$ and $m = 1$,

$$\Rightarrow \quad R_d = \frac{555.6\,nm}{75\,mm \times 1} = 7.408\,{}^{nm}\!/_{mm}$$

By using of a 5 μm slit (it means $W = 5\ \mu m$), the instrumental broadening will be equal to:

$$\Delta\lambda_{Instrumental} = 7.408\,\frac{nm}{mm} \times 5\,\mu m = 0.03704\,nm$$

B. Diffraction broadening measurement

As mentioned before, the reciprocal linear dispersion could be calculated by:

$$R_d = \frac{\Delta\lambda}{\Delta l} = \frac{d \times \cos(\beta)}{f \times m}$$

This equation can be written as follows:

$$\Delta\lambda = \Delta l.\frac{d\cos(\beta)}{f\,m} \tag{7}$$

The angular diffraction limit of an optical system can be used to calculation of the diffraction limit of grating.

$$\Delta\theta \equiv \frac{\Delta l}{f} = \frac{\lambda}{D} \tag{8}$$

Where D is a width of the diffraction grating. By combining the equations 7 and 8, the diffraction broadening becomes to

$$\Delta\lambda = \frac{\lambda}{D}\frac{d\cos(\beta)}{m} \tag{9}$$

For small diffraction angle $\cos(\beta) \approx 1$ and using of the total ruling number $N = D/d$, then

$$\Delta\lambda = \frac{\lambda}{m\,N} \tag{10}$$

The total ruling number of the spectroscopy system used on IR-T1 Tokamak, is:

$$N = 1800\,\frac{grooves}{mm} \times 75\,mm = 1.35 \times 10^5\,grooves$$

Thus, the diffraction broadening equation turns to

$$\Delta\lambda_{Diffraction} = \frac{\lambda}{1.35 \times 10^5}. \tag{11}$$

The broadenings caused by the diffraction limit for the selected impurities are listed in Table V. It is clearly seen that the instrumental broadenings is much more than the diffraction broadening.

C. Doppler broadening measurement

As mentioned above, the resolution broadening is:

$$\Delta\lambda^2_{Resolution} = \Delta\lambda^2_{Instrumental} + \Delta\lambda^2_{Diffraction}$$

so

$$\Delta\lambda_{Doppler} = \left[\Delta\lambda^2_{Measured} - \Delta\lambda^2_{Resolution}\right]^{1/2} \tag{12}$$

It means that the Doppler broadening can be obtained by measuring the total broadening of the spectrum emitted from the impurities. To calculate the total broadening, the Gaussian distribution function

TABLE V. The diffraction broadening measured for the selected impurities by equation 11.

Impurities (Multiplet Number)	Wavelength (nm)	$\Delta\lambda_{Diffraction}$ (nm)
N III (1)	409.731	0.00303
O V (4)	412.390	0.00305
Cr I (1)	425.434	0.00315
C II (6)	426.700	0.00316
Mo I (7)	427.726	0.00317
O II (5)	441.490	0.00327
Ar II (7)	443.018	0.00328
N II (15)	444.703	0.00329
Ni I (86)	446.264	0.00331
Ti I (42)	453.323	0.00336
C III (1)	464.742	0.00344
Mn I (21)	476.237	0.00353
O IV (9)	479.825	0.00355
W I (1)	484.382	0.00359
Fe I (15)	537.149	0.00398

in the MATLAB programming software is used. The results are shown in Figure 7. The Gaussian fitted curve is shown by dash line. The measured Doppler broadening for mentioned impurities are in Table VI.

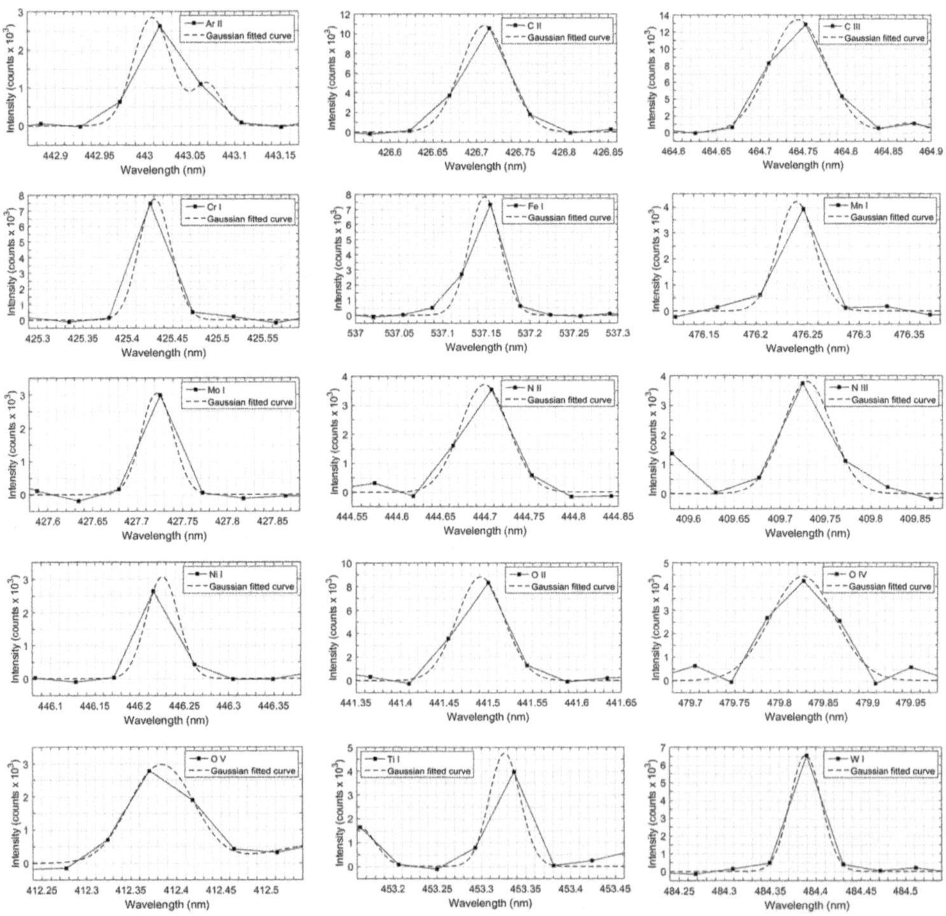

FIG. 7. The Gaussian fitted curves and the total broadening measured for the mentioned impurities.

TABLE VI. The Doppler broadening measured from equation 12.

Impurities (Multiplet Number)	Wavelength (nm)	$\Delta\lambda_{Measured}$ (nm)	$\Delta\lambda_{Resolution}$ (nm)	$\Delta\lambda_{Doppler}$ (nm)
N III (1)	409.731	0.0637	0.04003	0.04955
O V (4)	412.390	0.0838	0.04005	0.07361
Cr I (1)	425.434	0.0413	0.04015	0.00968
C II (6)	426.700	0.0645	0.04016	0.05047
Mo I (7)	427.726	0.0408	0.04017	0.00714
O II (5)	441.490	0.0631	0.04027	0.04858
Ar II (7)	443.018	0.0483	0.04028	0.02665
N II (15)	444.703	0.0644	0.04029	0.05024
Ni I (86)	446.264	0.0419	0.04031	0.01144
Ti I (42)	453.323	0.0415	0.04036	0.00966
C III (1)	464.742	0.0789	0.04044	0.06775
Mn I (21)	476.237	0.0423	0.04053	0.01211
O IV (9)	479.825	0.0861	0.04058	0.07594
W I (1)	484.382	0.0410	0.04059	0.00578
Fe I (15)	537.149	0.0426	0.04098	0.01164

V. ION TEMPERATURE MEASUREMENT

Ion temperature measurement is essential for each tokamak. To determine the formula for measuring the ion temperature, the one dimension Maxwellian distribution function is used.

$$f(v_x, T_i) = \left(\frac{m}{2\pi T_i}\right)^{1/2} e^{-\frac{mv_x^2}{2k_b T_i}} \tag{13}$$

Where m is ion mass, T_i is ion temperature, v_x is ion flow velocity along the X axis and k_b is Boltzmann constant. The full width at half maximum ($\Delta\lambda_{FWHM}$) for each peak can be calculated as:

$$I_0\, e^{-\frac{m\left(\frac{\Delta\lambda_{FWHM}}{2}\cdot\frac{c}{\lambda_0}\right)^2}{2k_b T_i}} = \frac{1}{2} I_0 \quad \Rightarrow \quad \frac{m\left(\frac{\Delta\lambda_{FWHM}}{2}\cdot\frac{c}{\lambda_0}\right)^2}{2k_b T_i} = \ln 2 \tag{14}$$

Where c is the speed of light and λ_0 is the wavelength of desired peak. So, the ion temperature could be measured as:

$$T_i = \frac{m c^2 \Delta\lambda_{FWHM}{}^2}{8 \ln 2 k_b \lambda_0{}^2} \tag{15}$$

TABLE VII. The ion temperature computed for selected impurities according to the Table VI and equation 16.

Impurities	λ_0 (nm)	M (amu)	Ti (eV)
N III	409.731	14	34.44
O V	412.390	16	85.75
Cr I	425.434	52	4.53
C II	426.700	12	28.24
Mo I	427.726	96	4.50
O II	441.490	16	32.59
Ar II	443.018	40	24.35
N II	444.703	14	30.06
Ni I	446.264	58.7	6.48
Ti I	453.323	47.9	3.66
C III	464.742	12	42.90
Mn I	476.237	55	5.98
O IV	479.825	16	67.41
W I	484.382	183.8	4.41
Fe I	537.149	55.8	4.40

It should be noted that $\Delta\lambda_{FWHM} = \Delta\lambda_{Doppler}$. By substituting constants in the equation 15, the ion temperature can be expressed by following equation:

$$T_i\,(eV) = M\,B\frac{\Delta\lambda_{Doppler}{}^2}{\lambda_0{}^2} \tag{16}$$

Where M is ion mass in amu (atomic mass unit) and $B = 1.68216831 \times 10^8$. The results of ion temperature measurement for mentioned impurities are shown in Table VII.

This is a fact that the plasma temperature will decrease by moving from the center toward the edge. So, it can be concluded that the rotational paths of impurities such as Ni I, Mi I, Cr I, Mo I, W I, Fe I and Ti I with low temperature are near the plasma edge. These results were partly predictable by considering the mass of metallic impurities. The results also showed that the maximum ion temperature in the IR-T1 tokamak plasma between 400-500 nm is related to the O V at 412.30 nm.

VI. CONCLUSION

A new spectrometer device for study on the spectrum emitted from plasma in the specified wavelength range is bought and installed on the IR-T1 tokamak. To shielding the spectrometer device from annoying magnetic fields and electromagnetic waves, a new leaden chamber is made. The plasma spectrum emitted from the IR-T1 tokamak between 400-500 nm was detected. For the first time, the main ions and impurities of plasma and their transitions between 400 - 550 nm, are identified. Detected spectral lines showed that the low rate of impurities such as N, O, Cr, C, Mo, Ar, Ni, Ti, Mn, O, W and Fe are existed inside the plasma. The Doppler broadening of the most intense peak of the impurities and their temperatures are measured in this experiment (for the first time). The results show that the ratio of ion temperature to electron temperature (T_i/T_e) at the plasma edge is almost 0.15-0.3. The methods in this paper can be used on all fusion devices.

ACKNOWLEDGMENTS

The authors would like to thank Mr. Reza Arvin, Mr. Syamak Mohammadi and Mr. Ali Nik Mohammadi as the IR-T1 tokamak team for their great efforts in carrying out the project.

Alipour, R., Meshkani, S., Elahi, A. S., and Ghoranneviss, M., "Investigation on the effect of pressure on turbulent transports of the IR-T1 Tokamak plasma," The European Physical Journal D **71**, 60 (2017).

Burrell, K. H., Gohil, P., Groebner, R. J., Kaplan, D. H., Robinson, J. I., and Solomon, W. M., "Improved charge-coupled device detectors for high-speed, charge exchange spectroscopy studies on the DIII-D tokamak," Review of Scientific Instruments **75**, 3455–3457 (2004).

Ghanbari, M. R., Ghoranneviss, M., Elahi, A. S., Tajdidzadeh, M., and Mohamadi, S., "Effect of limiter biasing on runaway electrons in tokamaks," Physica Scripta **83**, 055501 (2011).

Ghasemloo, M., Ghoranneviss, M., Salem, M. K., Arvin, R., Mohammadi, S., and Nik Mohammadi, A., "The effect of emissive biased limiter on the magnetohydrodynamic modes in the IR-T1 tokamak," Physics of Plasmas **20**, 032503 (2013).

Gomes, R. B., Varandas, C. A. F., Cabral, J. A. C., Sokolova, E., and Cortes, S. R., "High dispersion spectrometer for time resolved Doppler measurements of impurity lines emitted during ISTTOK tokamak discharges," Review of Scientific Instruments **74**, 2071–2074 (2003).

Ghoranneviss, M., Mohammadi, S., Elahi, A. S., and Arvin, R., "Tokamak edge plasma rotation in the presence of the biased electrode," Fusion Engineering and Design **88**, 94–99 (2013).

Hasanvand, N., Riazifar, M. R., Alipour, R., Meshkani, S., and Ghoranneviss, M., "Improving plasma confinement by controlling hard X-ray," Chinese Physics Letters **33**, 115202 (2016).

Hennequin, P., Honoré, C., Truc, A., Quéméneur, A., Fenzi-Bonizec, C., Bourdelle, C., Garbet, X., and Hoang, G. T., "Fluctuation spectra and velocity profile from Doppler backscattering on Tore Supra," Nuclear Fusion **46**, S771 (2006).

Hojabri, A., Ghoranneviss, M., Hajakbari, F., Salem, M. K., and Khorshid, P., "Effect of rotating magnetic helical field on the impurity radiation and the magnetic island in IR- T1 Tokamak," *32nd EPS*, Tarragona, Spain, ECA, 29C, P-4.064 (2005).

Kumagai, A., Kubo, H., Takenaga, H., Suzuki, S., Shimizu, K., Asakura, N., and Shimada, M., "Spatial variation of Dα line spectral profile and emission process in the divertor region of JT-60U," Plasma Physics and Controlled Fusion **42**, 529 (2000).

Meshkani, S., Ghoranneviss, M., and Alipour, R., "Study of particle transport in IR-T1 Tokamak," AIP Conference Proceedings **1824**, 030021 (2017).

Naydenkova, D., Stöckel, J., Weinzettl, V., Šesták, D., and Havlíček, J., "First spectroscopic measurements on the COMPASS tokamak," WDS **9**, 158–162 (2009).

Charlotte, E. M., "A multiplet table of astrophysical interest," Revised Edition, Part I-Table of Multiplets, Contributions from the Princeton University Observatory, 20, pp.1–110.

Onchi, T., Liu, Y., Dreval, M., McColl, D., Elgriw, S., Liu, D., Asai, T., Xiao, C., and Hirose, A., "Effects of compact torus injection on toroidal flow in the STOR-M tokamak," Plasma Physics and Controlled Fusion **55**, 035003 (2013).

Orlinkij, D. V. and Magyar, G., "Plasma diagnostics on large tokamaks," Nuclear Fusion **28**, 611 (1988).

Roudaki, F. S., Elahi, A. S., and Ghoranneviss, M., "Determination of electron energy distribution function in tokamak plasma," Journal of Fusion Energy **34**, 911–917 (2015).

Von Hellermann, M. G., Mandl, W., Summers, H. P., Weisen, H., Boileau, A., Morgan, P. D., Morsi, H., Koenig, R., Stamp, M. F., and Wolf, R., "Visible charge exchange spectroscopy at JET," Review of Scientific Instruments **61**, 3479–3486 (1990).

Weinzettl, V., Ďuran, I., Zajac, J., and Piffl, V., "Edge plasma biasing experiments on the CASTOR tokamak: Spectroscopic investigation and microwave measurements," Czechoslovak Journal of Physics **53**, 925–935 (2003).

Zweben, S. J. and Medley, S. S., "Visible imaging of edge fluctuations in the TFTR tokamak," Physics of Fluids B: Plasma Physics **1**, 2058–2065 (1989).

2

Tunable ferroelectric polarization of the bulk and free standing $ATiO_3$ (A = Ba and Pb) thin films via unit-cell thicknesses and strain engineering

A. Tariq and S. Nazir[a]
Department of Physics, University of Sargodha, 40100 Sargodha, Pakistan

Ab-Initio calculations are used to investigate the spontaneous polarization in the bulk and free standing $ATiO_3$ (A = Ba and Pb) thin films by increasing the corresponding unit-cell thickness in [001]-direction and by adjusting the bi-axial strain (both compressive and tensile) in *ab*-plane [110]. Our results exhibit that polarization gradually increases with the increase of unit-cell growth and compressive/tensile strain has a positive/negative impact on the spontaneous polarization for both bulk materials. For free standing thin films, it is found that polarization is strongly suppressed below a critical value of film thickness, due to depolarizing the intrinsic electric field induced by un-compensated dipoles at the surfaces. In $BaTiO_3$ thin film, 8.5/6 unit-cells (32/26 Å) are predicted to be the critical thickness for BaO-BaO/BaO-TiO_2 symmetric/asymmetric surface. In contrast, polarization gradually increases for TiO_2-TiO_2 symmetric surface upto 6.5 unit-cells and then raises sharply. In the case of $PbTiO_3$ thin film, 6.5 and 6 unit-cells (23 and 22 Å) are estimated the critical thicknesses for symmetric [(PbO-PbO) and (TiO_2-TiO_2)] and asymmetric (PbO-TiO_2) surfaces, respectively. Compressive and tensile strains display an increasing and decreasing trend for all thin film structures. However, the impact of compressive strain on the polarization is larger than tensile. Furthermore, we predict that asymmetric (AO-TiO_2) surfaces in both films show higher polarization as compared to symmetric [(AO-AO) and (TiO_2-TiO_2)] surfaces due to large octahedral distortions.

I. INTRODUCTION

Ferroelectricity is attributed to materials that show a spontaneous polarization in the absence of applied electric field.[1,2] In this respect, perovskite oxides with ABO_3 structures have offered a potential technological applications such as ferroelectric tunnel junctions (FTJ) as well as exhibit a giant electro-resistance effects,[3–5] dielectric capacitors,[6] sensors,[7] microelectronics,[8] non-volatile ferroelectric random access memory,[9] and electro-mechanical devices.[10–13] Ferroelectricity originates due to tetragonal distortion in the cubic perovskite structures and the fine balance between long range coulombic and short range covalent interactions.[14,15] Previously, ferroelectric properties have been investigated by means of structural parameters and strain-engineering in paraelectric materials.[16] In this respect, $BaTiO_3$ (BTO)[17] and $PbTiO_3$ (PTO)[15] are prototypical ferroelectric materials in tetragonal phases, which have been studied extensively from past few decades. BTO undergoes phase transitions from cubic to tetragonal, rhombohedral, and orthorhombic phases at 400 K, 278 K, and 200 K with respective space groups are $Pm\bar{3}m$, $P4mm$, and $Amm3$,[18] respectively. On the other hand, PTO has only one phase transition which is from cubic phase with space group $pm\bar{3}m$ to tetragonal phase ($P4mm$) at 760 K.[19] Many experimental[20–23] and theoretical[24–28] investigations have been

[a]safdar.nazir@uos.edu.pk

performed on the BTO and PTO in order to study their ferroelectric properties. The experimentally observed values of spontaneous polarization in the bulk form are 26 μCcm^{-2} and 75 μCcm^{-2} for BTO[29] and PTO,[15] respectively. To make these systems more suitable as a high performer ferroelectric candidates, it is therefore necessary to enhance their spontaneous polarization in bulk as well in thin film structures. In this way, unit-cell growth and strain engineering are an important parameters to tune the physical properties of the materials.

Thickness dependent ferroelectric properties have been studied experimentally,[30–33] where authors found that at nano-scale thin films make ferroelectric to paraelectric transitions due to high surface effects and depolarizing fields. Because accumulation of the charges on surfaces lead to the destabilization of ferroelectric distortions, which suppress the intrinsic internal electric field. Similarly, Junquera *et al.*[34] has demonstrated that BTO thin film between two metallic $SrRuO_3$ electrodes exhibits a critical thickness of 24 Å (2.4 nm), below which ferroelectric ordering vanish. On the other hand, Stachiotti *et al.*[35] has observed a critical thickness of 36 Å (3.6 nm). Very recently, it is also examine that BTO film becomes almost paraelectric when its thickness below to 44-48 Å (4.4-4.8 nm).[36] In contrast, due to formation of closure domain like structures in the ultra-thin film of PTO, ferroelectric state does not disappears[37] and Bnglun *et al.*[38] figured out the mechanism (Pb-O bond covalency) through which PTO thin films sustain their ferroelectric ordering.

Strain often has a significant influence on the physical properties of the materials as Kesong *et al.*[39] theoretically found that by applying appropriate strain on an ordinary band insulators, leads to a topological insulator. Hence, technologically strain dependent ferroelectric properties are very appealing as the materials properties can be tuned by growing on different rigid substrates (*i.e.*, by lattice mismatch technique)[40] that causes defects[41] in deposited materials, which results in geometry deformation and finally effects on the polarization.[42] As, it has been studied that several perovskite oxides show sensitivity to the strain, as an enhancement in polarization is observed in supper-lattice due to epitaxial strain which is induced by substrate.[43–45]

The main purpose of the present work is to enhance the spontaneous polarization of bulk and free standing thin films in BTO and PTO perovskite oxides by increasing the unit-cell thicknesses and applying the bi-axial (both compressive and tensile) strain in *ab*-plane [110] with in a reasonable range. In the case of free standing thin films, polarization is calculated for possible two symmetric [(AO-AO) and (TiO_2-TiO_2)] and one asymmetric (AO-TiO_2) surfaces. The critical thickness below which system make almost ferroelectric to paraelectric transition is discusses in detail for each configuration of thin film, which is different due to different surface terminations. Finally, the effects of both compressive and tensile strains are taken into account and results are explained on the basis of structural distortions.

II. COMPUTATIONAL METHODS

Present calculations employ on density functional theory, using a full-potential linearized augmented plane-wave method as implemented in the WIEN2K code.[46] The local density approximations (LDA) and generalized gradient approximations (GGA) are applied for the exchange-correlation functional. Fully relativistic effects are taken into account for the core states, while scalar relativistic approximation is used for the valence states. Therefore, the effect of spin-orbit coupling are neglected. We have fully optimized the crystal structure of the bulk and thin films by minimizing the atomic forces to 5 mR/a.u in (001)-direction. For the wave function expansion inside the atomic spheres, a maximum value of l_{max} = 12 is chosen and the plane-wave cutoff is set to $R_{mt}K_{max}$ = 5 with G_{max} = 12. The muffin-tin sphere radius are chosen as 2.0, 2.5, 1.55, and 1.40 for Ba, Pb, Ti, and O atoms, respectively. Moreover, a $9 \times 9 \times 1/7 \times 7 \times 1$ of k-mesh with 50/42 points within the irreducible wedge of the Brillouin zone is found to be well converged for the bulk/thin film case for atomic relaxation. A supercell of $1 \times 1 \times n$, where n = 1-5 and 1-9 is modeled for bulk and thin films, respectively. For thin film structures, a vacuum layer of 10 Å is adjusted along the [001] direction on both sides to resemble the actual epitaxial growth process of the materials and to avoid the interactions with other unit-cell. The symmetric structures [(AO-AO) and (TiO_2-TiO_2)] are constructed by AO and TiO_2 terminated surfaces on both sides. On the other hand, in asymmetric (AO-TiO_2) structure one surface consists of AO termination while other is terminated at TiO_2 layer in the same structure of thin film.

III. RESULTS AND DISCUSSION

A. Bulk properties

Calculated LDA and GGA lattice constants, atomic positions, and born effective charges of BTO and PTO are listed in Table I along with available experimental data and also previously reported theoretical values. From Table I, it is evident that our computed values are close to the experiment and also are in good agreement with the other theoretical works. However, the calculated values with LDA correlation is close to the experiments as compared to GGA. We found that the Ba-O and Ti-O bond lengths and O-Ti-O bond angles increases as the unit-cell thickness increases and atoms deviate from their centrosymmetric positions, which results in more distortions and causes high polarization. Because, it is found that structural relaxation has a significant effects on the chemical bonding, which latter causes a remarkable impact on the polarization.[19,47] To quantitatively illustrate, we calculated the spontaneous polarization (P) in each case as:[47]

$$P = \frac{e}{\Omega} \sum_{i=1}^{N} Z_i^* \delta z_i \tag{1}$$

where, e is the electronic charge (1.6×10^{-19} C), Ω is the total volume of the unit cell, Z_i^* is the born effective charge for each cation and δz_i is the relative displacement between the ith cation and O anion in the respective plane. The computed δz_{A-O}, δz_{Ti-O}, and P along with average O-Ti-O bond angles with respect to number of unit-cell thick for bulk BTO (left column) and PTO (right column) are plotted in Fig. 1. The calculated values of $\delta z_{Ba-O}/\delta z_{Ti-O}$ in one unit cell thick BTO are 0.0738/0.1259 Å and 0.0184/0.1214 Å for LDA and GGA correlations, respectively. Similarly, for the case of PTO, the calculated $\delta z_{Pb-O}/\delta z_{Ti-O}$ are 0.3608/0.3329 Å and 0.3809/0.3161 Å for LDA and GGA, respectively. Hence, the computed values of P in one unit-cell of bulk BTO/PTO are 26.33/75.71 μCcm^{-2} and 25.59/75.31 μCcm^{-2} for LDA and GGA scheme, respectively. Our results

TABLE I. Experimental (Expt.), DFT relaxed, and previously reported lattice parameters (Å), atomic positions of Ti/O atoms and born effective charges of $BaTiO_3$ and $PbTiO_3$ perovskite oxides. Both Ba and Pb atoms are positioned at (0,0,0).

	BTO				PTO			
	Exp.	Present work		Other Works	Exp.	Present work		Other Works
		LDA	GGA			LDA	GGA	
a	3.99,[a] 3.995[b]	3.870	4.111	3.943,[c] 3.966[d]	3.904,[e] 3.892[f]	3.787	4.021	3.904,[g] 3.834[h]
c	4.035,[a] 4.034[b]	3.914	4.156	3.994,[c] 4.035[d]	4.152,[e] 4.168[f]	4.027	4.277	4.151,[g] 4.302[h]
Ti(z)	0.522,[a] 0.514[b]	0.506	0.538	0.018[i]	0.540,[e] 0.542[f]	0.524	0.556	0.549,[g] 0.542[h]
O1(z)	−0.024,[a] −0.025[b]	0.021	0.025	−0.047[i]	0.612,[e] 0.629[f]	0.594	0.631	0.630,[g] 0.634[h]
O2(z)	0.489,[a] 0.488[b]	0.474	0.504	−0.023[i]	0.112,[f] 0.124[e]	0.109	0.115	0.125,[g] 0.134[h]
Z_A^*	–	2.87	2.960	2.90[i]	–	3.41	3.63	3.52[h]
Z_{Ti}^*	–	6.69	6.890	6.70[i]	–	5.03	5.34	5.18[h]
$Z_{O_1}^*$	–	−2.38	−2.45	−2.40[i]	–	−2.10	−2.22	−2.16[h]
$Z_{O_1}^*$	–	−4.76	−4.91	−4.80[i]	–	−4.25	−4.51	−4.38[h]

[a]Reference 22.
[b]Reference 23.
[c]Reference 24.
[d]Reference 25.
[e]Reference 20.
[f]Reference 21.
[g]Reference 27.
[h]Reference 28.
[i]Reference 26.

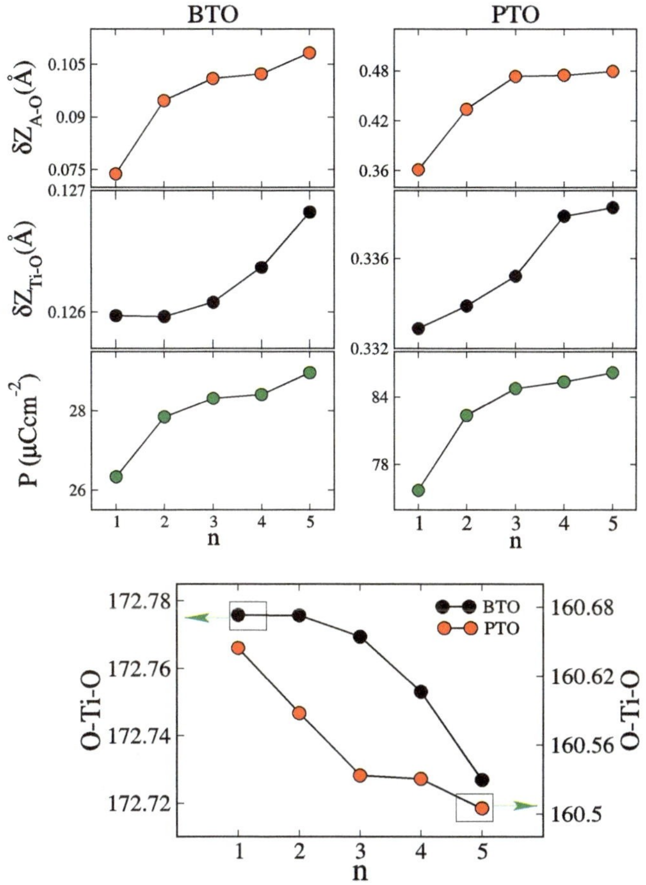

FIG. 1. Calculated average relative displacement δz_i of A and Ti cations with respect to O anions in the corresponding layers and spontaneous polarization (P) along [001] direction in $BaTiO_3$ (left column) and $PbTiO_3$ (right column) perovskite oxides with respect to corresponding unit-cell thicknesses (n). Along with computed average in-plane O-Ti-O bond angles (in degree) are plotted with respect to n for both compounds and arrow heads represent the corresponding material bond angles.

indicate that LDA calculated P values are closer to the experiment in both cases,[15,29] therefore for later study only LDA method is taken into account.

From Fig. 1, one can clearly see that the relative displacement δz_{A-O} (between A cations and O anions in the AO layers) and δz_{Ti-O} (between Ti and O atoms in the TiO_2 layers) are gradually increases with respect to increasing number of unit-cells in both cases, which results in higher P. Hence, both δz_{A-O} and δz_{Ti-O} play an important role in enhancing P. The increase in P can be understand from computed O-Ti-O bond angles (in degree) in Fig. 1. Our results indicate that bond angles decreases (more distorted) in both bulk materials as the number of unit-cells increases. The increase in structural distortion leads to the large changes in the relative displacements, which further effect on the spontaneous P. It is also noted that structures are more distorted within LDA correlation as compared to GGA, therefore P has a higher value in this case.

Next, the bi-axial strain effect is taken into account on P for both bulk BTO and PTO in *ab*-plane [110]. In the present study, bi-axial strain is applied up to 5 unit-cells for bulk systems which ranges from -5% to +5% in the original lattice constants. The calculated P with respect to bi-axial strain is plotted in Fig. 2(a) and (b) for bulk $(BTO)_n$ and $(PTO)_n$ structures, respectively, where n = 1, 3, and 5 represents the number of unit-cells. Our results clearly indicate that as the compressive strain increases, P also increases. Because, unit cell volume reduces and the structures show more distortion (O-Ti-O bond angles decreases from their 0% value). On the other hand, for tensile strains, volume increases and structures show less distortion (O-Ti-O bond angles increases from their 0% value), which results in decrease in P. Moreover, our results exhibit that

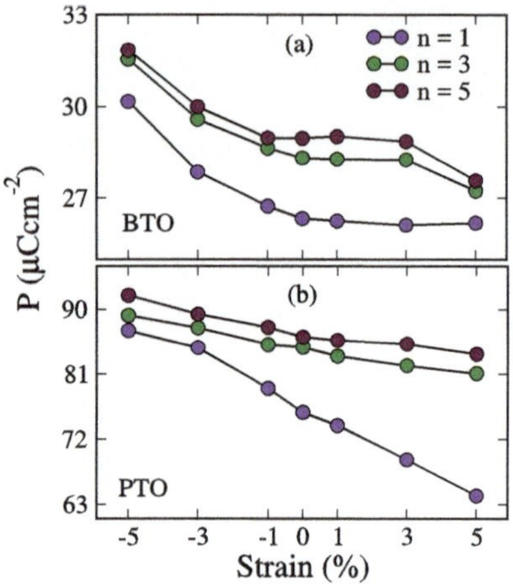

FIG. 2. Calculated spontaneous polarization (P) for (a) $(BaTiO_3)_n$ and (b) $(PbTiO_3)_n$ perovskite oxides, where $n = 1, 3$, and 5 with respect to bi-axial strain in ab-plane [110]. The −ve and +ve signs indicate compressive and tensile strains, respectively.

compressive strain shows high impact on P as compared to tensile strain which demands more experimental observations.

B. Free standing thin films

As BTO and PTO have wide range of applications at nano-scale, hence due to rapid development in electronic oxide industry, the minimizing the size of electronic devices has become an inevitable trend. Therefore, it is essential to form a good quality of epitaxial thin films for device applications. It is observed that when films are very thin, the surface effects become so strong and increase in surface tension, which can results in loss of ferroelectric ordering due to depolarizing intrinsic electric field induced by uncompensated dipoles.[48] Hence, we calculated P for two types of thin film structures such as symmetric [(AO-AO) and (TiO_2-TiO_2)] and asymmetric (AO-TiO_2) surfaces.

Calculated P for symmetric [(AO-AO) and (TiO_2-TiO_2)] and symmetric (AO-TiO_2) surfaces are plotted in Fig. 3 with respect to number of unit-cell thicknesses for BTO (left column) and PTO

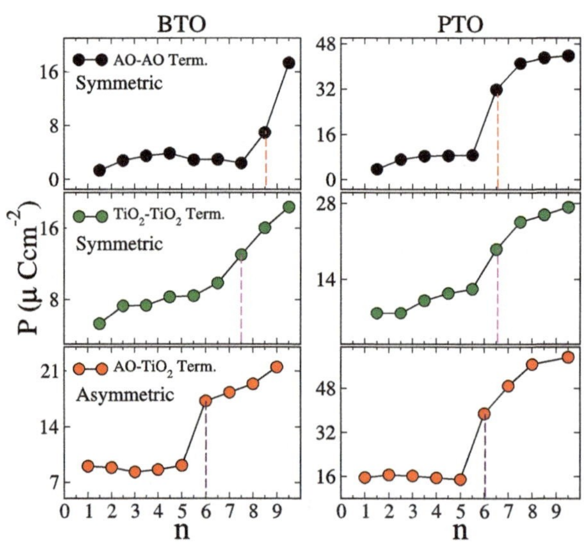

FIG. 3. Calculated spontaneous polarization (P) of $BaTiO_3$ (left column) and $PbTiO_3$ (right column) for symmetric [(AO-AO) and (TiO_2-TiO_2)] and asymmetric (AO-TiO_2) surfaces with respect to corresponding film thicknesses *(n)*. The vertical dashed lines represent the critical thickness in each surface structure.

(right column) thin films. From Fig. 3, one can clearly see that *P* is significantly suppressed (almost paraelectric) in symmetric (AO-AO) surfaces upto 7.5 unit-cells (27 Å) and 5.5 unit-cells (19 Å) for BTO and PTO thin films, respectively. However, when unit cell thicknesses reaches to 8.5 unit-cells (32 Å) and 6.5 unit-cells (26 Å) for BTO and PTO, respectively, *P* increases sharply. Hence, 32 Å and 26 Å film thicknesses are estimated the critical thicknesses for symmetric (AO-AO) surfaces, below which films are almost in paraelectric states. In the case of TiO_2-TiO_2 symmetric surfaces, *P* gradually increases with the increase of film thicknesses in both film structures (see Fig. 3). However, when film thicknesses approach to 7.5 unit-cells (28 Å) and 6.5 unit-cells (24 Å) for BTO and PTO, respectively, *P* significantly increases. Therefore, 28 Å and 24 Å can be considered the critical thicknesses for TiO_2-TiO_2 symmetric surfaces which is close to the experimentally observed value for BTO.[35] In the case of asymmetric (AO-TiO_2) surfaces, *P* remains constant with lower values of 8.43 μCcm^{-2}/12.26 μCcm^{-2} upto 5 unit-cells for BTO/PTO films (see Fig. 3). But a prominent increase in *P* is found at 6 unit-cells (22 Å and 23 Å), which is close to the previously calculated value in the case of BTO.[48] Hence, 22 Å and 23 Å are the critical thicknesses for BTO and PTO, respectively, for asymmetric (AO-TiO_2) surfaces. It is also noted that even for ultra-thin symmetric (TiO_2-TiO_2) and asymmetric (PbO-TiO_2) surfaces in PTO films, a reasonable *P* values resides, which is in accordance with previous work.[38] Moreover, our calculated values of *P* for asymmetric surfaces (AO-TiO_2) is higher as compared to symmetric [(AO-AO) and (TiO_2-TiO_2)] surfaces in both films. The different values of critical thicknesses depend on the surface terminations, which mainly in accordance with the octahedral distortions.

The remarkable changes in *P* can be attributed to the structural distortions. Therefore, we plotted average values of O-Ti-O bond angles (in degrees) with respect to unit-cell thicknesses in Fig. 4 for BTO (left column) and PTO (right column). Our results show that for symmetric (AO-AO) surfaces, O-Ti-O bond angle values are close to the perfect octahedral angle of 180°, which confirms the calculated *P* values in Fig. 3 where films are almost in paraelectric state for AO-AO symmetric surfaces in both film structures. However, further increase in the unit-cell thicknesses, bond angles show a prominent distortion which results in higher *P* at 9.5 unit-cells and 6.5 unit-cells for BTO and PTO, respectively. In contrast, for TiO_2-TiO_2 symmetric and AO-TiO_2 asymmetric surfaces, and bond angles are little bit distorted upto first few unit-cells of BTO and PTO thin films and they exhibit more distortion when critical thickness reaches in each case, see Fig. 4. The calculated average critical values of O-Ti-O bond angles (in degree) in BTO thin films at which ferroelectric

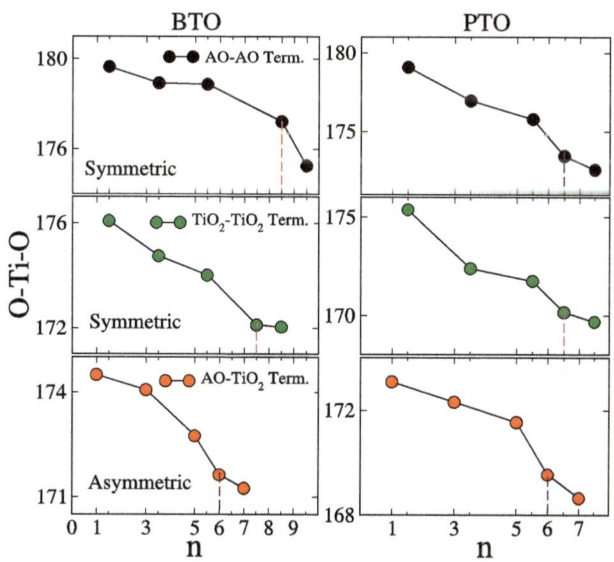

FIG. 4. Calculated average O-Ti-O bond angles (in degree) for symmetric [(AO-AO) and (TiO_2-TiO_2)] and asymmetric (AO-TiO_2) surfaces for $BaTiO_3$ (left column) and $PbTiO_3$ thin films with respect to corresponding film thicknesses. The vertical dashed lines represent the critical thickness in each surface structure.

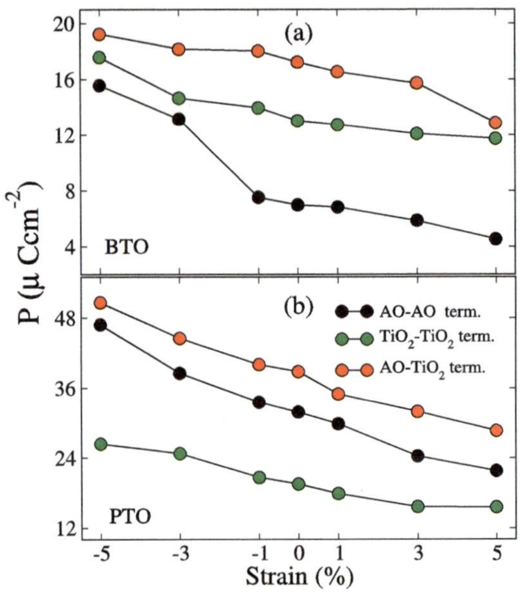

FIG. 5. Calculated spontaneous polarization (P) with respect to bi-axial (compressive and tensile) strain in the *ab*-plane [110] for symmetric [(AO-AO) and (TiO_2-TiO_2)] and asymmetric (AO-TiO_2) surfaces for the corresponding critical thicknesses in (a) $BaTiO_3$ and (b) $PbTiO_3$ thin films, respectively.

to almost paraelectric phase transition occurs for symmetric [(BaO-BaO)/(TiO_2-TiO_2)] and asymmetric (BaO-TiO_2) surfaces are 177°/172° and 171°, respectively. Similarly, in the case of PTO thin films, 173°/170° and 169° are found the critical bond angles for symmetric [(PbO-PbO)/(TiO_2-TiO_2)] and asymmetric (PbO-TiO_2) surfaces, respectively. Furthermore, it is also evident from Fig. 4 that O-Ti-O bond angles are more distorted for AO-TiO_2 asymmetric surfaces than the symmetric [(AO-AO) and (TiO_2-TiO_2)] surfaces in both thin film configurations, which results in higher value of P in asymmetric surfaces and also confirms the calculated values of P in Fig. 3.

Finally, the biaxial strain along [110] plane is applied at the estimated critical thicknesses for all types of thin film configurations. The calculated P for symmetric [(AO-AO) and (TiO_2-TiO_2)] and asymmetric (AO-TiO_2) surfaces at the corresponding critical thicknesses with respect to bi-axial strain in *ab*-plane [110] are plotted in Fig. 5(a) and (b) for BTO and PTO thin films, respectively. One can examine that when compressive strain increases from −1% to −5%, spontaneous P gradually increases. Because volume of the unit-cells reduce and structures show more distortions. In contrast, P decreases with the increase of tensile strain, due to suppression of local octahedral distortions. Moreover, we found that compressive strain has a large impact on P as compared to tensile strain as established in the bulk cases. Hence, these results indicate that P can be retain (due to higher octahedral distortions) even at the ultra-thin film thicknesses with the implementation of compressive strain inspite of higher surface effects. Because, the applied strain changes the volume of unit-cell and also effects on its chemical bonding, which consequently induced categorical changes in P.

IV. CONCLUSION

In summary, ferroelectric properties of two prototypical perovskite oxides $ATiO_3$ (A = Ba and Pb) have been investigated in bulk and free standing thin film structures with respect to unit-cells growth and strain engineering using density functional theory electronic structure calculations. For both bulk cases, polarization increases as the unit-cell thickness increases. Structures show larger and smaller distortion from their original value (0% strain) for compressive and tensile strains, respectively, which further enhances and diminish the polarization. Our calculations indicate that spontaneous polarization is suppresses below a certain critical value of film thicknesses and increases sharply after that for all thin film configurations (symmetric and asymmetric surfaces), which can be attributed to the surface effects that produced due to depolarizing fields induced by un-compensated surface dipoles. Hence, the present investigations demand for more comprehensive experimental characterization of ferroelectric properties of perovskite oxides for device applications.

ACKNOWLEDGMENTS

The computational work was supported by University of the Sargodha, Sargodha under ORIC grant No. UOS/ORIC/2016/55.

[1] X. Ren, Nat. Mater. **3**, 91 (2004).
[2] Y. Xu, *Ferroelectric Materials and Their Applications* (North-Holland, New York, 1991).
[3] X. S. Gao, J. M. Liu, K. Au, and J. Y. Dai, J. Appl. Phys. **101**, 142905 (2012).
[4] L. L. Tao and J. Wang, J. Appl. Phys. **119**, 224104 (2016).
[5] X. Lou, Y. Zheng, and B. Wang, J. Appl. Phys. **111**, 074102 (2012).
[6] U. Weber, G. Greuel, U. Boettger, S. Webe, D. Hennings, and R. Waser, J. Am. Soc. **84**, 759 (2001).
[7] D. Damjanovic, P. Muralt, and N. Setter, IEEE Sens. J. **1**, 191 (2001).
[8] J. F. Scott, Science **315**, 954 (2007).
[9] R. Ramesh, S. Aggarwal, and O. Auciello, Mater. Sci. Eng. **32**, 191 (2011).
[10] R. Ramesh, *Thin film Ferroelectric Materials and Devices* (Kluwer Academic, Boston, 1997).
[11] J. F. Scott, *Ferroelectric Memories* (Springer, Berlin, 2000).
[12] E. Lines and A. M. Glass, *Principles and Applications of Ferroelectrics and Related Materials* (Clarendon Press, Oxford, 1977).
[13] M. Dawber, K. M. Rabe, and J. F. Scott, Rev. Mod. Phys. **77**, 1098 (2005).
[14] R. Resta, M. Posternak, and A. Baldereschi, Phys. Rev. Lett. **70**, 1010 (1993).
[15] W. Zhong, R. D. King-Smith, and D. Vanderbilt, Phys. Rev. Lett. **72**, 3618 (1994).
[16] J. H. Haeni, P. Irvin, W. Chang, R. Uecker, P. Reiche, Y. L. Li, S. Choudhury, W. Tian, M. E. Hawley, B. Craigo, A. K. Tagantsev, X. Q. Pan, S. K. Streiffer, L. Q. Chen, S. W. Kirchoefer, J. Levy, and S. D. Schlom, Nature **430**, 758 (2004).
[17] R. W. Gray, U.S. Pat. No. 2, 486, 560 (1949).
[18] X. Deng, X. Wang, H. Wen, and Z. Gen, J. Am. Ceram. Soc. **89**, 1059 (2006).
[19] R. E. Cohen, Nature **358**, 136 (1992).
[20] A. M. Glazer and S. A. Mabud, Acta Cryst. B. **34**, 1060 (1978).
[21] H. Mestric, R. A. Eichel, T. Kloss, K. P. Dinse, S. Laubach, S. Laubach, P. C. Schmidt, K. A. Schonau, M. Knapp, and H. Ehrenberg, Phys. Rev. B **71**, 134109 (2005).
[22] G. H. Kwei, A. C. Lawson, S. J. L. Billinge, and S. W. Cheong, J. Phys. Chem. **97**, 2368 (1993).
[23] H. T. Evans, Acta Cryst. **14**, 1019 (1961).
[24] M. Fechner, S. Ostanin, and I. Mertig, Phys. Rev. B **77**, 094112 (2008).
[25] J. J. Wang, F. Y. Meng, Y. Q. Ma, M. X. Xu, and L. Q. Chen, J. Appl. Phys. **108**, 034107 (2010).
[26] M. Uludogan and T. Cagin, Turk J. Phys. **30**, 277 (2006).
[27] A. Garcia and D. Vanderbilt, Phys. Rev. B **54**, 3817 (1996).
[28] G. Saghi-Szabo, R. E. Cohen, and H. Krakauer, Phys. Rev Lett. **80**, 4321 (1998).
[29] H. H. Wieder, Phys. Rev. **99**, 1161 (1955).
[30] S. Rios, J. F. Scott, A. Lookman, J. McAneney, R. M. Bowman, and J. M. Gregg, J. Appl. Phys. **99**, 024107 (2006).
[31] Y. Bastani, T. S. Kempen, A. Roelofs, and N. B. Gharb, J. Appl. Phys. **109**, 014115 (2011).
[32] Y. L. Sang, B. Liu, and D. N. Fang, Comput. Mater. Sci. **44**, 404 (2008).
[33] Y. Y. Liu and J. Y. Li, Appl. Phys. Lett. **97**, 042905 (2008).
[34] J. Junquera and P. Ghosez, Nature **422**, 506 (2003).
[35] M. G. Stachiotti, Appl. Phys. Lett. **84**, 251 (2003).
[36] Y. Li, Y. Rong, H. Zhou, Z. Cheng, X. Wang, L. Li, and J. Zhu, J. Am. Cream. Soc. **98**, 2710 (2015).
[37] T. Shimada, S. Tomoda, and T. Kitamura, Phys. Rev. B **81**, 144116 (2010).
[38] B. Yin and S. Qu, J. Appl. Phys. **114**, 063703 (2013).
[39] K. Yang, W. Setyawan, S. Wang, M. B. Nardelli, and S. A. Curtarolo, Nat. Mater. **11**, 614 (2011).
[40] W. D. Nix, Metall. Trans. A **20**, 2217 (1989).
[41] H. Li, A. L. Roytburd, S. P. Alpay, T. D. Tran, L. Salamanca-Riba, and R. Ramesh, Appl. Phys. Lett. **78**, 2354 (2001).
[42] W. D. Nix and B. M. Clems, J. Mater. Res. **14**, 3467 (1999).
[43] J. B. Neaton and K. M. Rabe, J. Appl. Phys. **82**, 1586 (2003).
[44] C. Bungaro and K. M. Rabe, Phys. Rev. B **69**, 184101 (2004).
[45] O. Dieguez, K. M. Rabe, and D. Vanderbilt, Phys. Rev. B **72**, 144101 (2005).
[46] P. Blaha, K. Schwarz, G. K. H. Madsen, D. Kvasnicka, and J. Luitz, in *WIEN2K: An Augmented Plane Wave plus Local Orbitals Program for calculating Crystal Properties*, edited by K. Schwarz (Vienna University of Technology, Austria, 2011).
[47] S. Nazir, J. Cheng, and K. Yang, ACS Appl. Mater. Interfaces **8**, 390 (2016).
[48] Y. Su, H. Chen, J. J. Li, A. K. Soh, and G. J. Weng, J. Appl. Phys. **110**, 084108 (2011).

3

PIC-MCC analysis of electron multiplication in a cold-cathode Penning ion generator and its application to identify ignition voltage

H. Noori,[1,a] A. H. Ranjbar,[1] and M. Mahjour-Shafiei[2]
[1]*Faculty of Physics, Shahid Bahonar University of Kerman, Kerman 7616914111, Iran*
[2]*Department of Physics, University of Tehran, Tehran 1439955961, Iran*

A cold-cathode Penning ion generator (PIG) has been developed in our laboratory to study the interaction of charged particles with matter. The ignition voltage was measured in the presence of the axial magnetic field in the range of 460-580 G. The performed measurements with stainless steel cathodes were in argon gas at pressure of 4×10^{-2} mbar. A PIC-MCC (particle-in-cell, Monte Carlo collision) technique has been used to calculate the electron multiplication coefficient M for various strength of axial magnetic field and applied voltage. An approach based on the coefficient M and the experimental values of the secondary electron emission coefficient γ, was proposed to determine the ignition voltages, theoretically. Applying the values of secondary coefficient γ leads to the average value of $\gamma M(V, B)$ to be = 1.05 $\pm$ 0.03 at the ignition of the PIG which satisfies the proposed ignition criterion. Thus, the ion-induced secondary electrons emitted from the cathode have dominant contribution to self-sustaining of the discharge process in a PIG.

I. INTRODUCTION

A Penning ion generator (PIG), with cold or hot cathodes, provides a high-voltage and low-pressure plasma discharge. An axial magnetic field is applied for electron confinement between two cathodes and inside the hollow cylindrical anode. PIG (also named Penning ion source) is based on the gas discharge technique invented by Penning.[1] Some advantages such as compact design, simple geometry, high current density, low-pressure and long-time operation make it one of the most interesting and instrumental ion sources for both fundamental and applied researchers. This ion source has many purposes of applications in isotope mass separators and accelerators.[2–4] PIGs are, also, very practical and can be used in sputtering, evaporation of surfaces, fusion application, and neutron generators.[5–7]

The behavior of Penning gas discharge has been investigated both experimentally and numerically.[8] In the experimental part, PIG and Penning discharge characteristics have been studied for various parameters such as the anode length, magnetic field, working-gas pressure, and so on.[9–12] In particular, the effect of the magnetic field, such as, the effect of configuration of magnetic field on the extracted ion current has been considered in the literature.[13] In some studies to avoid the complexities associated with results interpretation, the constant and uniform magnetic fields have been considered.[11,14] The authors have, previously, taken into account the effect of the uniform axial magnetic field on the breakdown voltage of PIG and proposed a theoretical expression.[15]

The experimental results have shown that in Penning discharges, there is also a group of fast electrons so-called secondary electrons that are emitted from the cathode surface due to the

[a]Electronic mail: hadi-noori@uk.ac.ir

ion bombardment.[16] The number of those electrons is quantified by the secondary electron emission coefficient, γ, that is electron yield per ion falling on the cathode[17] and depends on the ion impact energy.[18] They are accelerated in the electric field between the anode and the cathode and acquire enough energy to ionize the neutral gas atoms for several times due to the inelastic impacts. Despite the fact that those electrons are a small fraction of the electrons presented in the discharge, they have a dominant contribution to the ionization and play a main role to self-sustaining of the discharge.[16] However, the contribution of the electrons resulting from ionizing collisions of the fast electrons with the gas atoms, to the ionization is not significant. Most of them are produced inside the anode cylinder and do not acquire sufficient energy from the weak electric field in that region.[16] The electron multiplication coefficient, M, is defined as the number of electrons produced for each secondary electron released from the cathode and it has been investigated in the uniform and non-uniform electric field.[19,20] In a simple case, the coefficient was obtained for the uniform electric field and in the absence of magnetic field, by well-known equation $M = \exp(\alpha d)$.[17]

A conventional theoretical method to study the breakdown characteristics of gas discharges is PIC-MCC (particle-in-cell, Monte-Carlo collision) technique.[21–23] Employing this method in the previous works, the breakdown has been, theoretically, identified based on the time variation of the number of ions and electrons.[24–26] In the present work, we also used the same method to track the electrons and their collisions inside the PIG. Based on the coefficient M and the experimental values of the coefficient γ, the ignition voltages for various strength of axial magnetic field has been reckoned while the energy dependency of γ was also considered. Due to the less number of moving charged particles presented in the simulation domain, the advantage of this method is a faster calculation of M relative to number-time plot of electrons and ions for a given applied voltage and a magnetic field.

II. MATERIAL AND METHODS

A. Experimental set-up

A cold-cathode PIG has been developed in our laboratory to study the interaction of charged particles with the matter. A schematic representation of lateral cross section of the electrodes of the generator is shown in Fig.1. The electrodes fabricated from unmagnetized stainless steel material which is categorized into the dirty metal as defined by Phelps et al.[18] The hollow cylindrical anode is fixed inside the chamber by a ceramic ring. The cathodes stands 30 mm apart and the anode length is

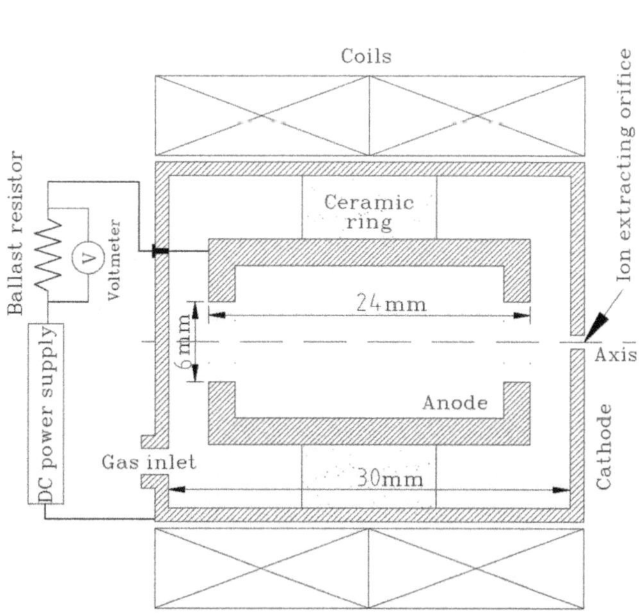

FIG. 1. Lateral cross section of the developed Penning ion generator.

24 mm. The discharge chamber is placed between two cylindrical solenoids powered by a switched-mode power supply to establish a uniform axial magnetic field with a homogeneity of no worse than 7% in the discharge region. The magnitude of the magnetic field can be varied between 460 G to 580 G.[15]

Difference voltage between the electrodes was applied using a variable DC voltage power supply connected to a fixed ballast resistor of 495 kΩ in series. The discharge current was determined by measuring the dropped voltage across the resistor. It provides an accuracy of 10^{-8} A for the obtained values of the current. A complete vacuum system including a diffusion pump and a rotary van pump, was used to evacuate the apparatus. Before the operation of the ion source, the system must be evacuated to about 5×10^{-5} mbar to remove the residual gases. The argon gas was, then, injected through a needle valve which has been used to adjust the gas pressure. The operation pressure was kept at 4×10^{-2} mbar during the measurements using a digital vacuum meter.

B. Theoretical bases

The dominant mechanisms to control the number of positive ions in PIG are: (1) the ion production due to the ionizing collisions of the ion-induced secondary electrons emitted from the cathodes, with the neutral atoms, (2) the deionization of the ions by the recombination of the electrons and the positive ions, and (3) the ion absorption by the cathodes. It should be noted that most of the electrons and ions are confined within the effective volume, $A'd'$, inside the anode. Thus, in the infinite-small time interval the ion number density, n_i, changes as[17]

$$A'd'dn_i = \gamma I_{i,c} M(V,B)dt - CA'd'n_i n_e dt - I_{i,c}dt, \tag{1}$$

where γ is the secondary electron emission coefficient, C is the constant of proportionality for the recombination, and n_e is the electron number density. $I_{i,c}$ is the number of the ions incident on the cathodes per unit time and it is given by

$$I_{i,c} = n_{i,c} A v_i, \tag{2}$$

where v_i is the ion velocity falling on the cathode, A is the area of the cathode, and $n_{i,c}$ is the density of the incident ions. Considering the ion collisions, then we have $v_i = \mu_i E$ where μ_i is the ion mobility and E is the electric field between the anode and cathodes. Assuming $n_{i,c} = Dn_i$ (D is the constant of proportionality) and $n_i = n_e$, therefore Eq. 1 becomes

$$dn_i = \frac{D\mu_i EA}{A'd'}[\gamma M(V,B) - 1]n_i dt - Cn_i^2 dt. \tag{3}$$

For the initial condition $n_i(t = 0) = n_0$, the solution of the equation

$$\frac{dn_i}{dt} = \frac{D\mu_i EA}{A'd'}[\gamma M(V,B) - 1]n_i - Cn_i^2, \tag{4}$$

gives a state of equilibrium after a long period with the ion density of

$$n_\infty = \left(\frac{D\mu_i EA}{CA'd'}\right)[\gamma M(V,B) - 1]\,. \tag{5}$$

The PIG can be ignited when

$$\gamma M(V,B) > 1 \tag{6}$$

This criterion has, previously, been applied to the breakdown between a simple geometry of two parallel-plate electrodes.[27]

C. Numerical simulation

In the discharge chamber, plasma has been formed due to ionizing electron impacts to the neutral atoms of operating gas. Secondary electrons emitted from the cathode surface due to ion bombardment are accelerated in the electric field between anode and cathodes where they obtain enough energy between successive collisions to ionize the atoms. The total number of ions which is produced by a secondary electron is defined as the electron multiplication coefficient, *M*, and it was calculated by the numerical simulations using the XOOPIC code.

That two-dimensional code is an object oriented Particle-in-cell (PIC) code, developed by Plasma Theory and Simulation Group at the University of California, Berkeley. The code tracks three types of particles: primary electrons, secondary electrons, and singly charged particles (Ar^+). It, also, considers the neutrals as motionless particles and the charged particles that are tracked under the forces of the electromagnetic fields and electron-neutral collisions.

Bearing in mind the fact that the magnetic field generated by the motion of charged particles within the discharge chamber is negligible, a constant and axial external magnetic field was applied to the chamber. In contrast, the electric field is a dynamic field due to the charged particles and the boundaries (electrodes) that could be calculated by the solution of electrostatic approximation of Maxwell equations. XOOPIC code considers elastic, excitation, and ionization collisions between electrons and neutral atoms as well as elastic and charge exchange collisions between ions and neutrals.

To determine the coefficient M a number of electrons, N_e, were initially seeded on the surface of the cathode in front of the anode apertures and the rest of the discharge chamber was considered free of charged particles. The neutral atoms of the background gas (Ar) were distributed uniformly with the density equivalent to the pressure of 4×10^{-2} mbar. Based on the electron cyclotron frequency and the electron collision frequency, time step was chosen to be 10^{-11} s. Over the time steps, the electrons were tracked in the presence of electric and magnetic fields inside the discharge chamber. As the boundary condition, an electron was removed from the simulation when it reaches to the anode. The ionizing electron-neutral collisions lead to production of positive ions before the electrons are absorbed to the anode or their energies fall below the ionization energy of the neutral atoms. The total number of positive ions produced due to the electron collisions, N_i, were obtained. Then according to the definition, the value of electron multiplication coefficient, $M(V, B) = N_i/N_e$, were achieved for various magnitudes of axial magnetic field and applied voltage.

III. RESULTS AND DISCUSSION

A. Experimental results

Table I gives the measured values of the ignition voltage for various axial magnetic fields. The uncertainty of the measured value of the voltages was about 10 V. The applied voltage was increased by a rate of 1 V/s while the current of the circuit was monitored. It was realised that at the ignitions the current suddenly increased from almost zero to 10^{-6} A. The results indicate that a stronger magnetic field leads to a lower ignition voltage because of the electron magnetic confinement. It decreases the diffusion losses of the electrons and therefore increases the number of ionizing collisions.

B. Numerical simulation results

The simulation results showed that the electron multiplication coefficient M depends extremely on the position of seeded primary electrons relative to the center of the cathode, r. Hence, before working out M, the ion distribution on the cathode surface as a function of r was obtained. The results for $V = 700$ V and $B = 500$ G are plotted in Fig. 2. This figure indicates that the lensing effect of the round apertures of the anode determines the behaviour of the ion distribution on the cathode surface.

TABLE I. Measured values of ignition voltage of the PIG for various magnitudes of axial magnetic field.

B (G)	460	480	500	520	540	560	580
V (V)	890	800	700	640	530	480	450

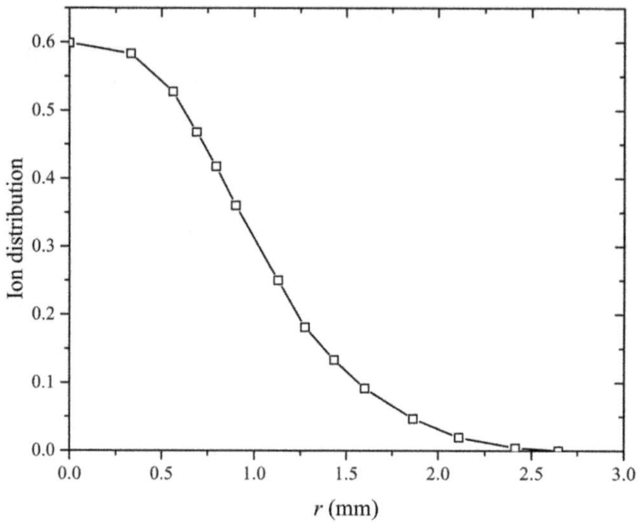

FIG. 2. Ion distribution on the cathode surface in terms of the distance from the axis for $V = 700$ V and $B = 500$ G using the XOOPIC code.

Thus, the primary electrons, as the ion-induced secondary emission electrons, are spread based on the ion distribution. The simulations resulted that this distribution is not significantly influenced by the change of the applied voltage from 100 to 900 V and the strength of magnetic field up to 800 G. The average values of the coefficient M were, subsequently, determined as a function of the axial magnetic field (B) for $V = 700$ V that are illustrated in Fig. 3. As it can be seen, the coefficient M is an increasing function of the magnetic field. This behaviour is due to the rise of the magnetic field confinement of the electrons more efficiently around the axis. This leads to the reduction of the electron losses and increasing the ionization frequency. It should be noted that M is a strictly increasing function of B in the range of $B = 0$ to 800 G.

Fig. 4 shows the dependency of M on the applied voltage (V) for various strength of magnetic fields from 460 G to 580 G. It can clearly be seen that increasing the voltage increases the multiplication coefficient to a certain voltage value then starts to decrease, slowly. On the left side of this value, the ionization of the neutral atoms increases as V is increased. But, further increase of the voltage

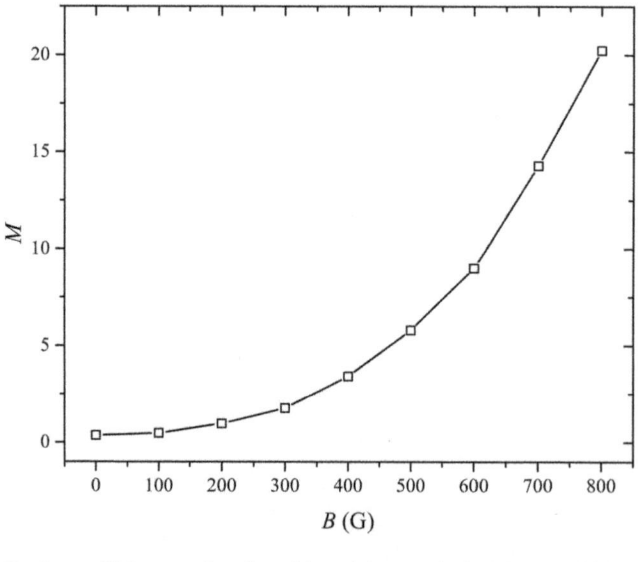

FIG. 3. Electron multiplication coefficient as a function of the axial magnetic field for $V = 700$ V using the XOOPIC code.

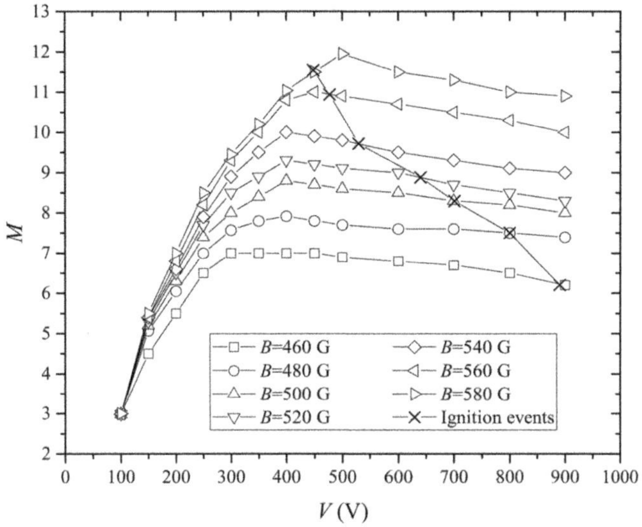

FIG. 4. Electron multiplication coefficient as a function of applied voltage using the XOOPIC code, open symbols; the experimental ignition voltages for each magnetic field, cross symbols.

rises the electron attractions to the anode apertures. This in turn counteracts the enhancing ionization. Morever, the ionization cross section of argon gas for the electron energy more than 50 eV, is a decreasing quantity.[28]

According to the criterion given in Eq. 6, the values of M should be greater than the inverse values of γ which depends on the energy of the incident ions on the cathodes. As shown in Fig. 4, the measured values of V and B for each ignition occurrence are marked (cross symbols) and the amounts of the coefficient M were accordingly obtained. Correspondingly, the values of the coefficient γ as a function of the ion energy is reported by Phelps et al. for dirty metals.[18] Eventually, the average energy of the incident ions on the cathode plate for various V were worked out using the numerical simulations. The results, M and γ^{-1}, are plotted in Fig. 5. It is clear that the criterion given in Eq. 6 is fulfilled at the PIG ignitions so that the average value of $\gamma M(V, B)$ equals to 1.05 ± 0.03. Inserting this value in Eq. 5 gives an approximate n_∞ of the order of 10^{10} cm^{-3} which is a reasonable result.[8]

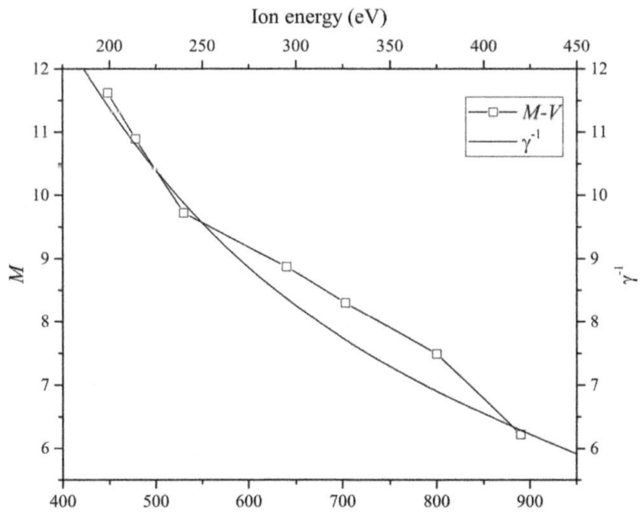

FIG. 5. Electron Multiplication coefficient as a function of ignition voltage, open symbol; the inverse values of the secondary coefficient γ reported in Ref.[18] versus the applied voltage and the equivalent energy of incident ions on the cathode, solid line.

In conclusion, the electron multiplication coefficient M is one of the main factors to control the ignition behaviour of PIGs. Hence, the effect of parameters such as magnetic field, pressure, gas type, and geometry parameters can be readily studied employing the suggested coefficient.

IV. CONCLUSION

The electron multiplication coefficient of a Penning ion generator (PIG) has, numerically, been obtained considering the effect of applied voltage and axial magnetic field. Based on a theoretical approach, it was shown that the ignition criterion of the PIG can be presented as $\gamma M(V, B) > 1$. The obtained values of M at the PIG ignitions for various strength of magnetic field, gives the average value of $\gamma M(V, B)$ to be 1.05 ± 0.03 which satisfies the criterion. Inserting this value in Eq. 5 gives an approximate n_{∞} of the order of 10^{10} cm^{-3} which is a reasonable result. Thus, the secondary electrons yield due to the ion bombardment of the cathode have dominant contribution to self-sustaining of the discharge in a PIG. It is in agreement with the experimental results reported in the literature. Employing the defined multiplication coefficient M leads to the reasonable results for the PIG ignition, so that it can be used to study the effects of parameters such as the gas pressure and the geometry parameters. Currently, a study is being worked out by the authors to explore a comprehensive picture of the coefficient M and its dependencies.

[1] F. M. Penning, Physica **4**, 71 (1937).
[2] A. T. Finkelstein, Rev. Sci. Instrum. **11**, 94 (1940).
[3] C. E. Anderson and K. W. Ehlers, Rev. Sci. Instrum. **27**, 809 (1956).
[4] G. Frech, W. Ensinger, H. Hueller, M. Hans, and G. K. Wolf, Nucl. Inst. & Meth. B **68**, 402 (1992).
[5] R. Keller, Rev. Sci. Instrum. **81**, 02B311 (2010).
[6] J. L. Rovey, Plasma Sources Sci. Technol. **17**, 035009 (2008).
[7] W. Liu, M. Li, K. Gao, and D. Gu, Nucl. Instr. Meth. Phys. Res. A **768**, 120 (2014).
[8] A. S. Dikalyuk and S. E. Kuratov, J. Phys.:Conf. series **815**, 012001 (2017).
[9] N. V. Mamedov, N. N. Shchitov, and I. A. Kanshin, Instrum. Exp. Tech. **59**, 870 (2016).
[10] R. P. Babertsyan, E. S. Badalyan, G. A. Egiazaryan, and E. I. Ter-Gevorgyan, Tech. Phys. The Russ. J. Appl. Phys. **43**, 1035 (1998).
[11] L. A. Zyul'kova and A. V. Kozyrev, Tech. Phys. Lett. **30**, 672 (2004).
[12] B. K. Das and A. Shyam, Rev. Sci. Instrum. **79**, 123305 (2008).
[13] B. K. Das, A. Shyam, R. Das, and A. D. P. Rao, Nucl. Instr. Meth. Phys. Res. A **669**, 19 (2012).
[14] R. P. Babertsyan, E. S. Badalyan, G. A. Egiazaryan, and E. I. Ter-Gevorgyan, Zh. Tekh. Fiz. **66**, 77 (1996).
[15] M. M. Shafiei, H. Noori, and A. H. Ranjbar, Rev. Sci. Instrum. **82**, 80 (2011).
[16] S. P. Nikulin, Tech. Phys. **43**, 795 (1998).
[17] E. Nasser, *Fundamental of gaseous ionization and plasma electronics* (John Wiley & Sons, Inc., 1971).
[18] A. V. Phelps and Z. Lj Petrović, Plasma Sources Sci. Technol. **8**, R21 (1999).
[19] L. Friedland, J. Phys. D: Appl. Phys. **7**, 2246 (1974).
[20] N. Sato and H. Tagashira, J. Phys. D: Appl. Phys. **18**, 2451 (1985).
[21] C. K. Birdsall, IEEE Trans. Plasma Sci. **19**, 65 (1991).
[22] J. Teunissen1 and U. Ebert, Plasma Sources Sci. Technol. **25**, 044005 (2016).
[23] M. Jiang, Y. Li, H. Wang, and C. Liu, Phys. Plasmas **24**, 102112 (2017).
[24] M. Radmilovic-Radjenovic, Z. Lj. Petrovic, G. N. Malovic, and D. Maric, Czech. J. Phys. **56**, B996 (2006).
[25] A. Derzsi, I. Korolov, E. Schünge, Z. Donkó, and J. Schulze, Plasma Sources Sci. Technol. **24**, 034002 (2015).
[26] D. Levko and L. L. Raja, J. Appl. Phys. **117**, 173303 (2015).
[27] H. Noori and A. H. Ranjbar, J. Appl. Phys. **112**, 1 (2012).
[28] M. R. Talukder, S. Bose, M. A. R. Patoary, A. K. F. Haque, M. A. Uddin, A. K. Basak, and M. Kando, Eur. Phys. J. D **46**, 281 (2008).

4

First-principle study on the electronic structure and optical property of new diluted magnetic semiconductor $(Y_{0.75}Sr_{0.25})$ $(Cu_{0.75}Mn_{0.25})SO$

Li Zhang,[1,a] Haoze Chen,[1] Shan Feng,[1] Linxian Li,[1] Yuke Li[2] and Jianye Chen[1]
[1]*Department of physics, China Jiliang University, Hangzhou 310018, China*
[2]*Department of physics, Hangzhou Normal University, Hangzhou 310018, China*

In this research, we investigated the electronic structure and optical property of new diluted magnetic semiconductor $(Y_{0.75}Sr_{0.25})$ $(Cu_{0.75}Mn_{0.25})SO$ with decoupled charge and spin doping by first-principle calculation. We also compared $(Y_{0.75}Sr_{0.25})$ $(Cu_{0.75}Mn_{0.25})SO$ with pure YCuSO, and found that $(Y_{0.75}Sr_{0.25})$ $(Cu_{0.75}Mn_{0.25})SO$ is still a direct semiconductor with a band gap 1.20eV. The important difference from YCuSO is that the DOS of $(Y_{0.75}Sr_{0.25})$ $(Cu_{0.75}Mn_{0.25})SO$ presents asymmetry around fermi surface, leading to obvious spin order and ferror-magnetism due to the p-d hybridization through Mn 3d and S 3p. In case of the optical properties (including reflectivity coefficient, absorption coefficient, the imaginary part and real part of complex dielectric constant), the intensity of all peaks for both pure and doped YCuSO tends to zero above 15.5eV. However, the value of all the peaks for $(Y_{0.75}Sr_{0.25})$ $(Cu_{0.75}Mn_{0.25})SO$ reduces to some extent due to the Sr and Mn dopant. In addition, the position of all the peaks for $(Y_{0.75}Sr_{0.25})$ $(Cu_{0.75}Mn_{0.25})SO$ have a "red shift", probably originating from the impurity states generated by the p-d hybridization through Mn 3d and S 3p. This will be beneficial for searching new 1111 phase DMS.

I. INTRODUCTION

Diluted magnetic semiconductor(DMS) mainly refers to new semiconductor formed by the magnetic ions in group 3d transition metal or group 4f rare earth metal partially substituting for non-magnetic ions in group III-V, II-VI and IV-VI compounds. It has triggered extensive research due to its charge and spin characters and potential to be applied in spintronics for storage, transmission and processing of information.[1–3] However, the typical systems, such as (Ga,Mn)As, which are based on III-V semiconductors face serious challenges. For example, substitution of divalent Mn^{2+} into trivalent Ga^{3+} sites not only has severely limited chemical solubility, but also leads to the difficulty of individually control of the spin and charge degrees. Another limitation is that (Ga,Mn)As specimen is only available as thin film resulting in considerable difficulty for discovering the mechanism of ferromagnetism.[4,5] Recently, new types of DMS Li(Zn,Mn)As with the Curie temperature up to 50 K[6] and $(Ba,K)(Zn,Mn)_2As_2$ with Tc up to 230 K,[7,8] based on structure of iron-based superconductors, has been reported with decoupled charge and spin, where its spin is doped via isovalent Mn^{2+} in place of Zn^{2+} while its charge is provided by off-stoichiometry of the Li concentration or K^+ instead of Ba^{2+}. These new DMSs have some good qualities. For example, they are bulk materials in favor of performing muon spin relaxation (μSR),[6,7,9] nuclear magnetic resonance (NMR)[10] and even neutron scattering.[11] Most importantly, they have diverse characters with similar structures:(Ba,K)

[a]lzhang@cjlu.edu.cn

$(Zn,Mn)_2As_2$ is a new DMS, (Ba,K) Fe_2As_2 is an iron-based superconductor, $BaMn_2As_2$ is an anti-ferromagnetic semiconductor and $BaZn_2As_2$ is a non-magnetic semiconductor, which takes advantage of designing and synthesizing multifunctional junctions.

The bulk form DMSs with decoupled charge and spin doping have been developed into four families according to the molecule formula and crystal structures: 111,[6] 122,[7] 1111[9] and 32522.[12] 1111-type DMS was initially reported by Ding et al in PRB 2013, which inspires discovery of many 1111-type DMS including LaCuSO,[13] and of course YCuSO. So far there are more than twenty samples synthesized, such as Li(Zn,Mn)P, (La,Ca)(Zn,Mn)SbO, (La,Ba)(Zn,Mn)AsO, (La,Sr)(Cu,Mn)SO, (Sr,Na)$(Cd,Mn)_2As_2$, and (A,Na)$(Zn,Mn)_2As_2$ (A= Ca, Sr)[14–17] and latest (Ba,K)F(Zn,Mn)As[18] and $Ba(Zn_{1-2x}Mn_xCu_x)_2As_2$.[19] Besides synthesizing samples, the research includes the investigation of properties by experimental measurements and theoretical calculations.[20–24] However, up to now there are only few articles about the first-principle study in the 122 phase of new DMS, and rarely on optical properties investigation.[25] In case of 1111 phase, LaCuSO is a wide band gap (3.1eV) p-type conductive oxysulfide semiconductor,[26] and it can become metallic with Sr doping. Sharp optical absorption and emission peaks near the band gap were observed in LaCuSO polycrystalline thin films at room temperature.[27,28] Thus LaCuSO is a promising candidate for optoelectronic application such as light-emitting devices. In $La_{1-x}Sr_xCu_{0.925}Mn_{0.075}SO$(x=0.075), it exhibits a ferromagnetic order with Tc around 200K,[16] which is among the highest Tc record of known 1111 phase bulk DMS materials up to now.

In order to lay a material foundation for further device application based on the new DMS materials, we need to investigate the optical properties of DMS besides electronic and magnetic properties and combine light, electricity and magnetism together. As a 1111-phase II-II-V based DMS, (Y,Sr)(Cu,Mn)SO, similar to LaCuSO, is an ideal prototype for theoretical study to elucidate the intrinsic physics of DMSs. In this paper, we investigate the electronic structure and optical properties by first-principle study on $(Y_{0.75}Sr_{0.25})$ $(Cu_{0.75}Mn_{0.25})$SO, and discuss the influence of the Mn and Sr co-doping with decoupled spin and charge doping compared the results with the parent YCuSO.

II. CALCULATION METHOD

Based on the density functional theory (DFT), the Perdew-Burke-Ernzerhof generalized gradient approximation (PBE-GGA)[29–32] was used in our calculations, implemented in the CASTEP module of Material Studio Suite.[33] The plane wave cut-off energy of 340 eV, energy convergence of 10^{-5} eV and force convergence criterion of 10^{-3} eV/Å were used for pure YCuSO phases in present calculations. The Monkhorst-Pack grid was used for the unit cell YCuSO with 4×4×2 k-points.

YCuSO phase adopts a tetragonal layered structure of the ZrCuSiAs-type with a space group P4/nmm, No.129, Z=2. a=b=3.99250 Å, c=8.55250 Å, $\alpha=\beta=\gamma=90°$, belong to 4mm point group.[34] This structure can be schematically described as a stacking of quasi two-dimensional $[Y_2O_2]$ and $[Cu_2S_2]$ blocks in the sequence -$[Y_2O_2]/[Cu_2S_2]/[Y_2O_2]/[Cu_2S_2]$. The atomic positions are Y:2c(1/4,1/4, 0.14219);Cu:2b(3/4,1/4,1/2);S:2c(1/4,1/4, 0.3117) and O:2a(3/4,1/4,0). The optimized lattice constants a=b=3.97118 Å and c=8.49978 Å are consistent with the experimental value, respectively. Electronic configuration of outer layer is Y:$4d^15s^2$, Cu:$3d^{10}4s^1$, S:$3s^23p^4$ and O:$2s^22p^4$.

$(Y_{0.75}Sr_{0.25})$ $(Cu_{0.75}Mn_{0.25})$SO, $(Y_{1-x}Sr_x)(Cu_{1-y}Mn_y)$SO systems were built via (Y^{3+}, Sr^{2+}) substitution and (Cu^{2+}, Mn^{2+}) substitution. Based on the similar method for YCuSO, the plane waves cut-off energy of 340 eV, energy convergence of 2.0×10^{-5} eV and force convergence criterion of 0.002 eV/Å were used in the calculation for $(Y_{0.75}Sr_{0.25})$ $(Cu_{0.75}Mn_{0.25})$SO. Since holes introduced by Sr dopants lead to metallic behavior in $(Y_{0.75}Sr_{0.25})$ $(Cu_{0.75}Mn_{0.25})$SO, the electronic structures and optical properties of supercells (Y,Sr)(Cu,Mn)SO are all achieved at the PBE-GGA level.

III. RESULTS AND DISCUSSION

A. Study of YCuSO

We first focus on the electronic structure of its parent compound YCuSO according to the structure of fig. 1(a). As Cu is a diamagnetic metal, we calculate the electronic structure without considering its spin. The DOS and band structure were depicted in fig. 1(b) and (c). In this paper, all the DOS and

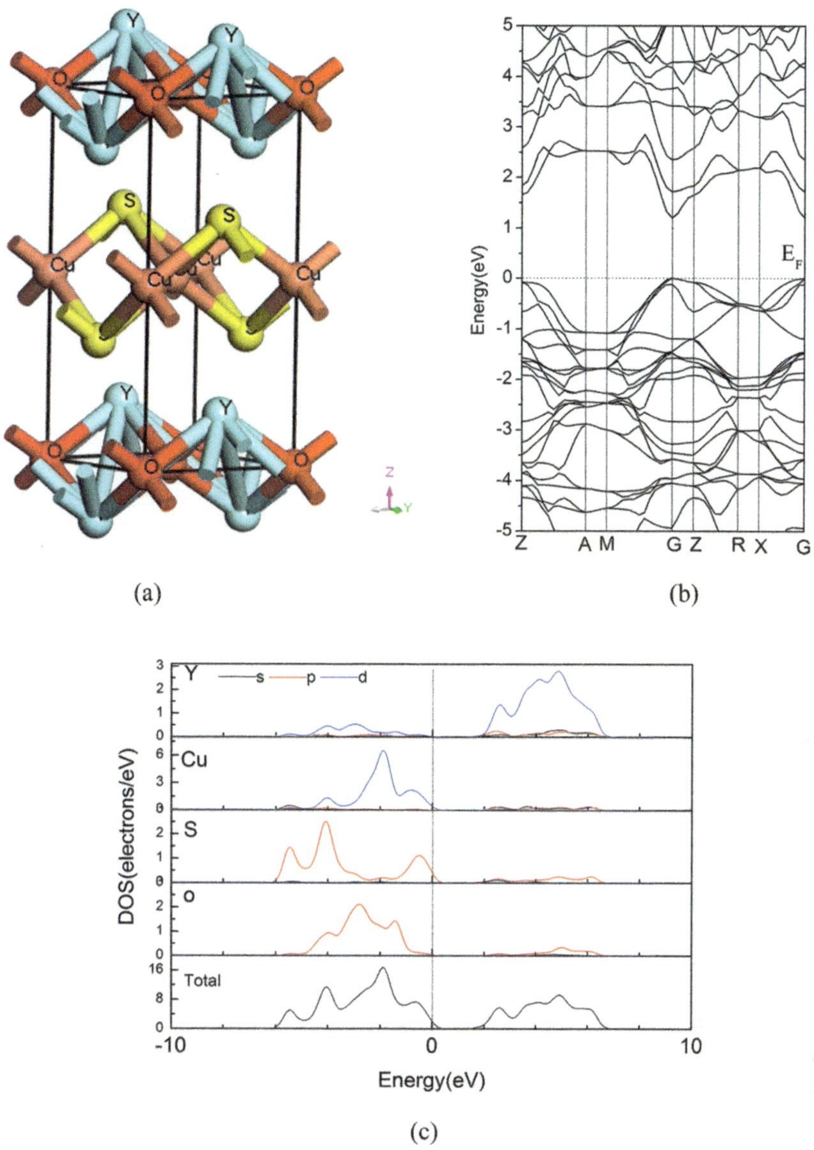

FIG. 1. (a)Crystal structure (b) Band structure and (c) DOS of $(Y_{0.75}Sr_{0.25})$ $(Cu_{0.75}Mn_{0.25})SO$.

band figures are plotted with the Fermi level being zero. Obviously, YCuSO is a direct semiconductor with the band gap 1.25eV at high symmetry Γ point in Brillouin zone, which is a little smaller than the band gap of LaCuSO, which is 1.7eV. Its optical band gap is just 55% of 3.1eV via reflectivity measurement,[28] and it is a little larger than the band gap of the latest reported fluoride parent DMS-BaFZnAs with a band gap 0.86eV from transport measurement.[18] The states range from -6.0 eV to 0 eV are mainly contributed by S3p, O2p and Cu3d orbital, which consist of valence bands. The top of valence bands contains primarily S3p state and Cu 3d, forming covalent Cu-S bonds around fermi surface. The Cu–S bonds in the CuS layers dominantly determine the electronic structure near the band gap. However, the conduction band is mainly composed of Y4d states with a gap 2eV above Ef. There are scarcely any Y states near the Fermi energy.

B. Superlattice structures of $Y_{0.75}Sr_{0.25}Cu_{0.75}Mn_{0.25}SO$

In order to calculate the electronic structure and study the optical property of $(Y_{0.75}Sr_{0.25})$ $(Cu_{0.75}Mn_{0.25})SO$, we constructed a 2×1×1 superlattice based on the structure of parent YCuSO.

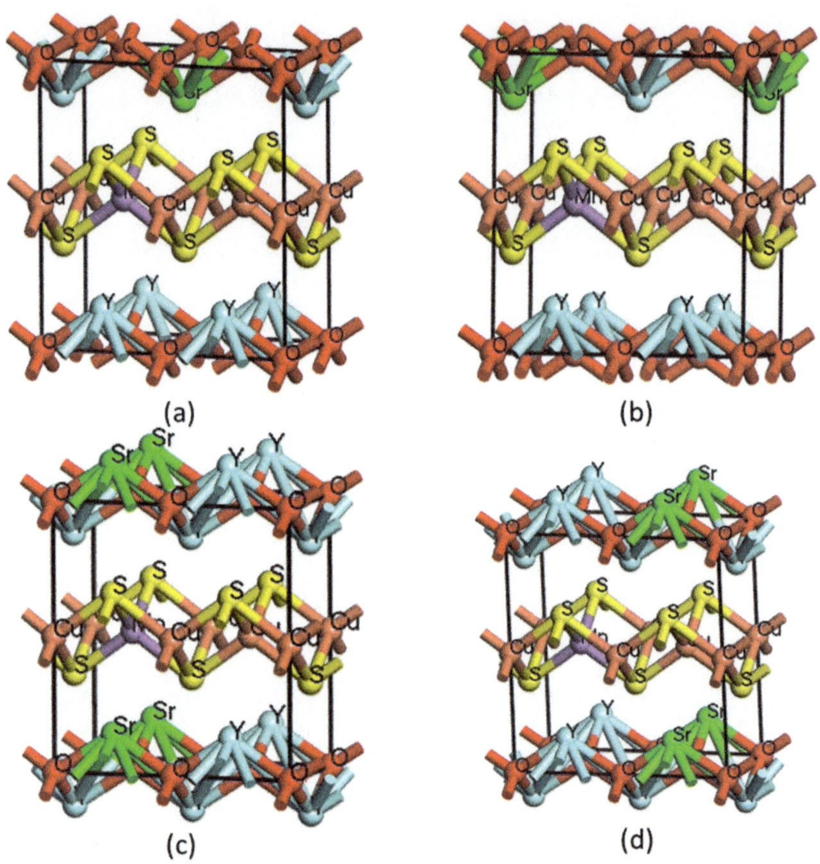

FIG. 2. Superlattice structures of $(Y_{0.75}Sr_{0.25})$ $(Cu_{0.75}Mn_{0.25})SO$. Different positions of Sr doped atom when Mn-doped atom is fixed in (a)Position 1, (b)Position 2, (c)Position 3 and (d)Position 4.

There are 16 atoms in the super lattice, including 4 atoms of Y, Cu, S and O respectively, without doping. When substituting a Mn atom for a S atom and simultaneously replacing a Y atom by a Sr atom, we get the $(Y_{0.75}Sr_{0.25})$ $(Cu_{0.75}Mn_{0.25})SO$ structure with 0.25 dopant of Mn and Sr. For simplicity, if we fix the Mn atom (denoted by the velvet ball), the Sr dopant atom (denoted by the green ball) probably has 4 positions, as shown in fig. 2. We calculate the energy of the whole super lattice under the 4 situations respectively and find out the lowest energy (-9374.7898eV) relates with the Sr atom in the position 1, which will be applied on following calculation.

C. Electronic structure of $Y_{0.75}Sr_{0.25}Cu_{0.75}Mn_{0.25}SO$

The band structure of $(Y_{0.75}Sr_{0.25})$ $(Cu_{0.75}Mn_{0.25})SO$ with spin up and down for comparison was plotted in fig. 3. Whether the spin is up or down, the bottom of conduct band and upper of valence band locate at high-symmetry point G in Brillouin zone, which indicates $(Y_{0.75}Sr_{0.25})$ $(Cu_{0.75}Mn_{0.25})SO$ is still a direct semiconductor. Taking account of spin, the number of valence bands below the fermi surface Ef increases in the spin up situation while increases a lot above the fermi surface in the spin down situation. The band gap is about 1.20eV in the spin up state while it is just 1eV in the spin down state. Compared with pure YCuSO, the band gap apparently reduces a lot after doping, particularly in the latter state. Compared with each other, in general the energy level around the Fermi surface moves down in the spin down state, leading to a smaller band gap.

The total density of states (DOS) and partial density of states (PDOS) for $(Y_{0.75}Sr_{0.25})$ $(Cu_{0.75}Mn_{0.25})SO$ are plotted in fig. 3(c). Overall, after Mn and Sr atoms are doped into the lattice, the top of valence bands contain primarily O2p, S3p state, Cu 3d and Mn 3d, forming Mn-S bonds besides Cu-S bands. The conduction bands are mainly composed of Y4d and Mn 3d band with

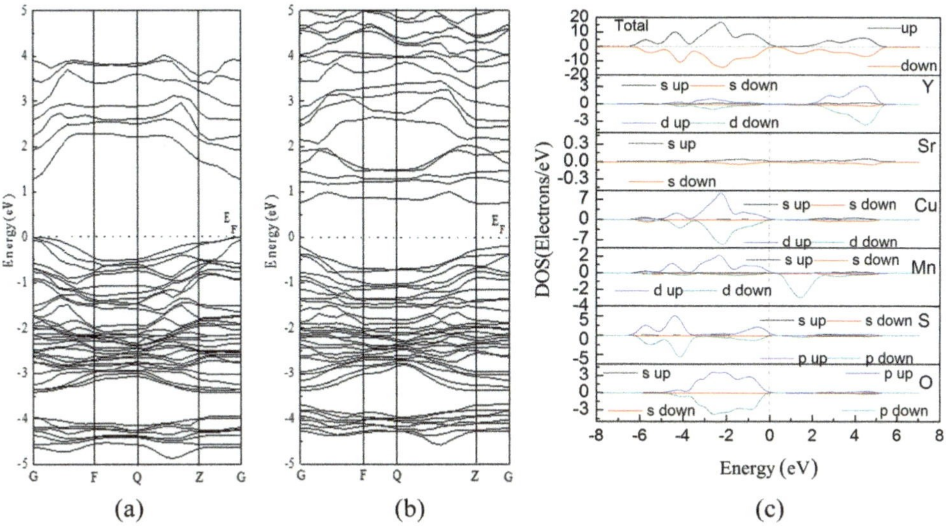

FIG. 3. (a)Band structures with spin up, (b) down and (c) Total density of states and partial density of states of superlattice($Y_{0.75}Sr_{0.25}$) ($Cu_{0.75}Mn_{0.25}$)SO.

a little gap of about 0. 5eV. Mn dopant plays an important role in the DOS around the fermi level. Below Ef from -6eV to 0eV, the asymmetry of PDOS of Mn irons in the up state dominate compared with that in the down state. At about -2eV, it reaches the maximum. Above Ef, it dominates in the down state. The largest PDOS locates at 1.8eV. The gap exists in the PDOS of Mn ion above Ef. The PDOS of S ion also contributes a few to the asymmetry of TDOS between spin up and spin down. Thus, it is the hybridization between Mn 3dand S 3p electrons that modifies the Fermi level, makes asymmetry for the TDOS and results in magnetic order in the compound.

Combining the DOS and band structures, ($Y_{0.75}Sr_{0.25}$) ($Cu_{0.75}Mn_{0.25}$)SO behaves as a DMS. Although there are few articles performing first-principle study in new DMS, we still find a few typical results for comparison. In 111 phase of Li($Zn_{0.875}Mn_{0.125}$)As, $Li_{1.1}$($Zn_{0.875}Mn_{0.125}$) As and $Li_{0.9}$($Zn_{0.875}Mn_{0.125}$)As,[25] Mn injectors make the sp-d hybridization among Mn3d, As4p and other orbital, leading to a spin polarized impurity band. Both excess and deficient of Li increase the half metallic character of the system, and the band gap changes from 1.643 eV to 1.516 eV and 1.810 eV corresponding to Li($Zn_{0.875}Mn_{0.125}$)As, $Li_{1.1}$($Zn_{0.875}Mn_{0.125}$)As and $Li_{0.9}$($Zn_{0.875}Mn_{0.125}$)As, respectively. Excess of Li could improve the Curie temperature and conductivity of the material. Therefore the separation of spin and charge injection mechanisms may be achieved in LiZnAs semiconductor. In 122 phase of (Ba,K)(Zn,Mn)$_2$As$_2$,[23] compound $BaZn_2As_2$ is a paramagnetic semiconductor with a calculated band gap 0.28 eV. Ba($Zn_{1-y}Mn_y$)$_2$ As$_2$ system shows anti-ferromagnetic behavior caused by the short-range super exchange interactions between localized spins of Mn 2p ions. Holes introduced by K dopants lead to metallic behavior in ($Ba_{1-x}K_x$)($Zn_{1-y}Mn_y$)$_2$As$_2$ system and drive the system toward ferromagnetism. The ferromagnetic coupling in the Mn-Mn pair is attributed to a long-range p-d exchange interaction mediated by itinerant holes, which is shown in J. K. Glasbrenner, I. Žutić, and I. I. Mazin.[22] It compared with the experimental results from $La_{1-x}Sr_xCu_{0.925}Mn_{0.075}SO$ (x=0.075,Tc=200K),[16] in which the Tc increases with the Sr doping level (x), proportional to the carrier density, indicating that the induction of hole-type charge carriers is crucial to the enhancement of FM order. We suggest the mechanism in ($Y_{0.75}Sr_{0.25}$) ($Cu_{0.75}Mn_{0.25}$)SO is the same as in ($Ba_{1-x}K_x$)($Zn_{1-y}Mn_y$)$_2$As$_2$. Just notice that holes introduced by the moderate Sr dopants lead to metallic behavior and with the moderate Mn dopants together drive the system toward ferromagnetism, and which is consistent with the carrier-induced origin of the ferromagnetism.

D. Optical properties of $Y_{0.75}Sr_{0.25}Cu_{0.75}Mn_{0.25}SO$

First-principle calculation in studying diluted magnetic superconductors plays a more and more important role of investigating optical property, such as complex dielectric constant, absorption coefficient and reflectivity coefficient. These optical parameters are connected with one another and can be deduced directly from the radiative-transition-rate theory and Kramers-Kroning dispersion relation as following equations.

$$\boldsymbol{\varepsilon}(\omega) = \boldsymbol{\varepsilon}_1(\omega) + \mathbf{i}\boldsymbol{\varepsilon}_2(\omega) \tag{1}$$

$$\boldsymbol{\varepsilon}_1 = 1 + \frac{2e}{\varepsilon_0 m^2} \sum\nolimits_{V.C} \int_{BZ} \frac{2dK}{(2\pi)^3} \cdot \frac{|\vec{a} \cdot M_{VC}(K)|^2}{[E_C(K) - E_V(K)]^2/\hbar^2 - \omega^2} \tag{2}$$

$$\boldsymbol{\varepsilon}_2 = \frac{\pi e^2}{\varepsilon_0 m^2 \omega^2} \sum\nolimits_{V.C} \frac{2dK}{(2\pi)^3} |\vec{a} \cdot M_{VC}(K)|^2 \delta [E_C(K) - E_V(K) - \hbar\omega] \tag{3}$$

$$\mathbf{I} = \frac{2\omega K}{c} \tag{4}$$

$$\mathbf{R} = \frac{(n-1)^2 + K^2}{(n+1)^2 + K^2} \tag{5}$$

In formula (1), $\boldsymbol{\varepsilon}$ is the complex dielectric constant, formula (2) and (3) exhibit the detailed information for $\boldsymbol{\varepsilon}_1$-the real part of complex dielectric constant and $\boldsymbol{\varepsilon}_2$-the imaginary part of it. (4) **I** represents the absorption coefficient and (5) **R** is on behalf of the reflectivity coefficient.

As can be seen from Fig. 4(a), the peaks of loss function for both pure and doped YCuSO are very distinct and sharp. The largest difference between them is the "red shift", which is the peak for pure YCuSO locates at 15eV while the peak of $(Y_{0.75}Sr_{0.25})$ $(Cu_{0.75}Mn_{0.25})$SO at 10.5eV, about 4eV difference. And the peak value for doped material reduces to about 1/2 of that for pure material. As for reflectivity the situation is similar, there is also 4eV difference between the position of main peaks, expressing the "red shift". But the two peaks are no longer as sharp and distinct in low-energy side. There are 3 peaks in the curve for YCuSO, 15eV, 5ev and 2.5eV respectively, in contrast to two peaks for doped YCuSO, 10.5ev and 3eV. Note that the value of the peaks cuts down a little for doped YCuSO (about 94% of YCuSO). Above 20eV (equal to 62nm wavelength), the value of reflectivity tends to zero.

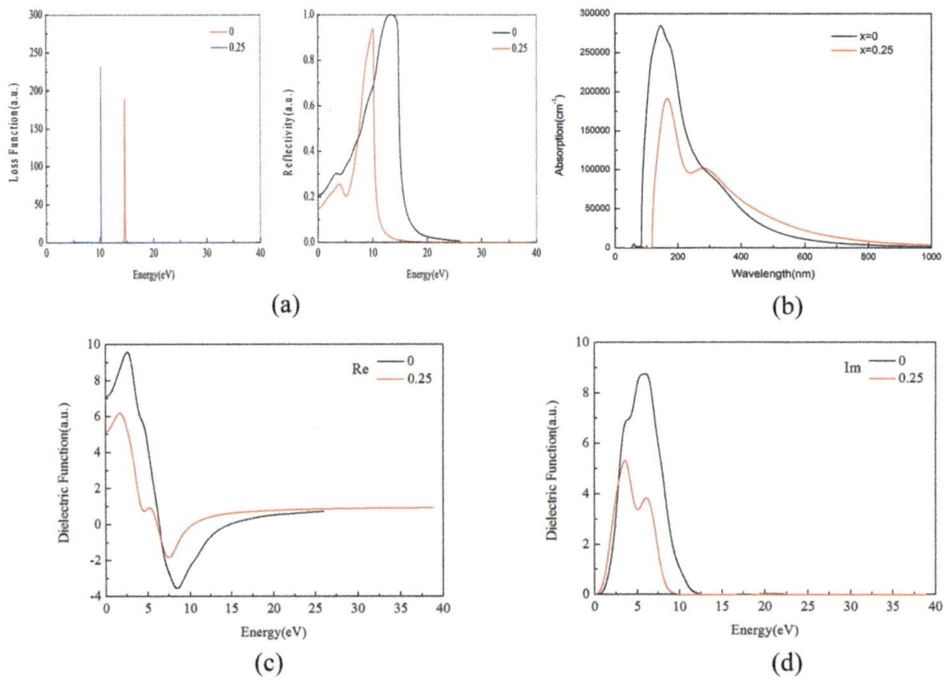

FIG. 4. Comparison of (a)energy loss and reflectivity (b)absorption (c)the real part of dieletric function and (d)the imaginary part of dieletric function of $(Y_{0.75}Sr_{0.25})$ $(Cu_{0.75}Mn_{0.25})$SO with YCuSO.

Fig. 4(b) is the comparison of the absorption of $(Y_{0.75}Sr_{0.25})$ $(Cu_{0.75}Mn_{0.25})SO$ with YCuSO. There also exits the "red shift" to some extent between the two compounds. In parent YCuSO, the largest peak situated at 160nm wavelength (7.76eV energy) while for the doped material, it is at 180nm wavelength (6.9eV energy). The intensity is about 2/3 ($1.75 \times 105cm^{-1}$) compared with the pure one ($2.75 \times 105cm^{-1}$). Clearly there is another peak at about 300nm wavelength (4.14eV energy) for doped one. In ultra-red range, the absorption has a tendency to zero. Apparently, the absorption is almost zero below the 80nm wavelength (above 15.5eV). K. Ueda, S. Inoue and H. Hosono[27] reported that there are two peaks in the optical absorption spectra of LaCuSO thin film, probably locating at about 3.2eV and 3.5eV, respectively. And the intensity of the main peak is as large as $1.5 \times 10^5 cm^{-1}$ at 6K decreasing with increasing temperature in contrast with the calculated absorption spectrum locating at 4.5eV and just 1/3 compared with the experimental one. It may indicate that doping wakes up another peak in pure YCuSO.

We described the difference of the imaginary part and real part of the complex dielectric index between pure and doped YCuSO in fig. 4(c) and (d), respectively. The dielectric index fluctuates in the lower energy area. The static dielectric index corresponding to 0eV reduces dramatically from 6.3 for pure YCuSO to 3.85 for doped YCuSO. Above 20eV the real part tends to zero and above 12.5eV the imaginary part equals to zero. In fig. 4(c), there are two peaks in the real part for pure YCuSO, one is at 3.5eV, the other is at 5eV. For doped YCuSO, a very little "red shift" is also observed. The value of the main peaks is about 4.8 for doped and 8.8 for pure YCuSO.

In fig. 4(d), the difference between the pure and doped YCuSO lies in the region of 0-12.5eV, i.e. above 99nm wavelength. We think the largest peak for pure YCuSO at 6.2eV is the superposition produced by two supposed peaks greater and less than 6.2eV. For doped YCuSO, the two peaks with almost the same height appear at 4eV and 6.2eV with distinct "red shift" compared with the position of the two supposed peaks. And the value of the peak for pure YCuSO is almost 2.5 times as large comparing to the doped YCuSO. In $Li^{1+}(Zn,Mn)As$ system, the peak of the imaginary part situates at 4.21eV and the position fluctuates a little with the changing of chemistry stoichiometry.[25] We suggest the peak around4eV is probably consistent with Mn ions.

Taking all the figs. 4a, 4b, 4c, and 4d into account, it concludes that the investigated optical properties tend to zero above 15.5eV, which explain the YCuSO system shows no change in extreme far-ultraviolet radiation. Most importantly, the relative intensity of all the peaks decreases and the position of all the peaks shows "red shift" when doped. The peak intensity decreases when doping attributes to the scattering from the Sr and Mn dopant. The red shift originates from the impurity band generated by the holes to compensate the charge imbalance due to Sr doping, which agrees with the decreasing band gap when doped.

IV. CONCLUSION

We have performed a study of the electronic structure and optical property on Sr and Mn separately doped YCuSO by density functional theory within the generalized gradient approximation. Parent YCuSO is a direct semiconductor with a gap of 1.25 eV. The topmost valence bands and the lowest conducting band are dominated by the S 2p-orbitals and the Cu3dorbitals. After Mn and Sr ions are separately doped into the lattice, we constructed the 2×1×1 superlattice for $Y_{0.75}Sr_{0.25}Cu_{0.75}Mn_{0.25}SO$ system. It is still the direct semiconductor with the band gap 1.20 eV in the spin up state while 1eV in the spin down state. The top of valence bands contains primarily O 2p, S3p state, Cu 3d and Mn 3d, forming Mn-S bond besides Cu-S bond. The conduction bands are mainly composed of Y 4d and O 2p orbitals. The magnetic order comes from the hybridization between Mn 3d and S 3p electron around fermi surface.

When it comes to the optical properties, above 15.5eV region, all the optical properties tend to zero. It suggests that in the extreme far-ultraviolet radiation, the YCuSO system shows no change. When doped the intensity of all the peaks decreases and the position of all the peaks presents "red shift". The intensity decreases owing to the scattering from the Sr and Mn dopant. The "red shift" probably originates from the impurity states generated by the holes produced by Sr doping, which is in agreement with the decreased band gap when doped. Future work is needed to gain deeper understanding of the magnetic and optical behaviors and to achieve higher Tc value of 1111 phase.

ACKNOWLEDGMENTS

This work was financially supported by the National Science Foundation of China (No. 61376094) and China Scholarship Council (No. 201408330028).

[1] H. Ohno, A. Shen, F. Matsukura, A. Oiwa, A. Endo, and Y. K. SandIye, Appl. Phys. Lett. **69**, 363 (1996).
[2] T. Dietl and H. Ohno, Rev. Mod. Phys. **86**, 187 (2014).
[3] T. Jungwirth, J. Sinova, J. Mašek, J. Kučera, and A. H. MacDonald, Rev. Mod. Phys. **78**, 809 (2006).
[4] T. Dietl, Nat. Mater. **9**, 965 (2010).
[5] M. Dobrowolska, K. Tivakornsasithorn, X. Liu, J. K. Furdyna, M. Berciu, K. M. Yu, and W. Waluliewicz, Nat. Mater. **11**, 444 (2012).
[6] Z. Deng, C. Q. Jin, Q. Q. Liu *et al.*, Nat. Commun. **2**, 422 (2011).
[7] K. Zhao, Z. Deng, X. C. Wang *et al.*, Nat. Comm. **4**, 1442 (2013).
[8] K. Zhao, B. Chen, G. Zhao *et al.*, Chinese Science Bulletin **59**, 2524 (2014).
[9] C. Ding, H. Man, C. Qin *et al.*, PRB **88**, 041102 (2013).
[10] C. Ding, C. Qin, H. Man *et al.*, PRB **88**, 041108 (2013).
[11] C. Ding, S. Guo, Y. Zhao *et al.*, JPCM **28**, 026003 (2016).
[12] H. Man, C. Qin, C. Ding *et al.*, EPL **105**, 67004 (2014).
[13] X. Yang, Y. Li, C. Shen *et al.*, APL **103**, 022410 (2013).
[14] C. Ding, X. Gong, H. Y. Man *et al.*, EPL **107**, 17004 (2014).
[15] Z. Deng, K. Zhao, B. Gu *et al.*, PRB **88**, 4365 (2013).
[16] X. Yang, Y. Li, P. Zhang *et al.*, Journal of Applied Physics **114**, 223905 (2013).
[17] S. Guo, Y. Zhao, X. Gong *et al.*, EPL **114**, 57008 (2016).
[18] H. Man, S. Guo, Y. Sui *et al.*, Scientific Repo. Rts. **5**(1), 5507 (2015).
[19] B. Chen, Z. Deng, W. Li *et al.*, Scientific Repo. Rts. **6**, 36578 (2016).
[20] H. Suzuki, Y. J. Uemura, C. Q. Jin, and A. Fujimori, PRB **92**, 235120 (2015).
[21] F. Sun, N. N. Li, B. J. Chen *et al.*, PRB **93**, 224403 (2016).
[22] J. K. Glasbrenner, I. Žutić, and I. I. Mazin, PRB **90**, 140403 (2014).
[23] J. T. Yang, S. J. Luo, Y. C. Xiong *et al.*, Solid State Sciences **46**, 102 (2015).
[24] J. Yang, S. Luo, Z. Cheng *et al.*, Materials Research Express **3**, 105903 (2016).
[25] W. Ai-Ling, W. Zhi-Min, W. Cong *et al.*, Acta Phys. Sin. **62**, 137101 (2015) (in chinese).
[26] V. V. Bannikov, I. R. Shein, and A. L. Ivanovskii, Solid State Sciences **14**(1), 89 (2012).
[27] K. Ueda, S. Inoue, and H. Hosono, APL **78**(16), 16 (2001).
[28] Y. Goto, M. Tanaki, Y. Okusa *et al.*, APL **105**, 2701 (2014).
[29] J. P. Perdew, S. Burke, and M. Ernzerhof, PRL **77**, 3865 (1996).
[30] K. G. Fürthmüller, J Comput. Mater. Sci. **6**, 15 (1996).
[31] P. E. Blöchl, PRB **50**, 17953 (1994).
[32] G. Kresse and D. Joubert, PRB **59**, 1758 (1999).
[33] M. D. Segall, P. J. D. Lindan, M. J. Probert *et al.*, J. Phys. Condens. Matter **14**, 2717 (2002).
[34] S. Feng, "First-principles study on the electronic structures and optical properties of $Y_{1-x}Sr_xCu_{1-y}Mn_ySO$ system and some experimental results," Master Dissertation (China Jiliang University, 2017).

5

Megahertz high voltage pulse generator suitable for capacitive load

Yu Xu,[1,2,3] Wei Chen,[1,3] Hao Liang,[1,3] Yu-Huai Li,[1,2] Fu-Tian Liang,[1,2,3] Qi Shen,[1,2] Sheng-Kai Liao,[1,2,a] and Cheng-Zhi Peng[1,2]

[1]*Department of Modern Physics and Hefei National Laboratory for Physical Sciences at Microscale, University of Science and Technology of China, Hefei 230026, China*

[2]*Chinese Academy of Sciences (CAS) Center for Excellence and Synergetic Innovation Center in Quantum Information and Quantum Physics, University of Science and Technology of China, Shanghai 201315, China*

[3]*State Key Laboratory of Particle Detection and Electronics, University of Science and Technology of China, Hefei 230026, China*

A high voltage pulse generator is presented to drive Pockels cell. The Pockels cell behaves like a capacitor which slows the rise/fall time of the pulse and restrains the repetition rate of the generator. To drive the Pockels cell applied in quantum communication system, it requires about 1 MHz repetition rate with the rise/fall time of the pulse less than 50 ns, adjustable amplitude up to 800 *V* and an adjustable duration. With the assistance of self-designed transformers, the circuits is simplified that a pair of high current radio frequency (RF) MOSFET drivers are employed to switch the power MOSFETs at a high speed, and the power MOSFETs shape the final output pulse with the requirements. From the tests, the generator can produce 800 *V* square pulses continously at 1 MHz rate with 46 *ns* in risetime and 31 *ns* in falltime when driving a 51 *pF* capacitive load. And the generator is now used to drive Pockels cell for encoding the polarization of photons.

I. INTRODUCTION

Linear electro-optic effect, known as the Pockels effect, is widely used for modulating amplitude, polarization or phase of photons. Owing to the ultra-fast response time, Pockels cells play critical roles in quantum information experiments, such as selecting measurement bases in quantum key distribution and quantum entanglement distribution,[1–3] arranging propagating path of pulses,[4,5] or performing Q-swishing in laser cavities.[6] For such applications, the rise/fall time and the repetition rate become key parameters for Pockels cells. Generally, higher repetition rate is strongly desired for quantum information experiments, as higher repetition rate corresponds to higher trial rate.

A Pockels cell is usually constituted of one or two electro-optic crystals, and modulated by an external electric field. For common electro-optic crystals such as KDP, KTP or BBO, the external electric field needs to be at the order of 100 kV/m to accumulate π-phase retardation per 100 millimeters. Achieving the high voltage together with the demands of a fast rise/fall time and high repetition rate is the main goal and also the difficulty when building drivers for Pockels cells for quantum information experiments.

Further more, the Pockels cell behaves like a capacitor, so the high voltage pulse generator specifically designed for a Pockels cell actually realizes fast charging and discharging of a capacitive load, which slows the rise/fall time of the pulse, releases extra heat and finally turns out to be one of the bottlenecks when raising the repetition rate.

---

[a]skliao@ustc.edu.cn

As an instrument that helps encode the polarization of the photon, regularly obtainable multi-kilohertz repetition rate Pockels cell drivers could not meet the requirements in the high trial rate quantum information experiments anymore. A higher repetition rate voltage square pulse generator with variable duration and amplitude is eagerly desired. And several key parameters need to be satisfied for this generator: pulse repetition rate over 1 MHz, voltage adjustable from 0 *V* to over 800 *V*, rise/fall time less than 50 *ns*, and driven capacitive load over 50 *pF*.

There are three main ways to realize a high voltage pulse generator, from the study. The first way is to use the vacuum tubes such as the krytron, thyratron, or planar triode to generate high voltage pulses.[7] Nanosecond rise/fall time could be achieved with such method. However, the devices are bulky and the circuit is complicated. The second way is based on avalanche transistor stack topology and avalanche Marx stack topology.[8] Although this method can generate high voltage pulses with rise time and fall time in the scale of nanoseconds or sub-nanoseconds, the repetition rate of up to 10 KHz is still relatively low.[9,10]

With the development of the semiconductor technologies, semiconductor devices, such as power MOSFET and the insulated gate bipolar transistor (IGBT) have very high potential for applications in generating high repetition rate, and high voltage pulses.[11] They have significant advantages over traditional vacuum tubes and avalanche transistors in performance, cost, and drive complexity. The push-pull configuration based on power MOSFET have both nanosecond rise and fall times,[12,13] this generator is adapted for capacitive load and has a powerful driving capability for the Pockels cell, but the repetiton rate is still low or not mentioned in these works.

In this paper, we designed a high voltage pulse generator that provides high voltage pulses with a variable amplitude from 0 to 800 *V* that works at a repetition rate of 1 MHz when driving a capacitive load equivalent to 51 pF. The widths of the high voltage pulses are adjustable from 300 *ns* to 700 *ns* with 46 *ns* rise time and 31 *ns* fall time. The generator is specifically designed for capacitive load, and it also has wide application prospects where high voltage pulse generators is demanded.

II. THE CIRCUITS

As illustrated in Fig. 1, the high voltage pulse generator contains two main parts: a gate driver circuit triggered by external input, and a pulse generator circuit with a high voltage power supply. The high voltage pulse generator circuit is used to generate high voltage pulses with fall time

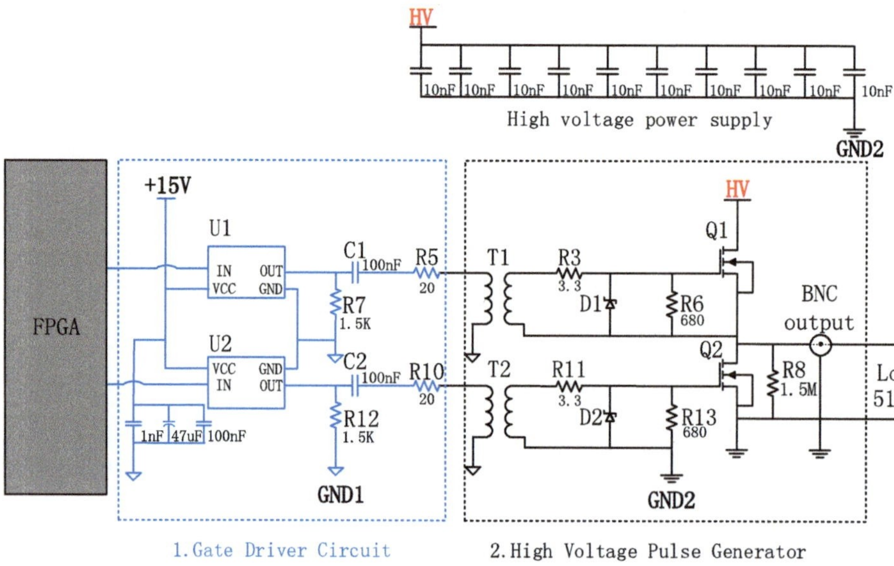

FIG. 1. Schematic diagram of high voltage generator circuit.

and rise time in ten-nanoseconds, which is achieved by quickly turning on and off the two MOSFETs Q1 and Q2 in sequence. There is parasitic or stray capacitances C_{gs} between gate (G) and source (S), C_{gd} between gate (G) and drain (D), and the input equivalent capacitance $C_{ISS} = C_{gd} + C_{gs}$. When the C_{ISS} is charged or discharged to gate threshold voltage V_{th}, the power MOSFETs turn on or off. Higher current is provided, more quickly the power MOSFETs switch. Thus, a gate driver circuit with high driving ability is employed to generate a pair of short trigger signals to achieve quick switching of the power MOSFETs Q1 and Q2.

In the following, two main parts of the design will be discussed in detail.

A. Gate driver circuit

The gate driver circuit mainly consists of two RF MOSFET drivers U1(2) and corresponding AC coupling circuits. The high current RF MOSFET drivers IXRFD630 is chosen because it can deliver 30 A of peak current which ensures the C_{ISS} can be charged to V_{th} in nanoseconds and turn on the power MOSFET Q1(2). The gate driver circuit requires 15 V DC power supply for IXRFD630. To supply instantaneous high current and filter out high frequency noise, three different package capacitors are located at the VCC pins of U1 and U2. Different packages of capacitors bring different parasitic inductances and lead to different impedance and ESR(equivalent series resistance) curves corresponding to frequencies.[14] When a combination of different ESR curves applied in the filtering net, better filtering performance may be achieved.

The output pin of U1 is directly connected to AC coupling capacitance C1 and resistance R5. Because the primary isolation transformer T1 can be equivalently regarded as an inductor L1, the H-trigger signals, which turns on the power MOSFET Q1, will pass through an RLC circuit. The RLC circuit is expected to operate in over-damped condition and the oscillation is highly suppressed, which requires $R5 \geq 2\sqrt{L1/C1}$. In the experiment, the value of the induction L1 and the coupling capacitor C1 is about 10 uH and 100 nF, and the over-damped condition requires $R5 \geq 20\Omega$.

B. Pulse generator circuit

The pulse generator circuit is the key part of the system, where the push-pull configuration is introduced to enhance the output driving ability and generate a high voltage pulse with ten-nanoseconds scale rise/fall time. The pulse generator circuit consists of three main parts: two pulse isolation transformers, a push-pull configuration circuit and the high voltage power supply.

The isolation transformer is a widely-used and effective tool in achieving current isolation. Because the high side power MOSFET Q1 of the push pull configuration is floating, a high repetition pulse isolation transformer is applied to achieve galvanic isolation between Q1 and the output of the gate drive circuit. However, the commercial isolated gate drivers (eg. ADUM1233, IX6R11) usually fail to meet the requirement in insulation voltage (800 V), while the transformer chips (eg. PA2531NL, PA2621INL) usually fail to meet the requirement in repetition rate (over 1 MHz).

In order to work through the challenge of both high insulation voltage and high repetition rate. We designed our own isolation transformers to meet the requirement of both high insulation voltage and high repetition rate. The NiZn (Nickel Zinc) family of ferrites is characterized by permeabilities $\mu \leq 2500$, high transmission line bandwidth, and high practical frequency. A ring-shaped NiZn ferrite was chosen as the core of the transformer. The primary and secondary windings whose break down voltage reaches 10 kV are braided in alternate directions around the ring-shaped core, and the turns ratio is 1:1.

The push-pull configuration circuit mainly contains two FAIRCHILED power MOSFETs FQA8N100C denoted by Q1 and Q2, whose breakdown voltage is 1000 V, with corresponded continuous drain current being 8 A and RDS(on) being 1.45 Ω. The input high voltage power supply is connected to the drain of Q1. The gate of Q1 is floating above ground and connected to the secondary of T1 through the damping resistor R3. The drain of Q2 is connected with the source of Q1, and the source of Q2 is connected to the ground. Resistor R4 and R3 with resistance 3.3Ω are used to dampen the gate drive ringing. The trigger timing diagram and the waveform of output high voltage pulse are shown in Fig. 2. When the H-trigger signal arrives at the gate terminal of the

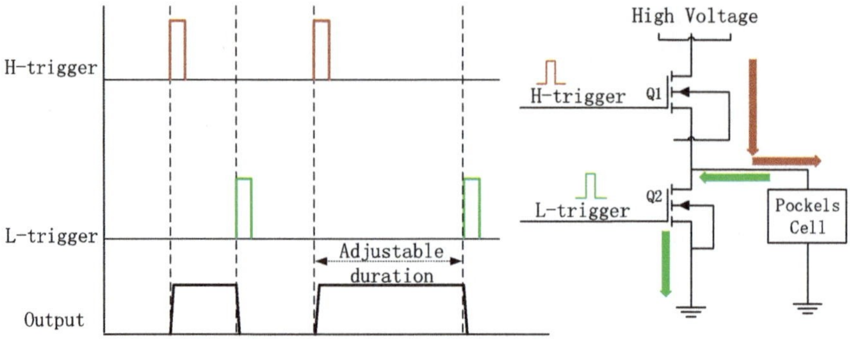

FIG. 2. The trigger timing diagram and the output high voltage pulse.

high side power MOSFET Q1, Q1 is turned on and charges the capacitive load to 800 *V* in 46 *ns*. Once the L-trigger signal arrives at the gate of the low side power MOSFET Q2, Q2 turns on, the drain of the Q2 is grounded and the capacitive Pockels cell is discharged to 0 *V* in 31 *ns* through Q2. The zener diode D1 is used to guarantee that the gate-source voltage of MOSFET does not exceed limitation. The resistor R6(680Ω) is to provide a low impedance channel to release the gate charge.

A high voltage power supply is used to provide high voltage up to 1 *kV* and can provide current up to 500 *mA* for the pulse generator circuit. In fact, a traditional high voltage power supply cannot provide sufficient instantaneous current. In this paper, a capacitor group consisting of ten high voltage ceramic capacitors with capacitance 10*nF* is utilized, which makes it possible to provide large enough instantaneous current for the pulse generator circuit. Because of the properties like fast rise-fall time, high repetition rate, and high instantaneous current, the following factors, such as ground splitting, high voltage trace width and components layouts should all be taken into careful consideration for electromagnetic interference (EMI) suppression and signal integrity. In the design, the PCB is partitioned into a drivers section and a pulse generator section, thus the ground is correspondingly split into two parts, which are connected together by the self-designed transformer, to prevent the EMI from influencing the drivers section.

III. TESTS

We test the high voltage pulse generator circuit loading with a 51 pF capacitor which is exactly the value of the equivalent capacitive load of the Pockels cell M350-50 from ConOptics. As illustrated in Fig. 3, the test setup is composed of 5 main instruments, including a megahertz high voltage pulse generator (a), an arbitrary generators AFG3252 (b) to provides one pair of TTL signals at repetition rate of 1 MHz in stead of the FPGA in the practical cases, a water circulation system (c) to cool the power MOSFETs and the RF MOSFET drivers, a Tektronix DPO4104 oscilloscope (d) to record the high voltage pulse waveform, and a power supply DH1722A-6 (e) to provide high voltage. The waveform of the output high voltage pulse is shown in Fig. 4. The 10% – 90% fall time is about 31 *ns* and the rise time is 46 *ns*, the amplitude of the high voltage pulses is 800 *V* and the repetition rate is 1 MHz.

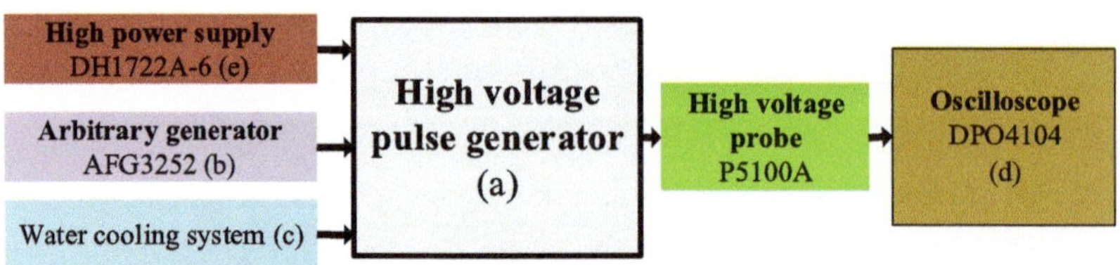

FIG. 3. The diagram of the test setup.

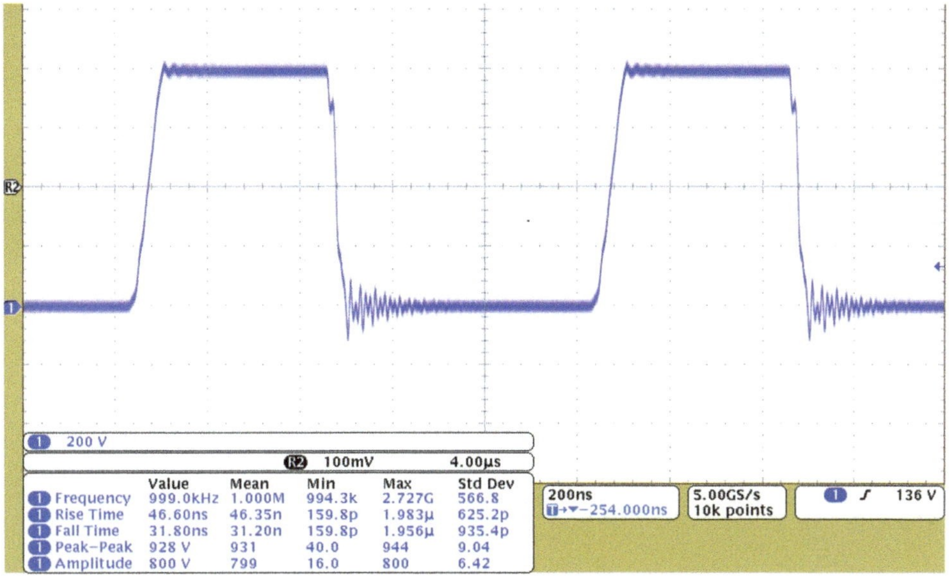

FIG. 4. Output voltage waveform of the push pull circuit.

IV. CONCLUSION

In this paper, the push pull configuration circuit is applied to the design of a high voltage pulse generator which is suitable for capacitive load. The high voltage pulse generator was designed and tested. All of the requirements, pulse repetition rate over 1 MHz, voltage adjustable from 0 *V* to 800 *V*, rise/fall time less than 50 *ns*, and driven capacitive load over 50 *pF*, are successfully met.

ACKNOWLEDGMENTS

This work has been supported by Chinese Academy of Sciences(CAS) Center for Excellence and Synergetic Innovation Center in Quantum Information and Quantum Physics, Shanghai Branch, University of Science and Technology of China, by the National Natural Science Foundation of China (under grant no.11405172), by CAS Key Technology Talent Program, by Shanghai Sailing Program.

[1] J. Yin, Y. Cao, Y.-H. Li, S.-K. Liao, L. Zhang, J.-G. Ren, W.-Q. Cai, W.-Y. Liu, B. Li, H. Dai *et al.*, "Satellite based entanglement distribution over 1200 kilometers," Science **356**, 1140–1144 (2017).

[2] M. Giustina, M. A. Versteegh, S. Wengerowsky, J. Handsteiner, A. Hochrainer, K. Phelan, F. Steinlechner, J. Kofler, J. Å. Larsson, C. Abellán *et al.*, "Significant-loophole-free test of bell's theorem with entangled photons," Physical Review Letters **115**, 250401 (2015).

[3] L. K. Shalm, E. Meyer-Scott, B. G. Christensen, P. Bierhorst, M. A. Wayne, M. J. Stevens, T. Gerrits, S. Glancy, D. R. Hamel, M. S. Allman *et al.*, "Strong loophole-free test of local realism," Physical Review Letters **115**, 250402 (2015).

[4] Y.-H. Li, Y. Cao, H. Dai, J. Lin, Z. Zhang, W. Chen, Y. Xu, J.-Y. Guan, S.-K. Liao, J. Yin *et al.*, "Experimental round-robin differential phase-shift quantum key distribution," Physical Review A **93**, 030302 (2016).

[5] H. Wang, Y. He, Y.-H. Li, Z.-E. Su, B. Li, H.-L. Huang, X. Ding, M.-C. Chen, C. Liu, J. Qin *et al.*, "High-efficiency multiphoton boson sampling," Nature Photonics **11**, 361–365 (2017).

[6] P. Kwiat, H. Weinfurter, T. Herzog, A. Zeilinger, and M. A. Kasevich, "Interaction-free measurement," Physical Review Letters **74**, 4763 (1995).

[7] W. M. Portnoy, G. J. Rohwein, and L. D. Roose, "Characterization of high power thyristors," *Pulsed Power Conference* 1, 336–341 (1995).

[8] Y. M. Daud and N. Bidin, "Nanoseconds switching for high voltage circuit using avalanche transistors," Applied Physics Research **1**, 25 (2009).

[9] R. Booth, E. S. Fulkerson, and D. C. Norman, "Driving pockels cells using avalanche transistor pulsers," *Pulsed Power Conference, 1997. Digest of Technical Papers* 2, 1341–1346 (1997).

[10] J. Li, Z. Zhao, Y. Sun, Y. Liu, Z. Ren, J. He, H. Cao, and M. Zheng, "A hybrid pulse combining topology utilizing the combination of modularized avalanche transistor marx circuits, direct pulse adding, and transmission line transformer," Review of Scientific Instruments **88**, 033507 (2017).

[11] G. Wang, J. Su, Z. Ding, X. Yuan, and Y. Pan, "A semiconductor opening switch based generator with pulse repetitive frequency of 4 mhz," Review of Scientific Instruments **84**, 125102 (2013).
[12] M. T. Bernius and A. Chutjian, "Improved high-voltage, high-frequency square-wave generator," Review of Scientific Instruments **61**, 925–927 (1990).
[13] Z. Tan, X. Feng, and X. Long, "Nanosecond square high voltage pulse generator for electro-optic switch," Review of Scientific Instruments **82**, 075102 (2011).
[14] J. Cain, "Comparison of multilayer ceramic and tantalum capacitors," AVX Technical Publication, AVX Corporation.

6

Charged particle reflection by a planar artificially structured boundary with electrostatic plugging

R. M. Hedlof and C. A. Ordonez[a]
Department of Physics, University of North Texas, Denton, Texas 76203, USA

A classical trajectory Monte Carlo simulation is used to investigate an artificially structured boundary for confinement and control of charged particles. The artificially structured boundary considered here incorporates a planar sequence of conducting wires, where adjacent wires carry current in opposite directions. Such a configuration creates a sequence of magnetic cusps and was studied previously [C. A. Ordonez, J. Appl. Phys. **106**, 024905 (2009)]. The effect of introducing a sequence of electrodes for electrostatic plugging of the cusps is investigated. The results of the simulations are used to identify regions of parameter space in which particle losses through the cusps may be negligible in the single particle limit. A trap based on a cylindrical generalization of the artificially structured boundary presented here may lead to a method for confining non-neutral and partially neutralized plasmas along the edge, such that the bulk of a confined plasma is effectively free of externally applied electromagnetic fields.

I. INTRODUCTION

An artificially structured boundary (ASB) creates a static electromagnetic field that is short in range compared to the size of the ASB, only affecting the trajectories of particles that are relatively close to the ASB. In the context of the present work, an ASB is a configuration of electrodes and permanent magnets or electromagnets that produces a spatially periodic arrangement of electrostatically plugged magnetic cusps. The dimensions of an ASB are chosen such that the spatial period of the applied electromagnetic field is much smaller than the dimensions of a charged-particle confinement volume that is effectively free of the applied electromagnetic field, except along the edge of the confinement volume.

There are many configurations that can be considered as ASBs. In a planar geometry, for example, a configuration could be created by a sequence of long wires, with adjacent wires carrying current in opposite directions.[1] Such a configuration is referred to as a 'picket-fence' configuration. Another possible configuration can be constructed by arranging a large number of magnetic dipoles (e.g., superconducting coils, cylindrical permanent magnets with axial or radial magnetization, or solenoids) in a grid pattern, with adjacent magnets having oppositely signed poles facing the particle confinement region. Trapping configurations that can be constructed from ASBs include electrostatically plugged spherical and cylindrical high-order magnetic multipole traps. Research involving charged particle reflection from various ASBs that produce static, spatially periodic, electric or magnetic fields has been reported previously.[1–4] An ASB with electrostatic plugging, but much more experimentally difficult to realize than the ASB considered here, was studied and proposed as a high purity ion trap/source in Ref. 5. In the context of fusion related applications, various magnetic cusp geometries with electrostatic plugging have been discussed in Ref. 6.

[a]Author to whom correspondence should be addressed: cao@unt.edu

The production of cold antihydrogen for high precision spectroscopy and gravity experiments relies heavily on an ability to confine and control positron and antiproton plasmas.[7–11] The ALPHA and ATRAP experiments simultaneously synthesize and confine antihydrogen in Penning traps that are located within neutral atom traps.[7,8] In Penning traps, the presence of a strong magnetic field in the region where antihydrogen is formed reduces the three-body recombination rate by roughly an order of magnitude as compared to the rate when there is no magnetic field present.[12,13] In addition, positron space charge and collisions between antiprotons during the mixing process may have a significant detrimental effect on the number of antihydrogen atoms with kinetic energy suitable for further experimentation.[14,15] The difficulties associated with creating antihydrogen suitable for gravity experiments in a nested Penning trap, along with the possibility of increasing the three-body recombination rate, are motivations for developing an alternative positron and antiproton confinement scheme.

The confinement of a positive ion or positron plasma within the space charge of an edge confined electron plasma was investigated from a theoretical perspective in Ref. 16. In Ref. 16 it was shown that if the properties of the two species can be tailored properly, there will be a region near the trap center in which the two species will overlap. A trap that employs an ASB to confine a positron plasma along its edge might be able to confine an antiproton plasma within the electrostatic well created by the space charge of the positrons. A region of overlap of the two species may result in mixing the positron and antiproton plasmas in a region of space where strong magnetic fields are not present.

In Ref. 16, the exact structure of the reflecting boundary used to confine an electron plasma along its edge was not considered. Here, the possibility of using an electrostatically plugged picket-fence configuration as a reflecting boundary is investigated via computer simulation. The present work is an extension of that which was reported in Ref. 1, where charged particle interaction with an ASB employing a picket-fence magnetic field without electrostatic plugging was investigated. It was found that charged particles could be effectively reflected by the picket-fence magnetic field, provided the particles were directed toward the ASB at grazing angles of incidence.[1]

A conceptual illustration of the ASB considered here is shown in Fig. 1. The magnetic field of the ASB is produced by a planar arrangement of current carrying wires, whereby adjacent wires

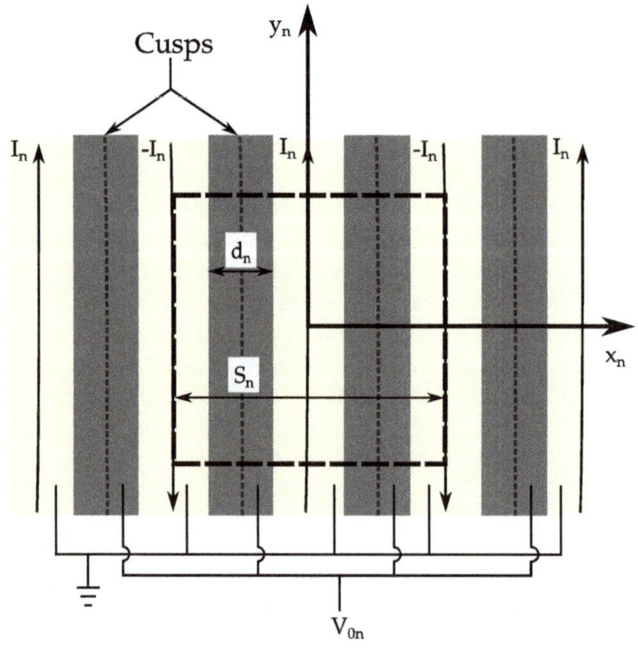

FIG. 1. A conceptual illustration of a small portion of the ASB. The dotted square represents the boundary of the simulation volume. The dark gray rectangles represent the plugging electrodes biased to a normalized potential V_{0n}, while the yellow rectangles represent a conducting background material that is held at zero potential. Current carrying wires are represented by the vertical arrows, where the direction of each of the arrows indicates the direction in which the normalized current I_n flows.

carry current in opposite directions. The wires are infinitely long in the y dimension, and wires with current running in the same direction are separated by a distance S. Such an arrangement creates a magnetostatic field that is periodic in the x dimension with a spatial period equal to S and that has a sequence of line cusps at each mid-plane between adjacent wires. As shown in Fig. 1, a series of plugging electrodes (dark gray strips) that are biased to a potential V_0 and have a width d are centered in each of the line cusps to enhance confinement. The plugging electrodes are considered to be flush with a planar conducting background material that is held at zero potential. The ASB is treated as infinite in extent in the x and y dimensions and as located at $z = 0$. Here, x, y, and z are Cartesian coordinates.

In Sec. II, a classical trajectory Monte Carlo simulation is presented that is used to track charged particle motion in the electromagnetic field of the ASB. In Sec. III A, the magnetic confinement properties of the ASB are discussed, and in Sec. III B, the effect of electrostatic plugging is investigated. A set of parameters that optimizes the reflection properties of the ASB is found. Section IV provides a discussion and concluding remarks.

II. CLASSICAL TRAJECTORY MONTE CARLO SIMULATION

A computer program has been developed that is used to calculate the electromagnetic field of the configuration and to conduct classical trajectory Monte Carlo (CTMC) simulations. A normalization procedure is carried out in such a way that all parameters used in the simulations are dimensionless quantities. Hereafter, normalized quantities are denoted by the same symbol as their unnormalized counterparts, except with a subscript n attached. Parameters used for normalization are the spatial period of the magnetic field S, the initial kinetic energy of a particle K_0, the particle's charge q, and the particle's mass m. Each of these parameters has a normalized value of unity, $S_n = K_{0n} = q_n = m_n = 1$. The normalized time, magnetic field, electric field, and electrostatic potential are given by $t_n = \sqrt{K_0/mS^2}\, t$, $\boldsymbol{B}_n = (qS/\sqrt{mK_0})\,\boldsymbol{B}$, $\boldsymbol{E}_n = (qS/K_0)\,\boldsymbol{E}$, $\phi_n = (q/K_0)\,\phi$, respectively. A particle's normalized position, velocity, and acceleration are $\boldsymbol{r}_n = \boldsymbol{r}/S$, $\boldsymbol{v}_n = \sqrt{m/K_0}\,\boldsymbol{v}$, and $\boldsymbol{a}_n = (mS/K_0)\,\boldsymbol{a}$.

Two charged particle sources are considered in this work: a normally incident monoenergetic beam and charged particles incident with a monoenergetic isotropic velocity distribution. The motion of all simulated particles is constrained to take place within a simulation volume defined by $-S_n/2 \le x_n \le S_n/2$, $-S_n/2 \le y_n \le S_n/2$, and $0 \le z_n \le 10S_n$. The initial Cartesian coordinates (x_{0n}, y_{0n}, z_{0n}) for each simulated particle are considered to be equally likely to be located at any point in the $z_n = z_{0n}$ plane within the simulation volume. The value of z_{0n} is chosen sufficiently large so that the effect of the electromagnetic field on a trajectory is initially negligible for all parameters considered in the simulations. The expressions used for the initial coordinates are

$$x_{0n} = \frac{S_n}{2} R_x,$$
$$y_{0n} = \frac{S_n}{2} R_y,$$
$$z_{0n} = 9.9, \qquad (1)$$

where R_x and R_y are random real numbers equally likely to have any value between -1 and 1.

Let $(v_{0n}, \theta_0, \phi_0)$ represent spherical coordinates in velocity space associated with each particle's initial velocity $\boldsymbol{v}_{0n}$. The particles have a normalized initial speed $v_{0n} = \sqrt{2}$, and the Cartesian initial velocity components are given by

$$v_{0xn} = \sqrt{2}\sin\theta_0\cos\phi_0,$$
$$v_{0yn} = \sqrt{2}\sin\theta_0\sin\phi_0,$$
$$v_{0zn} = \sqrt{2}\cos\theta_0. \qquad (2)$$

The values for θ_0 and ϕ_0 are sampled from a velocity space probability density function $f(\boldsymbol{v}_n)$ appropriate for describing the source of charged particles being considered. For simulations involving a beam of perpendicularly incident particles, $\theta_0 = \pi$ is used in Eqs. (2), giving $v_{0xn} = v_{0yn} = 0$ and $v_{0zn} = -\sqrt{2}$. Sampling expressions used for θ_0 and ϕ_0 in simulations involving monoenergetic particles that follow an isotropic velocity distribution are

$$\theta_0 = \arccos(-R_\theta),$$
$$\phi_0 = 2\pi R_\phi, \tag{3}$$

where R_θ and R_ϕ are random real numbers that are equally likely to have any value between 0 and 1. Equations (3) are arrived at by noting that the isotropic monoenergetic velocity distribution is separable in spherical coordinates and may be written as $f(v_{0n}, \theta_0, \phi_0) = f_0\delta(v_{0n} - \sqrt{2})f_\theta(\theta_0)f_\phi(\phi_0)$, where f_0 is a normalization constant, δ is the Dirac delta function, $f_\theta(\theta_0) = \sin\theta_0$, and $f_\phi(\phi_0) = 1$. The sampling expression for θ_0 is then obtained by solving

$$R_\theta = \frac{\displaystyle\int_{\theta_l}^{\theta_0} \sin\theta \, d\theta}{\displaystyle\int_{\theta_l}^{\theta_u} \sin\theta \, d\theta} \tag{4}$$

for θ_0. Here, $\theta_l = \pi/2$ and $\theta_u = \pi$ are the lower and upper bounds on θ_0, which are chosen such that all particles have a z-component of their initial velocity directed toward the ASB.

Single particle trajectories are simulated by numerically integrating the normalized equations of motion,

$$\dot{\boldsymbol{r}}_n(t_n) = \boldsymbol{v}_n(t_n),$$
$$\dot{\boldsymbol{v}}_n(t_n) = \boldsymbol{E}_n + \boldsymbol{v}_n(t_n) \times \boldsymbol{B}_n. \tag{5}$$

The simulation is written in C++ and uses a 4th order implicit L-stable (m,k)-method with adaptive timestep [Intel(R) ODE Library, 2008]. Periodic boundary conditions are used for particles that reach the $x_n = \pm S_n/2$ or $y_n = \pm S_n/2$ planes. If a particle reaches the $z_n = 0$ plane, the particle is considered to be transmitted through the ASB. If a particle reaches the $z_n = z_{0n}$ plane at a time $t_n > 0$, it is considered to be reflected. All trajectories are numerically integrated until each particle is reflected or transmitted.

Due to the short-range nature of the electromagnetic field, it is essential to implement a maximum timestep on the numerical integration. Otherwise, the adaptive algorithm may make the step size so large while particles are in regions where the electromagnetic field is negligible, that the particles interact with the field by a reduced amount, or, in the worst case scenario, some of the particles skip over the ASB altogether between timesteps. Additionally, to correctly track charged particles in the presence of a magnetic field, the timestep should be sufficiently small so as to resolve the cyclotron period of each particle. To satisfy these conditions, the maximum timestep in all simulations is specified as

$$\Delta t_{n,\max} = \min\left(\frac{\Delta s_{\max}}{v_n}, \frac{c_\tau}{\omega_{cn}}\right), \tag{6}$$

where min is a function returning the smallest value among its arguments, $\Delta s_{\max}$ is the maximum normalized distance the particle is allowed to move between timesteps if the electromagnetic field is negligible, v_n is the normalized speed of the particle at the time t_n, c_τ is a numerical factor, and $\omega_{cn} = q_n B_n / m_n = B_n$ is the normalized cyclotron angular frequency. To ensure that the cyclotron motion of each particle is resolved properly, a value of $c_\tau = 0.1$ is used for all simulations. The value of $\Delta s_{\max}$ is chosen based on the z coordinate of the particle:

$$\Delta s_{\max} = \begin{cases} 0.3, & z_n(t_n) > 2 \\ 0.1, & 1 < z_n(t_n) \le 2 \\ 0.05, & 0 \le z_n(t_n) \le 1. \end{cases} \tag{7}$$

For example, all particles would take at least 20 steps moving from $z_n = 1$ to $z_n = 0$.

The magnetic field of the ASB is modeled as being created by a series of infinitely long, infinitesimally thin, parallel wires that carry current in the $\pm y$ direction as shown in Fig. 1. The wires span the entire $z = 0$ plane, with the ith wire located at $x_i = i\,S/2$, where $i = 0, \pm 1, \pm 2, \ldots$, and S is the spatial period of the magnetostatic field. Such an array of wires produces an unnormalized magnetic field given by[1]

$$\boldsymbol{B}(x, y, z) = 2B_m\boldsymbol{\beta}\left(\frac{x}{S}, \frac{y}{S}, \frac{z}{S}\right), \tag{8}$$

where

$$\boldsymbol{\beta}(x_n, y_n, z_n) = -\frac{\cos(2\pi x_n)\sinh(2\pi z_n)}{\cos(4\pi x_n) - \cosh(4\pi z_n)}\hat{\boldsymbol{\imath}} + \frac{\sin(2\pi x_n)\cosh(2\pi z_n)}{\cos(4\pi x_n) - \cosh(4\pi z_n)}\hat{\boldsymbol{k}}, \tag{9}$$

$B_m = \mu_0 I/S$ is the magnitude of the magnetic field in the center of each cusp, μ_0 is the permeability of free space, I is the magnitude of the current in each wire, and $x_n = x/S$, $y_n = y/S$, and $z_n = z/S$ are the normalized Cartesian coordinates with associated unit vectors $\hat{\boldsymbol{\imath}}, \hat{\boldsymbol{\jmath}}$, and $\hat{\boldsymbol{k}}$. In terms of normalized parameters, the magnetic field is

$$\boldsymbol{B}_n(x_n, y_n, z_n) = \frac{2\sqrt{2}\,\mathrm{sgn}(q)}{r_{cn}}\boldsymbol{\beta}(x_n, y_n, z_n), \tag{10}$$

where sgn(q) is the sign of charge of the particles incident on the ASB, and $r_{cn} = \sqrt{2mK_0}/(|q|B_m S)$ is an artificial parameter introduced to reduce the parameter space. Physically, r_{cn} is the normalized cyclotron radius of a charged particle (with mass m, charge q, and kinetic energy K_0) moving perpendicular to a uniform magnetic field of magnitude B_m. By varying a single parameter r_{cn}, information about how the simulation results depend on the strength and spatial period of the magnetic field, as well as the mass, initial kinetic energy, and charge magnitude of the incident particles can be ascertained. Figure 2(a) shows a plot of the magnetic field in the $y_n = 0$ plane. Figure 2(b) shows the ratio of the magnitude of the normalized magnetic field in the $x_n = 0.25$ plane $B_n(0.25, y_n, z_n)$ to the magnitude of the normalized magnetic field in the cusp $B_{mn} = B_n(0.25, y_n, 0)$ as a function of z_n. The magnitude of the normalized magnetic field in the cusps at $x_n = 0.25$ drops by 99.6% as z_n is increased from $z_n = 0$ to $z_n = 1$. For example, if $r_{cn} = 10^{-2}$, $B_n(0.25, y_n, 1.0) = 0.53$ and $B_n(0.25, y_n, 0) = 140$.

The normalized electrostatic potential in each simulation is calculated within the first octant of the Cartesian coordinate system depicted in Fig. 1. To calculate the electrostatic potential, the Laplace equation with normalized coordinates,

$$\nabla^2\phi_n = 0, \tag{11}$$

is solved subject to Neumann boundary conditions in the x_n and y_n dimensions

$$\begin{aligned} \partial_x\phi_n\Big|_{x_n=0} &= \partial_x\phi_n\Big|_{x_n=S_n/2} = 0, \\ \partial_y\phi_n\Big|_{y_n=0} &= \partial_y\phi_n\Big|_{y_n=S_n/2} = 0, \end{aligned} \tag{12}$$

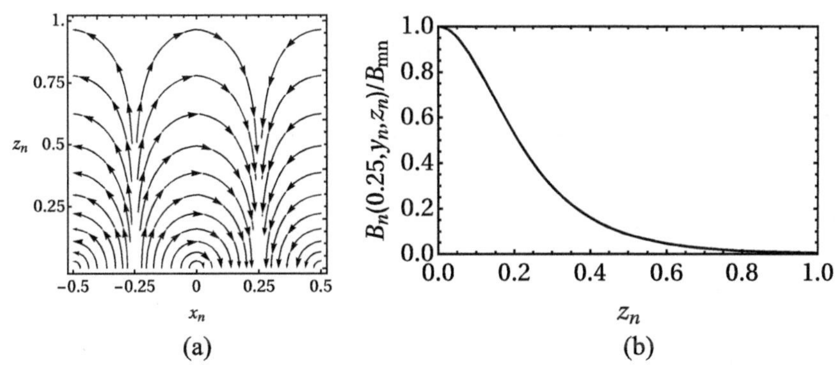

FIG. 2. (a) Magnetic field lines in the $y_n = 0$ plane and (b) the relative strength of the normalized magnetic field in the $x_n = 0.25$ plane.

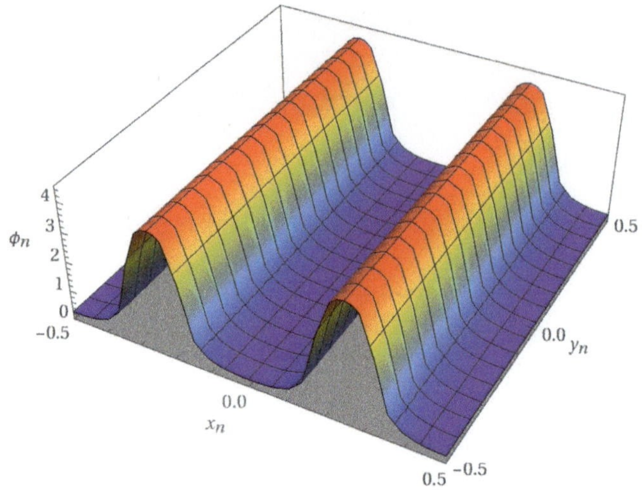

FIG. 3. Normalized electrostatic potential in the $z_n = 0.025$ plane.

and Dirichlet and Neumann boundary conditions on the $z_n = 0$ and $z_n = 10S_n$ planes, respectively,

$$\phi_n\Big|_{z_n=0} = V_n(x_n, y_n),$$
$$\partial_z\phi_n\Big|_{z_n=10S_n} = 0. \quad (13)$$

The Dirichlet boundary conditions on the $z_n = 0$ plane are given by

$$V_n(x_n, y_n) = \begin{cases} V_{0n}, & \frac{1}{2}\left(\frac{S_n}{2} - d_n\right) \leq x_n \leq \frac{1}{2}\left(\frac{S_n}{2} + d_n\right) \\ 0, & \text{otherwise.} \end{cases} \quad (14)$$

A three-dimensional Cartesian mesh is defined in the first octant of the simulation volume, and the Intel(R) Math Kernel Library [2017 Update 3 for Linux*] fast Poisson solver routines are used to numerically solve for the electrostatic potential on the grid. The number of grid points used in the x_n, y_n, and z_n dimensions is $n_x = 100$, $n_y = 100$, and $n_z = 2000$, respectively. After the numerical solution for ϕ_n is obtained on the grid in the first octant, the normalized electrostatic potential is obtained at non-grid points via a tri-linear interpolation algorithm, and the normalized electric field is calculated via $\mathbf{E}_n(x_n, y_n, z_n) = -\nabla\phi_n(x_n, y_n, z_n)$. The normalized electric field and electrostatic potential can then be calculated at any other point in the simulation volume by a linear coordinate transformation. Figure 3 shows $\phi_n(x_n, y_n, 0.025)$ obtained with $V_{0n} = 5.0$ and $d_n = 0.15$.

Let N_p denote the number of simulated particles, and P_R denote the percentage of those particles that were reflected by the ASB. For all simulations in which a charged particle beam is incident on the ASB, a value of $N_p = 100,000$ is used. For all simulations in which particles with an isotropic velocity distribution are incident on the ASB, a value of $N_p = 1,000,000$ is used. Given a set of input parameters, the simulation solves Laplace's equation with appropriate boundary conditions, integrates N_p particle trajectories, and outputs the total number of particles that have been reflected, transmitted, and any additional data that is specified (e.g., particle coordinates at each time step, magnetic/electric field values, etc.). The total energy of each particle in the simulations is conserved to within 0.5 %.

III. RESULTS

A. Magnetic confinement

The magnetic confinement properties of the ASB is evaluated by varying the parameter r_{cn} and setting $V_{0n} = 0$. The initial phase space quantities are sampled as described in Sec. II. Figure 4 shows

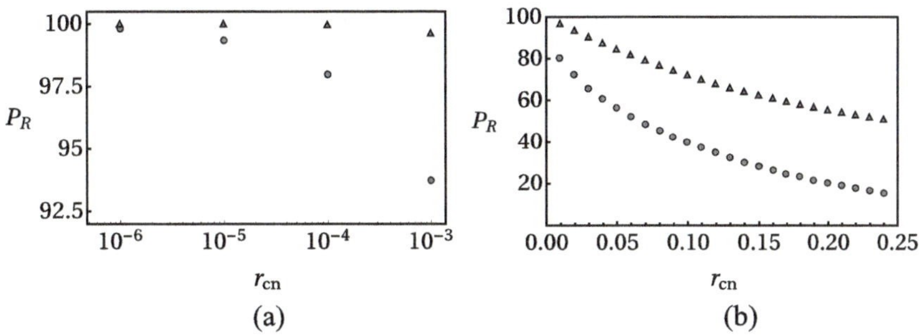

FIG. 4. Dependence of the percentage of reflected particles on r_{cn} for (a) $10^{-6} \leq r_{cn} \leq 10^{-3}$ and (b) $10^{-2} \leq r_{cn} \leq 0.25$ for a normally incident beam (circles) and an isotropic velocity distribution (triangles).

the percentage of incident particles that are reflected by the magnetic field of the ASB as a function of r_{cn} when V_{0n} is held at ground potential. There is no dependence on the value of d_n, because the electrodes are held at ground potential. The circles show $P_R(r_{cn})$ for a monoenergetic beam of positively charged particles that is normally incident on the ASB, while the triangles show $P_R(r_{cn})$ for a monoenergetic isotropic initial velocity distribution. The results indicate that when $r_{cn} = 10^{-6}$ over 99.8% of all incident particles are reflected for both a beam and an isotropic velocity distribution of incident particles. However, for all values of r_{cn} considered here, some particles are transmitted by the ASB when no electrostatic plugging is present. Figure 4 indicates that when a beam of charged particles is normally incident on the ASB without electrostatic plugging fewer particles are reflected as compared to an isotropic velocity distribution.

The area between adjacent wires where particles can travel within the magnetic field of the ASB at $z_n = 0$ is referred to as the transmission region, and the width of the transmission region is referred to as the transmission width W_{Tn}. Figure 5 shows the distribution of transmitted particles as they cross the $z_n = 0$ plane for (a) $r_{cn} = 10^{-4}$ and (b) $r_{cn} = 10^{-2}$ for a beam. Note that different plot ranges are used for each of the two plots. As r_{cn} increases, the area between adjacent wires where particles can travel within the magnetic field of the ASB increases. Figure 6 shows the normalized transmission width W_{Tn} as a function of r_{cn} for a beam. As r_{cn} decreases, the normalized transmission width decreases, and fewer particles are transmitted by the ASB.

Recalling that $r_{cn} = \sqrt{2mK_0}/(|q|B_mS)$: As the mass or initial kinetic energy of the particles increases, r_{cn} increases, and more particles are transmitted through the ASB. Alternatively, as the charge magnitude of the particles, or the strength or spatial period of the magnetic field are increased, r_{cn} decreases, and more particles are reflected by the ASB.

B. Effect of electrostatic plugging

A parameter scan was conducted with a beam of particles normally incident on the ASB to evaluate the effectiveness of electrostatic plugging. Parameter values for V_{0n} and r_{cn} used in the

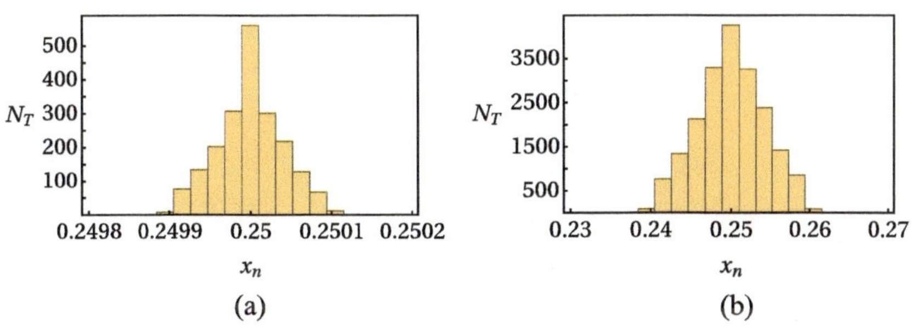

FIG. 5. Distribution of transmitted particles from a beam in the $z_n = 0$ plane for (a) $r_{cn} = 10^{-4}$ and (b) $r_{cn} = 10^{-2}$, with $V_{0n} = 0$. N_T is the number of transmitted particles per bin of width $\Delta_{xn} = 2.1 \times 10^{-5}$ for (a) and $\Delta_{xn} = 2.1 \times 10^{-3}$ for (b).

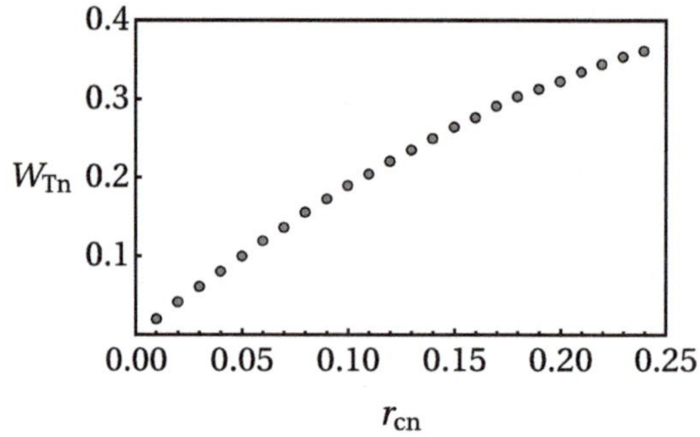

FIG. 6. Normalized transmission width for a beam of incident particles as a function of r_{cn} when $V_{0n} = 0$. The transmission width is defined as the maximum separation in the x dimension between any two transmitted particles within the same cusp at the $z_n = 0$ plane.

parameter scan are in the ranges $1.0 \leq V_{0n} \leq 15.0$ and $0.01 \leq r_{cn} \leq 0.25$ in increments of $\Delta V_{0n} = 1.0$ and $\Delta r_{cn} = 0.01$, respectively. Figure 7 shows $P_R(r_{cn}, V_{0n})$ for (a) $d_n = 0.15$ and (b) $d_n = 0.25$. The flat, red portions of the plots are regions in parameter space where 100% of all incident particles were reflected. For the simulations with $d_n = 0.15$, 100% reflection was found to occur when $2 \leq V_{0n} \leq 15$ and $0 \leq r_{cn} \leq 0.1$. For $d_n = 0.25$, 100% reflection was found to occur when $3 \leq V_{0n} \leq 15$ and $0 \leq r_{cn} \leq 0.1$. For the larger value of d_n, a larger plugging voltage V_{0n} is required to reflect 100% of all incident particles.

An optimized parameter value for V_{0n} is defined here to result in 100% reflection for the widest range of r_{cn} values. The optimized values of V_{0n} for various d_n values are shown in Table I for both a beam and a monoenergetic isotropic velocity distribution. The values for V_{0n} are obtained by determining the value that results in 100% reflection for $0 < r_{cn} \leq r_{cn,\max}$, where $r_{cn,\max}$ is the largest value of r_{cn} yielding $P_R = 100\%$ for a given d_n. As the value of d_n increases, the value of $r_{cn,\max}$ increases until $d_n = 0.45$, at which point a lower value of $r_{cn,\max}$ is found for both the beam and isotropic velocity distribution. While the ASB reflects 100% of all particles for various parameters with each velocity distribution, the value of $r_{cn,\max}$ is lower for a given value of d_n when the incident particle velocities are sampled from the isotropic velocity distribution.

In Table I, the row in which $r_{cn,\max}$ has a maximum value for both particle sources provides two sets of optimal normalized parameter values for charged particle reflection from an electrostatically plugged picket-fence ASB. The optimal normalized parameter values are ($d_n = 0.4$, $V_{0n} = 5.0$, and $r_{cn,\max} = 0.18$) for a normally incident beam of monoenergetic particles and ($d_n = 0.4$, $V_{0n} = 6.0$, $r_{cn,\max} = 0.09$) for monoenergetic particles that are incident with an isotropic velocity distribution. The normalized parameters d_n, V_{0n}, and r_{cn} are related to their unnormalized counterparts by $d = d_n S$,

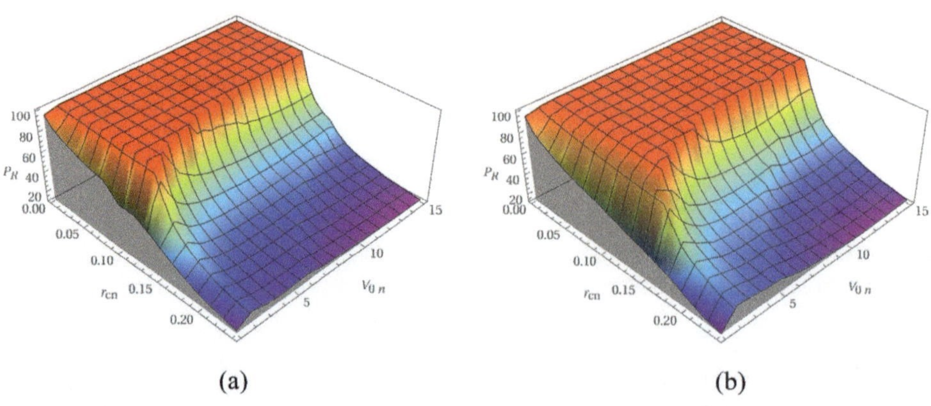

FIG. 7. Effect of electrostatic plugging for a normally incident beam of particles. (a) *dn* = 0:15 and (b) *dn* = 0:25.

TABLE I. Optimized V_{0n} values.

	Beam		Isotropic	
d_n	V_{0n}	$r_{cn,\max}$	V_{0n}	$r_{cn,\max}$
0.05	5.0 – 8.0	0.10	8.0 – 15.0	0.03
0.10	3.0	0.11	7.0 – 11.0, 14.0, 15.0	0.04
0.15	2.0	0.13	9.0, 12.0, 13.0	0.05
0.20	2.0	0.14	2.0 – 8.0, 10.0 – 15.0	0.05
0.25	3.0	0.14	3.0 – 9.0	0.06
0.30	3.0	0.16	3.0, 4.0	0.08
0.35	4.0	0.16	4.0	0.09
0.40	5.0	0.18	6.0	0.09
0.45	10.0, 11.0	0.17	13.0	0.08

$V_0 = V_{0n}K_0/q$, and $B_m = \sqrt{2mK_0}/(|q|r_{cn}S)$. The simulation results indicate that 100% reflection occurs for $0 < r_{cn} \leq r_{cn,\max}$.

IV. DISCUSSION AND CONCLUDING REMARKS

A classical trajectory Monte Carlo simulation has been used to find two sets of optimal normalized parameter values for charged particle reflection from a picket-fence ASB with electrostatic plugging. The two sets of optimal normalized parameter values were defined to be associated with the largest values of $r_{cn,\,max}$ for the two sources of charged particles considered. By way of example, radial plasma confinement within a cylindrical trap in the large radius limit is now considered. The confined plasma is assumed to be a non-drifting positron plasma with an isotropic Maxwellian velocity distribution and with a density low enough for a single-particle treatment to be suitable. The second set of optimal normalized parameter values are used to write the following conditions in terms of unnormalized parameters: $d = 0.4S$, $V_0 = 6K_{0,\,\max}/q$, $B_{m,\min} = \sqrt{2mK_{0,\max}}/(0.09|q|S)$, where $B_{m,\min}$ is the minimum magnetic field strength within the magnetic cusps for a particle with a kinetic energy that is less than $K_{0,\max}$ to be expected to remain confined radially in the single particle limit. Suppose that the value of $K_{0,\max}$ is chosen to be ten times larger than the average kinetic energy of a positron, $K_{0,\max} = 15kT$. For confinement of a positron plasma having a temperature of 40 K, with a trap that has a spatial period $S = 1$ cm, the repelling electrodes would have a width of $d = 0.4$ cm and an applied voltage of $V_0 = 0.31$ V. The magnets would have to produce a minimum magnetic field strength within the cusps of $B_{m,\min} = 8.5$ gauss.

In vacuum, the trap would have an average electric potential away from the ASB equal to the area weighted average of the potentials along the boundary, $0.8V_0$. Thus, the difference in electric potential between a repelling electrode at potential V_0 and a location many spatial periods away from the ASB would be approximately $V_0 - 0.8V_0 = 0.2V_0$. Such a difference represents a change in particle potential energy which is much larger than the temperature in energy units: $\Delta U = 0.2qV_0 = 18kT$. Consequently, in the single-particle limit, if a plasma particle travels from a location many spatial periods away from the ASB directly into the center of a magnetic cusp such that the magnetic field doesn't affect it, the plasma particle must have a kinetic energy more than 18 times larger than the plasma temperature in energy units to reach the repelling electrode.

The work presented here represents an extension to the research reported in Ref. 1, which considered only a picket-fence ASB without electrostatic plugging. In Ref. 1, it was reported that charged particles could be effectively reflected by the ASB, provided that the particles were directed toward the ASB at sufficiently small angles of incidence transverse to the wires that produce the magnetic field. In the work reported here, charged particle reflection at all angles of incidence is found possible under certain conditions, when the magnetic cusps are electrostatically plugged.

The confinement of a positive plasma species within the space charge of an edge confined electron plasma was investigated from a theoretical perspective in Ref. 16. The exact structure of the reflecting boundary used to confine an electron plasma along its edge was not considered in Ref. 16. A charged particle trap based on a cylindrical generalization of the planar ASB configuration considered here may lead to a method for confining a non-neutral plasma along the plasma's edge. Such a plasma would be effectively free of externally applied electromagnetic fields, and, under certain conditions, the electric field produced by the plasma's space charge may serve to confine a second oppositely charged plasma species.

In nested Penning traps used for antihydrogen production, the positron space charge can have a significant detrimental effect on the number of antihydrogen atoms with kinetic energy suitable for further experimentation.[14,15] In an ASB-based trap used for antihydrogen production, the positron space charge would have a beneficial effect by producing an electric field that serves to confine the antiprotons. With the development of a sufficiently cold antihydrogen source, various possibilities exist for conducting antimatter gravity experiments, such as those described in Refs. 17–20.

In conclusion, simulation studies of an electrostatically plugged picket-fence ASB have been presented, and the conditions under which the system can effectively reflect charged particles have been reported. The ASB creates a short range electromagnetic field, and might lead to a method for confining cryogenic plasmas in a region of space largely free of external electromagnetic fields. If the ASB is used to confine a positron plasma along its edge, the space charge of the positrons may provide a means to confine antiprotons in a three-dimensional electrostatic well. The mixing of the two plasmas in a region of space devoid of a strong magnetic field may increase the three-body recombination rate and may lead to an increase in the number of cold antihydrogen atoms suitable for experimentation such as for an antimatter gravity experiment.

ACKNOWLEDGMENTS

The authors would like to thank Ryan Lane and Ryan Phillips for helpful discussions. Computational resources were provided by UNT's High Performance Computing Services. This material is based upon work supported by the National Science Foundation under Grant No. PHY-1500427 and by the Department of Energy under Grant No. DE-FG02-06ER54883.

[1] C. A. Ordonez, "Charged particle reflection from an artificially structured boundary that produces a spatially periodic magnetostatic field," Journal of Applied Physics **106**, 024905 (2009).
[2] C. A. Ordonez, "Charged particle transport through a periodic electrostatic potential having a small spatial period," Journal of Applied Physics **104**, 054903 (2008).
[3] C. A. Ordonez, "Effect of a periodic electrostatic potential on magnetized particle transport," Physics of Plasmas **15**, 114507 (2008).
[4] R. M. Hedlof and C. A. Ordonez, "Artificially structured boundary as a charged particle beam deflector shield," Physics Procedia **66**, 95 (2015).
[5] J. L. Pacheco, C. A. Ordonez, and D. L. Weathers, "Artificially structured boundary for a high purity ion trap or ion source," Nuclear Instruments and Methods in Physics Research B **332**, 312 (2014).
[6] T. J. Dolan, "Magnetic electrostatic plasma confinement," Plasma Physics and Controlled Fusion **36**, 1539 (1994).
[7] M. Ahmadi, B. X. R. Alves, C. J. Baker, W. Bertsche, E. Butler, A. Capra, C. Carruth, C. L. Cesar, M. Charlton, S. Cohen, R. Collister, S. Eriksson, A. Evans, N. Evetts, J. Fajans, T. Friesen, M. C. Fujiwara, D. R. Gill, A. Gutierrez, J. S. Hangst, W. N. Hardy, M. E. Hayden, C. A. Isaac, A. Ishida, M. A. Johnson, S. A. Jones, S. Jonsell, L. Kurchaninov, N. Madsen, M. Mathers, D. Maxwell, J. T. K. McKenna, S. Menary, J. M. Michan, T. Momose, J. J. Munich, P. Nolan, K. Olchanski, A. Olin, P. Pusa, C. O. Rasmussen, F. Robicheaux, R. L. Sacramento, M. Sameed, E. Sarid, D. M. Silveira, S. Stracka, G. Stutter, C. So, T. D. Tharp, J. E. Thompson, R. I. Thompson, D. P. van der Werf, and J. S. Wurtele, "Antihydrogen accumulation for fundamental symmetry tests," Nature Communications **8**, 681 (2017).
[8] G. Gabrielse, R. Kalra, W. S. Kolthammer, R. McConnell, P. Richerme, D. Grzonka, W. Oelert, T. Sefzick, M. Zielinski, D. W. Fitzakerley, M. C. George, E. A. Hessels, C. H. Storry, M. Weel, A. Müllers, and J. Walz, "Trapped antihydrogen in its ground state," Physical Review Letters **108**, 113002 (2012).
[9] N. Kuroda, S. Ulmer, D. J. Murtagh, S. Van Gorp, Y. Nagata, M. Diermaier, S. Federmann, M. Leali, C. Malbrunot, V. Mascagna, O. Massiczek, K. Michishio, T. Mizutani, A. Mohri, H. Nagahama, M. Ohtsuka, B. Radics, S. Sakurai, C. Sauerzopf, K. Suzuki, M. Tajima, H. A. Torii, L. Venturelli, B. Wünschek, J. Zmeskal, N. Zurlo, H. Higaki, Y. Kanai, E. Lodi Rizzini, Y. Nagashima, Y. Matsuda, E. Widmann, and Y. Yamazaki, "A source of antihydrogen for in-flight hyperfine spectroscopy," Nature Communications **5**, 3089 (2014).
[10] A. Kellerbauer, M. Amoretti, A. S. Belov, G. Bonomi, I. Boscolo, R. S. Brusa, M. Buchner, V. M. Byakov, L. Cabaret, C. Canali, C. Carraro, F. Castelli, S. Cialdi, M. de Combarieu, D. Comparat, G. Consolati, N. Djourelov, M. Doser, G. Drobychev, A. Dupasquier, G. Ferrari, P. Forget, L. Formaro, A. Gervasini, M. G. Giammarchi, S. N. Gninenko, G. Gribakin, S. D. Hogan, M. Jacquey, V. Lagomarsino, G. Manuzio, S. Mariazzi, V. A. Matveev, J. O. Meier, F. Merkt, P. Nedelec, M. K. Oberthaler, P. Pari, M. Prevedelli, F. Quasso, A. Rotondi, D. Sillou, S. V. Stepanov, H. H. Stroke, G. Testera,

G. M. Tino, G. Trenec, A. Vairo, J. Vigue, H. Walters, U. Warring, S. Zavatarelli, and D. S. Zvezhinskij, "Proposed antimatter gravity measurement with an antihydrogen beam," Nuclear Instruments and Methods in Physics Research Section B **266**, 351 (2008).
[11] P. Indelicato, G. Chardin, P. Grandemange, D. Lunney, V. Manea, A. Badertscher, P. Crivelli, A. Curioni, A. Marchionni, B. Rossi, A. Rubbia, V. Nesvizhevsky, D. Brook-Roberge, P. Comini, P. Debu, P. Dupre, L. Liszkay, B. Mansoulie, P. Perez, J.-M. Rey, B. Reymond, N. Ruiz, Y. Sacquin, B. Vallage, F. Biraben, P. Clade, A. Douillet, G. Dufour, S. Guellati, L. Hilico, A. Lambrecht, R. Guerout, J.-P. Karr, F. Nez, S. Reynaud, C. I. Szabo, V.-Q. Tran, J. Trapateau, A. Mohri, Y. Yamazaki, M. Charlton, S. Eriksson, N. Madsen, D. P. van der Werf, N. Kuroda, H. Torii, Y. Nagashima, F. Schmidt-Kaler, J. Walz, S. Wolf, P.-A. Hervieux, G. Manfredi, A. Voronin, P. Froelich, S. Wronka, and M. Staszczak, "The gbar project, or how does antimatter fall?," Hyperfine Interactions **228**, 141 (2014).
[12] M. E. Glinsky and T. M. O'Neil, "Guiding center atoms: Three-body recombination in a strongly magnetized plasma," Physics of Fluids B **3**, 1279 (1991).
[13] F. Robicheaux, "Three-body recombination for electrons in a strong magnetic field: Magnetic moment," Physical Review A **73**, 033401 (2006).
[14] C. A. Ordonez, "Effect of positron space charge on operation of an antihydrogen trap," Physical Review E **76**, 017402 (2007).
[15] C. A. Ordonez and D. L. Weathers, "Two-species mixing in a nested penning trap for antihydrogen trapping," Physics of Plasmas **15**, 083504 (2008).
[16] J. L. Pacheco, C. A. Ordonez, and D. L. Weathers, "Space-charge-based electrostatic plasma confinement involving relaxed plasma species," Physics of Plasmas **19**, 102510 (2012).
[17] A. H. Treacher, R. M. Hedlof, and C. A. Ordonez, "Multiple aperture-based antihydrogen parallel plate gravity experiment," Physics Procedia **66**, 180 (2015).
[18] J. R. Rocha, R. M. Hedlof, and C. A. Ordonez, "Aperture-based antihydrogen gravity experiment: Parallel plate geometry," AIP Advances **3**, 102129 (2013).
[19] R. M. Hedlof and C. A. Ordonez, "Simulation of an antihydrogen gravity experiment utilizing multiple apertures," AIP Conference Proceedings **1525**, 102 (2013).
[20] C. A. Ordonez and R. M. Hedlof, "Simulation of an aperture-based antihydrogen gravity experiment," AIP Advances **2**, 012176 (2012).

7

Laboratory simulation of the formation of an ionospheric depletion using Keda Space Plasma EXperiment (KSPEX)

Pengcheng Yu,[1] Yu Liu,[2,a] Jinxiang Cao,[1] Jiuhou Lei,[2] Zhongkai Zhang,[1] and Xiao Zhang[1]
[1]*Department of Modern Physics, University of Science and Technology of China, Hefei, Anhui 230026, China*
[2]*CAS Key Laboratory of Geospace Environment, School of Earth and Space Science, University of Science and Technology of China, Hefei, Anhui 230026, China*

In the work, the formation of an ionospheric depletion was simulated in a controlled laboratory plasma. The experiment was performed by releasing chemical substance sulfur hexafluoride (SF_6) into the pure argon discharge plasma. Results indicate that the plasma parameters change significantly after release of chemicals. The electron density is nearly depleted due to the sulfur hexafluoride-electron attachment reaction; and the electron temperature and space potential experience an increase due to the decrease of the electron density. Compared to the traditional active release experiments, the laboratory scheme can be more efficient, high repetition rate and simpler measurement of the varying plasma parameter after chemical releasing. Therefore, it can effective building the bridge between the theoretical work and real space observation.

I. INTRODUCTION

The ionosphere is an ionization area of the earth's atmosphere and there are a large number of free electrons which contributes significantly for the formation of atmospheric electricity and earth's magnetosphere, and the transmission of the radio in the ionosphere too.[1] However, the researchers observe that the ionosphere was disturbed seriously when the launch of the rockets. This is attributed to the neutral molecules (molecular hydrogen and water vapor) and the electronegativity chemicals (sulfur hexafluoride) are released which can subsequently reduce the electron density in the ionosphere.[2,3] Moreover, disturbance up to 1000 km in radius was observed in the ionospheric F layer during launching of Skylab1.[4] These ionospheric disturbance have a great influence on radio wave propagation and satellite communications. In order to overcome the drawbacks, it is important to investigate the formation and formation of man-made ionospheric hole. In the early time, artificially release of chemical substance was adopted to study physical processes of the ionospheric depletion by using of the space shuttles and rockets including of CRRESS, SIMPLEX, and CARE projects.[5] Although active release experiments have advantages such as the large space scale and the real environment of ionosphere, there are also have some drawbacks to limit the research. For example, the space experiment is expensive in artificially releasing chemicals in the real ionosphere, and many microcosmic physical process is lack of study, so many process needed by means of numerical simulation research. Compared to the space experiment, the advantages of the laboratory studies of chemical releasing are easy repeatability; experimental parameters can be controlled; the plasma diagnosis is accurately and high-resolution; the type of the plasma source is variable and the cost of the experiment is low and so on.

---

[a]Electronic mail: yliu001@ustc.edu.cn

Liu et al develop a method to study the formation of the ionospheric depletion in a laboratory plasma,[6] and boundary layer processes including of the excitation of electron-ion hybrid instability has been studied detailedly.[7–9] However, the plasma device using in that experiment is quiet small in size. Therefore, the formation of the plasma parameters cannot fully studied. In order to more clear and effective observe the changes of electron density, we used a much larger plasma device named as Keda Space Plasma EXperiment (KSPEX) to model an ionospheric depletion in laboratory. An inductively coupled plasmas (ICPS) have been adopted as the plasma source to make a large area uniform ambient plasma.[10–12] Therefore, it is accessible to simulate an artificially-created ionospheric depletion and without boundary effect of the device using this experimental design reported here.

In this work, the experiment was performed by a large vacuum chamber which is detail introduced in Sec. II, and the plasma parameter was measured by a quad-probe which was arranged at a two dimensions movable probe holder, which is driven by a stepper motor. Therefore, the quad-probe can give a two dimensions distribution of plasma parameters such as electron density, space potential and electron temperature, in prior and after the chemical releasing. The results suggest that releasing chemical substance not only reduce the electron densities but also change the effective electron temperature and space potential in ICPs discharge. Our experiment provides a simple and efficient approach for investigating by artificial releasing the chemical substance in the ionosphere. This scheme can be effective building the bridge between theoretical work and real space observation.[13]

II. EXPERIMENTAL SETUP AND DIAGNOSTIC TECHNIQUES

A. Experimental setup

A schematic diagram of the apparatus employed for this study is presented in Fig.1, which named the Keda Space Plasma EXperiment (KSPEX). This discharge device has already been described in previously.[14]

The whole vacuum chamber is made of the stainless steel (diameter 0.5 m and length 2.73 m) which is composed of three cylindrical chamber. On either side of the device are two source chambers which was 50 cm in diameter and 41.8 cm in length; the central of the vacuum chamber is experimental chamber which was 50 cm in diameter and 177.4 cm in length. All experiments were carried out in the experimental chamber, the working gas is pure argon and the releasing gas is sulfur hexafluoride. The background pressure was maintained at the order of 10^{-5} Pa by using two turbo molecular pumps (600L/s) which are backed by two mechanical pumps (8L/s). In order to effective improve the plasma density and uniformity, the vacuum chamber surface was surrounded by 6 electromagnet coils with water cooling which can produce a uniform axial magnetic field. An internal cavity which made of glass was in the side of the reactor with a 10 cm inner diameter and a 30 cm depth in the plasma source chamber. The coil was made of copper, which was embedded into the internal cavity and surrounded by the vacuum chamber. A 13.56MHz radio-frequency power generator and an automatic matching network supply power to the coil. In our experiment, the measurements, performed using a

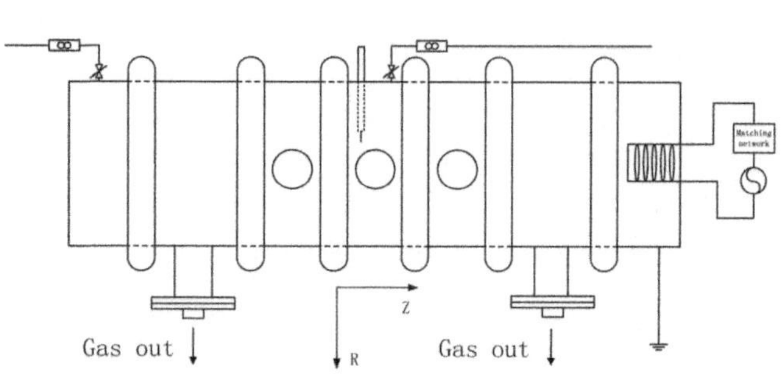

FIG. 1. The schematic diagram of the experimental device.

quad-probe, which all of the probes have the same size (1-mm-diameter and 10-mm-length). The DC voltage of the double probe were 100V. The quad-probe was arranged at a two dimensions movable probe holder, which is driven by a stepper motor. Through the stepper motor, the quad-probe can varied from -50 mm to 350 mm of the Z axis (more and more close to the plasma source and the 150 mm is the central of the vacuum chamber); and varied from 0mm to 400 mm of the R axis in a vacuum chamber (the 0 mm is the central of the vacuum chamber). Therefore, the quad-probe can give a two-dimension distribution of plasma parameters such as electron density, space potential and electron temperature in prior and after the chemical releasing. We designed the collection circuit of the quad-probe and the I-V curve of the probe were collected using a data acquisition system (DAQ) which have 600 KHz scan rate. The collection circuit of the quad-probe was fixed and sealed in an aluminum specimen box in order to makes the system stable mechanically and resistance of the static, electromagnetic interference.

B. Diagnostic techniques

In our experiment, we adopt a simple method for obtaining the plasma parameter through make use of the "quad-probe" which all the probes have same size at different electrostatic potentials. The quad-probe is consists of a double probe and two floating probes which are used to measure the floating potential immediately. Compared to the single probe, the quad-probe have those advantages as following: first, the current flowing into the probe is so small, which does not exceed the ion saturation current, so that the disturbance of the probe recording current in the plasma can be ignored. Second, the quad-probe can work in the plasma without a reference electrode. Third, the quad-probe have two floating probes which can measure the plasma floating potential at the same time. In addition, this system has some practical advantages over the other methods when applied to magneto plasmas.[15] In addition, the measurement of electron density was calibrated by a Langmuir probe, and the results show well agreement with that of the quad-probe. The calculation method of density, electron temperature and plasma potential from quad-probe can be found as following.

According to the circuit analysis of the double probe, the effective electron temperature formula can be derived by[16]

$$2\exp(\frac{eV_f}{kT_e}) = \exp(\frac{eV_1}{kT_e})\left[1+\exp(\frac{-eV}{kT_e})\right], \tag{1}$$

where V_1 is the positively biased electrode potential and V_f is the floating potential. If the applied potential V is satisfied the following conditions $V>>kT_e$ in the experiment, the Eq. (1) become

$$\frac{kT_e}{e} = \frac{1}{In2}\left(V_1 - V_f\right), \tag{2}$$

According to the Eq. (2), the effective electron temperature can be known immediately when the V_f have been measured by the floating probe. After that, the space potential can be obtained by use of the effective electron temperature and the floating potential.[17]

$$\phi_p = \phi_f + \eta T_e, \tag{3}$$

Where η is the coefficient of the effective electron temperature which is only related to the particle types. In our experiment, the discharge gas is pure argon, so the η is equal to 4.8 for argon.

After that, the current of the double probe I_D is

$$I_D = I_i\left\{\exp\left[\frac{e\left(V_1 - V_f\right)}{kT_e}\right] - 1\right\}, \tag{4}$$

In the condition of the collision-less plasma, the ion saturation current which be measured by the double probe is

$$I_i = \kappa e n_e A c_s, \tag{5}$$

where A is the physical area of the probe, n_e is the electron number density in a quasi-neutral plasma and $c_s = \sqrt{kT_e/m_i}$ is the Bohm velocity. Then the Eq. (4) is given by

$$n_e = \frac{I_D}{\kappa e A c_s \left\{ \exp\left[\frac{e(V_1 - V_f)}{kT_e}\right] - 1 \right\}}, \tag{6}$$

In our work, the applied potential $V >> kT_e$, so that the Eq. (6) become

$$n_e = \frac{I_D}{\kappa e A c_s}, \tag{7}$$

The value of κ can be obtained by make use of the energy conservation equation in the collision-less plasma

$$\frac{1}{2} m_i C_s^2 = e\Delta V_p, \tag{8}$$

where m_i is the ion mass, and ΔV_p is the difference potential value between the sheath boundary and the plasma. According to the Boltzmann relation, the unperturbed ion density n_s at the sheath boundary can be given by,

$$n_s = n_e \exp\left(-e\Delta V_p/kT_e\right) = n_e \kappa \approx 0.6 n_e, \tag{9}$$

Hence, Replacing κ in Eq. (7), the electron densities is,

$$n_e = \frac{I_i}{0.6 e A \sqrt{kT_e/m_i}}, \tag{10}$$

In order to effective improve the plasma density and uniformity in the vacuum experimental chamber, the additional magnetic field was added to constrictive plasma in our experiment. The magnetic field is produced by the 6 electromagnet coils which is controlled by the three magnetic field power supply. As we all know, the additional magnetic maybe cause some influences on data which is measured by the probe, so that the experimental data need to be careful analyze when the magnetic field is "very strong" or "strong".[18,19] However, the effect of the magnetic field to the ion current collecting can be neglect when the magnetic field is "weak". But how to define the strength of the magnetic field. The judgment criteria is the ratio of the ion or electron larmor radius to the radius of the probe. When the larmor radius of the ion and electron is much smaller than the radius of the probe ($r_P >> \rho_e; r_P >> \rho_i$), the magnetic can be defined "very strong". The "strong" magnetic field is the larmor radius of the electron much smaller than the radius of probe but the larmor radius of the ion much larger than probe ($r_P >> \rho_e; \rho_i >> r_P$). When the electron larmor radius ρ_e and ion larmor radius ρ_i are infinite or at least sufficiently large than the radius of the probe, the ion collection current could be considered unaffected by any magnetic field,[20,21] and this theory has already been verified in a magnetized alkali vapor plasma.[22] The larmor radius of the ion can be calculated by the equation as follow:

$$\rho_e = \frac{m_e v}{Be}, \tag{11}$$

Where m_e is the mass of the electron, the v is the hot electron velocities and the magnetic field intensity is B.

The larmor radius of the ion can also use the Eq. (11) after we replace the mass of the ion m_i to the mass of the electron m_e. Therefore, when the larmor radius of electron ρ_e and the larmor radius of ion ρ_i are much large than the probe radius r_p in the experiment, the magnetic field can be thought "week" and the influence of the magnetic field is not need to consider to the ion current collecting by the probe.

III. RESULTS AND DISCUSSION

A. Electron density

In our experimental environment, each magnetic field power can supply 20V, 40A to the electromagnet coils to create a uniform axial magnetic field of 160G. The ratio of the electron plasma frequency (ω_{pe}) to electron gyrofrequency (ω_{ce}), ω_{pe}/ω_{ce}, is 0.13~0.45 which are all within the valid ranges of 0.1~10.[23–25] Therefore, through control the parameters of plasma, the detailed physical process of an ionospheric hole boundary layer can be studied in the laboratory. The larmor radius of the electron or ion can be calculated by using Eq. (11). The result presents that the larmor radius of the electron is much larger than the radius of probe ($\rho_e = 6.5mm >> r_P = 0.1mm$) and the larmor radius of the ion is much larger than the radius of probe too ($\rho_i >> r_P$). Therefore, the effect of the magnetic field to the ion current collecting can be neglect and the Eq. (10) not need to corrected in our experimental environment. The pressure of the pure argon about 0.41pa is controlled by one of the mass flow meter (100sccm), and the release of SF_6 is controlled by another mass flow meter (30sccm). Relative to the pressure of pure argon, 30% chemical substance SF_6 are released into the plasma when the input power generator of the ICP source is kept at 300 W. In our experiment, the location of the quad-probe in the vacuum chamber is varied from 50mm to 250 mm of the Z axis and 0mm to 180 mm of the R axis in a vacuum chamber.

Figure 2 presents the formation of the electron density as a function of the distance to R axis and Z axis which is measured by the quad-probe. Figure 2(a) presents the pre-existing electron density while discharge gas is the pure argon and Figure 2(b) indicates the change of the electron density after releasing SF_6. According to the Figure 2(a), the electron densities decrease from $5\times10^{10}cm^{-3}$ to $1\times10^{10}cm^{-3}$. The maximum of the electron densities locate at a small area where is 200mm to 250mm in Z axis, and 0mm to 60mm in R axis; moreover, the minimum of the electron densities locate at 50mm to 250mm in Z axis and 120mm to 180mm in R axis. The electron density increases with the decrease of the distance to the source chamber in Z axis and decreases with the increase of the distance to the central of the vacuum chamber in R axis. From Fig 2(b), the electron density decreases from $7.5\times10^9cm^{-3}$ to $4\times10^9cm^{-3}$. The electron densities decreases to nearly 85% after the SF_6 release into the plasma region. The results present that the release of the SF_6 into the pure argon plasma discharge, can cause the decrease of electron density. The chemical reaction of releasing SF_6 is complex, and the whole of the reaction equations are sixteen.[26] According to the composition of the atoms and molecules in our vacuum experimental chamber, the main reason of the decrease electron density is SF_6-electron attachment,[27,28] and the two main processes are

$$SF_6 + e \rightarrow F + SF_5^-, \quad K_3 \cong 1.1 \times 10^{-7} cm^3 s^{-1}$$

$$SF_6 + e \rightarrow SF_6^-, \quad K_3 \cong 6.0 \times 10^{-8} cm^3 s^{-1}$$

Through the chemical reaction formula and the reaction rate, the releasing SF_6 can a certain extent reduce the density of the electrons in the plasma.

B. Effective electron temperature

In the experiment, the electrons obey the Maxwellian distribution in the plasma regions. If the temperature of plasma can be associated with the electron, the effective electron temperature can

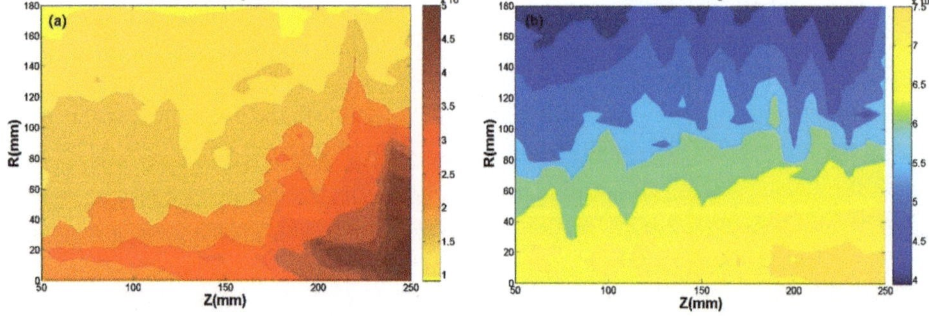

FIG. 2. The electron density of the plasma is measured by the quad-probe (a) the results of pure A_r; (b) the results of after releasing SF_6.

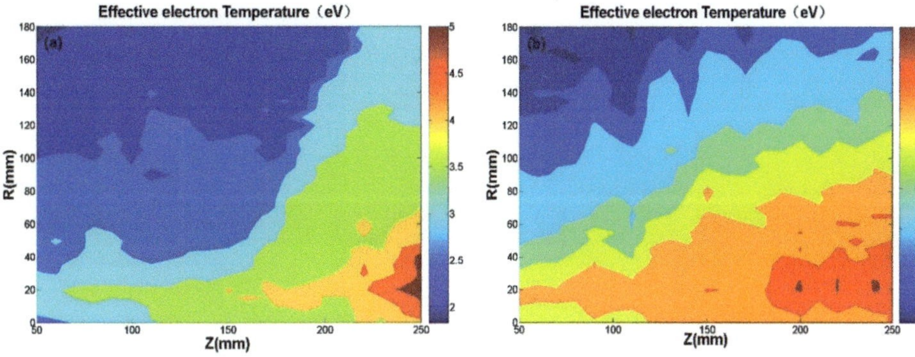

FIG. 3. The electron temperature of the plasma is measured by the quad-probe (a) the results of pure A_r; (b) the results of after SF_6 release.

be denoted by T_e. The electron temperature is an important parameter of the plasma which can determine the ambipolar diffusion coefficient,[29] and it is also crucial for the calculation of some other parameters such as Debye length, electron thermal velocity, ion sound velocity and so on. After the floating potential obtained by the two floating probe, the effective electron temperature can be calculated from Eq. (2).

As shown in Fig. 3, the effective electron temperature was measured by a quad-probe as a function of the distance. Figure 3(a) presents the pre-existing effective electron temperature. The electron temperature decreases with the increasing of the distance to the source chamber and the central of the vacuum chamber, from $5eV$ to $2eV$. The effective electron temperature increases with the decreasing of the distance to the source chamber in Z axis; and decreases with the increasing of the distance to the central of the vacuum chamber in R axis. The formation of the effective electron temperature by the distance is almost the same as the electron density. Figure 3(b) indicates that the change of the electron temperature after releasing SF_6, the peak of the effective electron temperature can reach $6eV$ and the nadir of the effective electron temperature is $2.5eV$. The result indicates that the change of the electron temperature is not obvious, and only have a slight increase after releasing SF_6. Compared to the Figure 3(a) and 3(b), the electron temperature was observed to increase about 25% after the releasing SF_6. That is may be the low energy electrons more easily attached by SF_6, and the rest of the high energy electron is more than the low energy electron in the vacuum chamber. Therefore, the effective electron temperature have a slight increase.

C. Plasma potential

Through the effective electron temperature obtained by the probes, we can further use the formula (3) to get the space potential fluctuation. The plasma potential is another crucial parameter which governs the electric field in the plasma. The trajectories of the electron and ion can be subsequently obtained by potential drop. Additionally, the electron densities can also be calculated by the electron saturation current.[30]

Figure 4 presents the formation of space potential over a large range by controlling the distance of R axis and Z axis. Figures 4(a) presents the pre-existing space potential while discharge gas is the pure argon, and the plasma potential decreases from $18V$ to $11V$, the maximum of the space potential locate at 170mm to 250mm in Z axis and 0mm to 50mm in R axis; and the minimum of the electron densities is located at 50mm to 200mm in Z axis and 100mm to 180mm in R axis. The space potential increases with the decrease of the distance to the source chamber in Z axis; and decreases with the increase of the distance to the central of the vacuum chamber in R axis. Figures 4(b) presents the space potential after release SF_6, it can be seen that the space potential decreases from $23V$ to $16V$. Compared to figure 4(a) and 4(b), it is clear that the space potential obviously increased after release SF_6. That is because of the electron is absorbed by the SF_6, which can reduce nearly 85% of the electron density after the release of SF_6. Therefore, reduction of the electron density would lead to the increase of the space potential. The suddenly change of the plasma potential can lead to the

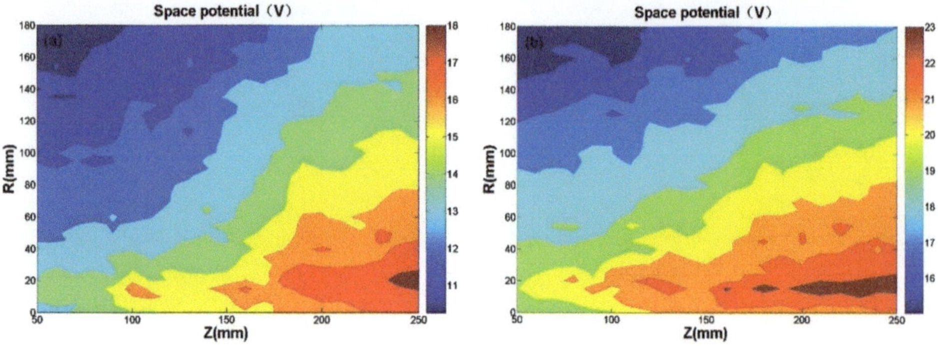

FIG. 4. The plasma potential of the plasma (a) the results of pure A_r; (b) the results of after SF6 release.

generation of inhomogeneous electric field in the boundary layer of an ionospheric depletion, which subsequently supply free energy for many plasma instabilities and waves.[7,8]

IV. CONCLUSIONS

In this study, we introduced a laboratory method for investigating ionosphere depletion by artificial releasing of chemical substance SF_6. The formation of the plasma parameters were measured by varying the position of the probe and releasing chemical substance. Results illustrate that the plasma parameters all have some changes after releasing SF_6. The electron densities was reduced about 85% due to attachment reaction, and the electron temperature have a slight increase after release of SF_6. Moreover, the space potential increased because of the electron densities is reduced, which could result inhomogeneous electric fields in the ionospheric depletion. Our scheme provides a simple and efficient approach for investigating artificial releasing the chemical substance in the ionosphere. It can effective building the bridge between theoretical work and real space observation. This experimental results also might be used to explain the other researches such as electronic quenching and so on.

ACKNOWLEDGMENTS

This project was supported by the National Natural Science Foundation of China (41604132), and the Fundamental Research Funds for the Central Universities.

[1] M. C. Kelley, *The Earth's Ionosphere: Electrodynamics and Plasma Physics [J]*. 2009.
[2] H. G. Booker, "A local reduction of F-region ionization due to missile transit [J]," Journal of Geophysical Research **66**(4), 1073–1079, https://doi.org/10.1029/JZ066i004p01073 (1961).
[3] J. E. Jackson, H. A. Whale, and S. J. Bauer, "Local ionospheric disturbance created by a burning rocket [J]," Journal of Geophysical Research **67**(5), 2059–2061, https://doi.org/10.1029/JZ067i005p02059 (1962).
[4] M. Mendillo, "The effect of rocket launches on the ionosphere [J]," Advances in Space Research **1**(2), 275–290 (1981).
[5] P. A. Bernhardt, J. O. Ballenthin, J. L. Baumgardner, A. Bhatt, I. D. Boyd, J. M. Burt, R. G. Caton, A. Coster, P. J. Erickson *et al.*, "Ground and space-based measurement of rocket engine burns in the ionosphere [J]," IEEE Transactions on Plasma Science **40**(5), 1267–1286 (2012).
[6] Y. Liu, J. Cao, J. Wang, Z. Zheng, L. Xu, and Y. Du, "Experimentally investigate ionospheric depletion chemicals in artificially created ionosphere [J]," Physics of Plasmas **19**(9), 092901 (2012).
[7] Y. Liu, J. X. Cao, L. Xu, X. Zhang, P. Wang, J. Wang, Y. Du, and Z. Zheng, "Coherent structure generated in the boundary layer of a laboratory-created ionospheric depletion [J]," Geophysical Research Letters **41**(5), 1413–1419, https://doi.org/10.1002/2014GL059211 (2014).
[8] Y. Liu, J. X. Cao, L. Xu, X. Zhang, P. Wang, J. Wang, Y. Du, and Z. Zheng, "Laboratory investigation of the boundary layer processes of artificially created ionospheric depletion [J]," Journal of Geophysical Research: Space Physics **119**(5), 4134–4145, https://doi.org/10.1002/2014JA019868 (2014).
[9] X. Zhang, J.-x. Cao, Y. Liu, P.-c. Yu, and Z.-k. Zhang, "Laboratory experiments in the argon plasma perturbed by injections of the electronegative gases [J]," AIP Advances **6**(7), 075304 (2016).
[10] J. Hopwood, C. R. Guarnieri, S. J. Whitehair, and J. J. Cuomo, "Langmuir probe measurements of a radio frequency induction plasma [J]," Journal of Vacuum Science & Technology A: Vacuum, Surfaces, and Films **11**(1), 152–156 (1993).
[11] E. Statnic and V. Tanach, "Investigation of the electrical discharge parameters in electrodeless inductive lamps with a re-entrant coupler and magnetic core [J]," Plasma Sources Science and Technology **15**(3), 465 (2006).

[12] A. J. Lichtenberg, V. Vahedi, M. A. Lieberman, and T. Rognlien, "Modeling electronegative plasma discharges [J]," Journal of Applied Physics **75**(5), 2339–2347 (1994).
[13] M. E. Koepke, "Interrelated laboratory and space plasma experiments [J]," Reviews of Geophysics **46**(3), https://doi.org/10.1029/2005RG000168 (2008).
[14] Y. Liu, Z. Zhang, J. Lei, J. Cao, P. Yu, X. Zhang, L. Xu, and Y. Zhao, "Design and construction of Keda Space Plasma Experiment (KSPEX) for the investigation of the boundary layer processes of ionospheric depletions [J]," Review of Scientific Instruments **87**(9), 093504 (2016).
[15] S. L. Chen and T. Sekiguchi, "Instantaneous direct-display system of plasma parameters by means of triple probe [J]," Journal of Applied Physics **36**(8), 2363–2375 (1965).
[16] D. M. Manos, "Characterization of laboratory plasmas with probes [J]," Journal of Vacuum Science & Technology A: Vacuum, Surfaces, and Films **3**(3), 1059–1066 (1985).
[17] N. Hershkowitz, "How Langmuir probes work [J]," *Plasma Diagnostics* 1, 113–183.1 (1989).
[18] F. F. Chen, C. Etievant, and D. Mosher, "Measurement of low plasma densities in a magnetic field [J]," The Physics of Fluids **11**(4), 811–821 (1968).
[19] I. G. Brown, A. B. Compher, and W. B. Kunkel, "Response of a Langmuir probe in a strong magnetic field [J]," The Physics of Fluids **14**(7), 1377–1383 (1971).
[20] J. R. Sanmartin, "Theory of a probe in a strong magnetic field [J]," The Physics of Fluids **13**(1), 103–116 (1970).
[21] I. M. Cohen, "Saturation currents to Langmuir probes in a collision-dominated plasma with uniform magnetic field [J]," The Physics of Fluids **12**(11), 2356–2361 (1969).
[22] F. F. Chen, C. Etievant, and D. Mosher, "Measurement of low plasma densities in a magnetic field [J]," The Physics of Fluids **11**(4), 811–821 (1968).
[23] C. C. Linand and L. A. Segel, *Mathematics Applied to Deterministic Problems in the Natural Sciences* Macmillan Publishing Co [J]. Inc. New York Google Scholar, 1974.
[24] P. J. Baum and A. Bratenahl, "The laboratory magnetosphere [J]," Geophysical Research Letters **9**(4), 435–438, https://doi.org/10.1029/GL009i004p00435 (1982).
[25] E. M. Tejero, W. E. Amatucci, G. Ganguli, C. D. Cothran, Crabtree, and E. Thomas, Jr., "Spontaneous electromagnetic emission from a strongly localized plasma flow," Physical Review Letters **106**(18), 185001 (2011).
[26] A. K. M. F. Hussain, "Coherent structures and turbulence [J]," Journal of Fluid Mechanics **173**, 303–356 (1986).
[27] P. A. Bernhardt, "A critical comparison of ionospheric depletion chemicals [J]," Journal of Geophysical Research: Space Physics **92**(A5), 4617–4628, https://doi.org/10.1029/JA092iA05p04617 (1987).
[28] Y. Johnson Charles, W. Sjolander Gary, S. Oran Elaine, R. Young Theodore, A. Bernhardt Paul, and V. da Rosa Aldo, "F region above Kauai: Measurement, model, modification," Journal of Geophysical Research: Space Physics **85**(A8), 4205–4213, https://doi.org/10.1029/JA085iA08p04205 (1980).
[29] E. O. Johnson and L. Malter, "A floating double probe method for measurements in gas discharges [J]," Physical Review **80**(1), 58 (1950).
[30] F. F. Chen, Langmuir probe diagnostics[C]//IEEE-ICOPS Meeting, Jeju, Korea. 2003: 2–6.

8

Origin of femtosecond laser induced periodic nanostructure on diamond

A. Abdelmalek,[1,a] B. Sotillo,[2] Z. Bedrane,[1] V. Bharadwaj,[2] S. Pietralunga,[2] R. Ramponi,[2] E.-H. Amara,[3] and S. M. Eaton[2]
[1]*Physics Department, Theoretical Physics Laboratory, Tlemcen University, 13000 Tlemcen, Algeria*
[2]*IFN-CNR and Dipartimento di Fisica, Politecnico di Milano, 20133 Milano, Italy*
[3]*Laser Material Processing Team, Centre for Advanced Technologies Development, CDTA, PO. Box 17, Baba-Hassen 16303, Algiers, Algeria*

We study the evolution of periodic nanostructures formed on the surface of diamond by femtosecond laser irradiation delivering 230 fs pulses at 1030 nm and 515 nm wavelengths with a repetition rate of 250 kHz. Using scanning electron microscopy, we observe a change in the periodicity of the nanostructures by varying the number of pulses overlapping in the laser focal volume. We simulate the evolution of the period of the high spatial frequency laser induced periodic surface structures at the two wavelengths as a function of number of pulses, accounting for the change in the optical properties of diamond via a generalized plasmonic model. We propose a hypothesis that describes the origin of the nanostructures and the principal role of plasmonic excitation in their formation during multipulse femtosecond laser irradiation.

I. INTRODUCTION

Laser–induced periodic surface structures (LIPSS) have attracted considerable attention by researchers, as indicated by the growth in the number of papers published on the topic in recent years.[1] LIPSS can enhance the absorption[2–5] at the surface of materials and can therefore open new avenues for technological applications such as micro–solar cells devices. LIPSS formation is a universal phenomenon and has been observed in dielectrics, semiconductors, metals and polymers.[1,6] Generally, two types of ripples can be observed during the irradiation by multi-pulse femtosecond laser, low spatial frequency LIPSS (LSFL) and high spatial frequency LIPSS (HSFL).

Derrien *et al.*[10] used an above ablation threshold fluence where the critical density of plasmonic excitation is achieved with the first femtosecond laser pulse. They observed ablation in the middle of the laser spot surrounded by LSFL and at the edge where the fluence is low, they found the HSFL. So we suggest that this repartition of the two kinds of LIPSS observed within the same volume spot laser are the result of different physical mechanisms.

LSFL formation for the majority of materials is well explained by the interference between the incident laser wave and the surface plasmon-polariton (SPP) excitation,[1,7,8] which involves a periodic spatial modulation of the energy deposited on the irradiated surface.[9,10] This interaction leads to the formation of nanostructures with periods close to or less than the laser wavelength λ_L. There is still some controversy in the literature regarding the physical mechanisms underlying the formation of HSFL. The physical models proposed include interference,[10,11] second harmonic generation (SHG)[12,13] and surface plasmon polariton excitation.[14–18]

---

[a]Electronic mail: ahmed7abdelmalek13@gmail.com

TABLE I. Data from literature for the observed LIPSS periods in different materials.

Wavelength (nm)	target	Dielectric environment	HSFL period (nm)	Ref.
800	Si	Water	400–150	15
800	I-b diamond	Air	200–50	19
800	DLC[a]	Air	170–60	14

[a]Diamond–Like Carbon.

The interference model depends necessarily on the angle of the incident laser, but it has been shown that the incident angle has no effect in the case of HSFL formation.[8,12] The SHG model is based on the formula: $\Lambda = \frac{\lambda_L}{2n}$ and so the nanostructure period depends strictly on the wavelength. In contrast, there are several experimental works where a change in the period of the HSFL has been observed by varying the pulse number at the same wavelength (Tab. I).

Miyazaki *et al.*[15] and Miyaji *et al.*[14,16,17] proposed a model for HSFL formation where they consider the effect of the number of pulses. They proved experimentally that the formation of nanostructures was due to a non-thermal process, where the random nanoablation is initiated by the disorder (non-thermal melting) during the multipulse femtosecond laser irradiation, followed by plasmonic excitation. In this model, a large number of laser pulses at low energy are necessary to generate HSFL type.[13,15] Forster *et al.*[19] showed a 50 nm period obtained due to a phase transition from crystalline diamond to a graphitic phase, along with 200 nm nanostructures consisting of crystalline diamond (see Tab. I). Clearly, a more sophisticated model is needed to provide insight into the formation of HSFL.

In this paper, we investigate the change in diamond surface morphology induced by ultrafast laser pulses when varying the scanning speed (i.e. number of pulses per spot size diameter) and the laser power (fluence) at 1030 nm and 515 nm wavelength. We illustrate the advantage of using plasmonic excitation in modeling the HSFL formation. We propose a general plasmonic model to follow the evolution of LIPSS formation and the changing diamond optical properties as a function of the electron-holes plasma excitation. Our model demonstrates the importance of plasmonic excitation in the dynamic change of non-thermal melting as an origin of HSFL formation.

II. EXPERIMENTAL SETUP

The samples used in this work were polished single crystal CVD diamond 7.5 mm × 7.5 mm × 0.5 mm. Their optical properties are discussed later in the theoretical model. To irradiate the diamond we used a Light Conversion Pharos amplified femtosecond laser delivering 230 fs pulses at 1030 nm and 515 nm wavelengths with a repetition rate of 250 kHz. The laser was incident normal to the target surface and focused with a 0.42 NA (50×) microscope objective, with a spot size of approximately 1.6 μm and 0.8 μm at 1030 nm and 515 nm wavelengths, respectively. To investigate the creation of surface nanostructures we varied the pulse number by changing the scanning speed, v, through the relationship: $N = \frac{2\omega_0 R_r}{v}$ where R_r is the repetition rate, $2\omega_0$ is the spot size and N the number of pulses delivered per spot diameter.[20] The laser polarization was parallel to the scanning direction, and all the experiments were carried out in air. After the irradiation, we analyzed the change in morphology of diamond surfaces using a scanning electron microscope (SEM).

III. THEORETICAL MODEL

We propose a generalized plasmonic model to study the development of periodical nanostructures on diamond surface during the irradiation with multipulse femtosecond laser at wavelengths of 1030 nm and 515 nm. We consider air as the dielectric environment. The plasmonic model has shown its validity in the case of metals.[35] Applying this model to semiconductors and dielectrics such as diamond is justified by the fact that a thin layer which has metallic character is induced during multipulse femtosecond laser irradiation.[27] In this paper, we call it “pseudo-metal layer”. For this reason, the surface plasmon polariton SPP excitation is used to calculate and explain the formation of periodic nanostructures in diamond. The SPP dispersion relation is:[21]

$$\omega^2 = c^2 k_{sp}^2 \frac{(\varepsilon_{pm} + \varepsilon_d)}{\varepsilon_{pm}\varepsilon_d}. \tag{1}$$

where ε_{pm} is the dielectric function of pseudo-metal layer and $k_{sp} = k_{sp1} + ik_{sp2}$ is the plasmon propagation number with $k_{sp1} = \Re k_{sp} = \frac{2\pi}{\lambda_{sp}} = \frac{\pi}{\Lambda}$ (Ref. 18) where λ_{sp} is the plasmon wavelength and Λ is the period of nanostructures and ε_d is the dielectric function of the material surrounding the pseudo-metal layer. This dielectric function is approximated as $\varepsilon_d = m\varepsilon_{di} + (1 - m)\varepsilon_{air}$ where m represents the factor of mixed state depending implicitly on the pulse number and it varies from 0 and 1, $\varepsilon_{air} = 1$ and $\varepsilon_{di} = 5.72$ and 5.90 at 1030 nm and 515 nm wavelengths, respectively, according to the relationship $n^2 = 1 + \frac{0.3306\lambda^2}{\lambda^2 - 175^2} + \frac{4.3356\lambda^2}{\lambda^2 - 106^2}$ (λ in nm) with $\varepsilon_{di} = n^2$ (Ref. 22). The dielectric function of the pseudo-metal layer is given by the Drude model:[23]

$$\varepsilon_{pm} = \varepsilon_{pm1} + i\varepsilon_{pm2} = 1 + (\varepsilon_{di} - 1)(1 - \frac{n_{eh}}{n_0}) - \frac{\omega_p^2}{\omega^2}\frac{1}{1 + \frac{i}{\omega\tau_{ee}}}. \tag{2}$$

with

$$\varepsilon_{pm1} = 1 + (\varepsilon_{di} - 1)(1 - \frac{n_{eh}}{n_0}) - \frac{\omega_p^2}{\omega^2}\frac{1}{1 + \frac{1}{\omega^2\tau_{ee}}}. \tag{3}$$

$$\varepsilon_{pm2} = \frac{1}{\omega\tau_{ee}}\frac{\omega_p^2}{\omega^2}\frac{1}{1 + \frac{1}{\omega^2\tau_{ee}}}. \tag{4}$$

where n_0 is the electron concentration in the valence band (we consider a value of 10^{23} cm^{-3}), τ_{ee} the electron-electron collision time (1 fs), ω is laser frequency and $\omega_p = \sqrt{\frac{n_{eh}e^2}{\varepsilon_0 m_{opt}^* m_e}}$ is plasma frequency, with $\varepsilon_0 = 8.85 \times 10^{-12}$ F.m and the optical effective mass m_{opt}^* estimated via the relationship[23] $m_{opt}^* = (m_e^{*-1} + m_h^{*-1})^{-1} = 0.30$ with $m_e^* = 3(m_{le}^{-1} + m_{te}^{-1} + m_{te}^{-1})^{-1}$ and $m_h^* = \frac{(m_{hh}^{\frac{3}{2}} + m_{lh}^{\frac{3}{2}})}{(m_{hh}^{\frac{1}{2}} + m_{lh}^{\frac{1}{2}})}$ (Ref. 24) being m_{le}, m_{te}, m_{hh}, m_{lh} being equal to 1.4, 0.36, 1.1, 0.3 (Refs. 25 and 26), respectively, for our single crystal CVD diamond.

M. Straub *et al.*[18] determined the groove period of Si after constructing the numerical formulas for k_{sp1} and k_{sp2} using the critical density of plasmon excitation n_{cr} for the purpose to simplify the calculations, such as $k_{sp1}(n_{cr}) = 0$. In our case is $n_{cr} = 2.32 \times 10^{21}$ cm^{-3} and 7.58×10^{21} cm^{-3} at λ_L = 1030 nm, 515 nm, respectively, but their formulas become valid only for the properties of Si and the wavelength used. In this paper, without employing the critical density n_{cr}, we will build the general formulas of k_{sp1} and k_{sp2} adapted for all semiconductor and dielectric materials and for all wavelengths.

According to Eq. (1)

$$k_{sp}^2 = \frac{\omega^2}{c^2}\frac{\varepsilon_{pm}\varepsilon_d}{\varepsilon_{pm} + \varepsilon_d}. \tag{5}$$

And

$$k_{sp}^2 = (k_{sp1} + ik_{sp2})^2 = k_{sp1}^2 - k_{sp2}^2 + i2k_{sp1}k_{sp2}. \tag{6}$$

$$\frac{1}{k_{sp1}^2} = \frac{c^2}{\omega^2}(\frac{1}{\varepsilon_d} + \frac{1}{\varepsilon_{pm}}). \tag{7}$$

$$\frac{1}{k_{sp1}^2} = \frac{c^2}{\omega^2}(\frac{\varepsilon_{pm1}^2 + \varepsilon_{pm2}^2 + \varepsilon_d\varepsilon_{pm1} - i\varepsilon_d\varepsilon_{pm2}}{\varepsilon_d(\varepsilon_{pm1}^2 + \varepsilon_{pm2}^2)}). \tag{8}$$

$$k_{sp1}^2 = \frac{\omega^2}{c^2}(\frac{\varepsilon_d(\varepsilon_{pm1}^2 + \varepsilon_{pm2}^2)}{\varepsilon_{pm1}^2 + \varepsilon_{pm2}^2 + \varepsilon_d\varepsilon_{pm1} - i\varepsilon_d\varepsilon_{pm2}}). \tag{9}$$

We set:

$$\alpha = \varepsilon_{pm1}^2 + \varepsilon_{pm2}^2 + \varepsilon_d \varepsilon_{pm1}, \quad \beta = \varepsilon_d \varepsilon_{pm2} \text{ and } \gamma = \varepsilon_d (\varepsilon_{pm1}^2 + \varepsilon_{pm2}^2)$$

Which implies:

$$k_{sp1}^2 = \frac{\omega^2}{c^2}\left(\frac{\gamma}{\alpha - i\beta}\right) = \frac{\omega^2}{c^2}\left(\frac{\gamma(\alpha + i\beta)}{\alpha^2 + \beta^2}\right). \tag{10}$$

We also have:

$$k_{sp1}^2 - k_{sp2}^2 = \frac{\omega^2}{c^2}\frac{\gamma\alpha}{\alpha^2 + \beta^2}. \tag{11}$$

$$2k_{sp1}k_{sp2} = \frac{\omega^2}{c^2}\frac{\gamma\beta}{\alpha^2 + \beta^2}. \tag{12}$$

$$(12) \longrightarrow k_{sp2}^2 = \left(\frac{\omega^2}{2c^2 k_{sp1}}\right)^2 \frac{\gamma^2\beta^2}{(\alpha^2 + \beta^2)^2}. \tag{13}$$

Replacing Eq. (13) in Eq. (11), we obtain:

$$4k_{sp1}^4 - 4\frac{\omega^2}{c^2}\frac{\gamma\alpha}{(\alpha^2 + \beta^2)}k_{sp1}^2 - \left(\frac{\omega^2}{c^2}\right)^2 \frac{\gamma^2\beta^2}{(\alpha^2 + \beta^2)^2} = 0. \tag{14}$$

The solution of Eq. (14) is:

$$k_{sp1} = \frac{1}{\sqrt{2}}\frac{\omega}{c}\left[\frac{\gamma\alpha}{(\alpha^2 + \beta^2)} + \frac{\gamma}{(\alpha^2 + \beta^2)^{\frac{1}{2}}}\right]^{\frac{1}{2}}. \tag{15}$$

Therefore, from Eq. (12) we obtain:

$$k_{sp2} = \frac{\omega^2}{2c^2 k_{sp1}}\frac{\gamma\beta}{(\alpha^2 + \beta^2)}. \tag{16}$$

This model was simulated using MATLAB software.

IV. RESULTS AND DISCUSSION

Diamond surface excitation by femtosecond laser pulses builds up a thin pseudo-metal layer[27] due to the non-thermal melting (disorder) which takes place at sub-vibrational time scale, since the pulse duration of the femtosecond laser is lower than the electron-phonon relaxation time.[27–31] This change of states affects optical properties of diamond such as reflectivity R, as shown in Fig. 1, with $R = \frac{(n-1)^2+k^2}{(n+1)^2+k^2}$, where $n = \frac{1}{\sqrt{2}}\left[\varepsilon_{pm1} + (\varepsilon_{pm1}^2 + \varepsilon_{pm2}^2)^{\frac{1}{2}}\right]^{\frac{1}{2}}$ is the real part of the refractive index and $k = \frac{\varepsilon_{pm2}}{2n}$ is the absorption coefficient.[32] We observe an increase of the reflectivity as a function

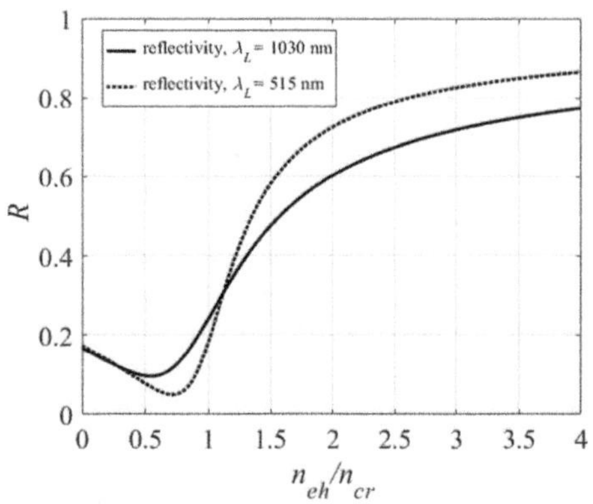

FIG. 1. The reflectivity of the pseudo-metal layer in diamond as a function of the electron-hole plasma density at both wavelengths.

of the excited electron-hole plasma density n_{eh} during the multipulse femtosecond laser interaction, where the reflectivity R is calculated for normal incident radiation. This indicates that the diamond surface takes a metallic character with high free electron density. We can remark that the reflectivity at 515 nm is higher than 1030 nm due to the difference in electron density excited (note that x-axis is different for each plot because $n_{cr}(1030 \text{ nm}) \neq n_{cr}(515 \text{ nm})$).

Under our experimental conditions, we have observed morphological changes of the femtosecond laser irradiated diamond surface with the variation of the scanning speed υ (i.e. number of pulses N) at a fixed fluence F, always keeping the laser polarization parallel to the translation direction of the sample. Figure 2 shows SEM images taken on the irradiated diamond surface for the two selected wavelengths. According to these images, the femtosecond laser is inducing nanostructures with sub-wavelength period (HSFL) perpendicular to the polarization of the laser beam. Previous results suggest that random nanoablation would result from the lattice disorder and intense near-field.[14,16] However, we have observed in experiment an ordered nanostructure and therefore we suggest that a periodic process is responsible for the ordering of the nanoablation. Moreover, we observed in Fig. 2 that the period of the nanostructure decreases when the number of pulse increases. With each incident femtosecond laser pulse, the density of excited electrons increases, resulting in a decrease of the real part of dielectric function of the irradiated layer, allowing for plasmonic excitation. Therefore, we suggest that the formation of HSFL can be attributed initially to the excitation of SPP in the surface layer. From our simulations, we show the importance of the plasmonic excitation on the formation of the HSFL, and also shed insight into other physical processes involved. It can also be observed in Fig. 2 that there is ablation of the material at the center of the micro-grooves, where the peak fluence of the Gaussian beam is highest. This effect is more noticeable for lower scan speeds, as heat accumulation is more pronounced for a greater number of pulses interacting within the focal volume in the pseudo-metal layer.[20] Despite the ablation at the center of the tracks, nanostructures can be observed at the sides of the micro-grooves, indicating that the periodic nanostructures are produced at fluences slightly below the ablation threshold.[11]

Figure 3 shows SEM images of the HSFL formed with fluences ranging from 7.96, 8.36 and 9.15 J/cm^2 (average laser power 20, 21 and 23 mW) at 1030 nm wavelength. The periods of the nanostructures decreases slightly with increasing fluence, owing to a strong disorder of the lattice

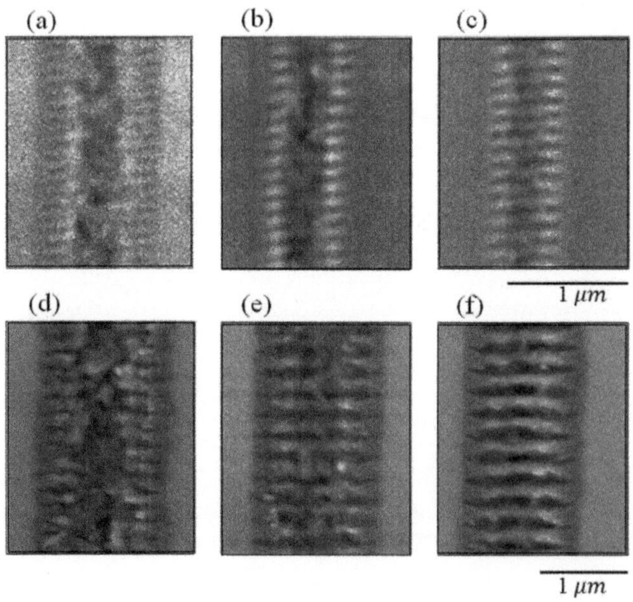

FIG. 2. SEM images of diamond surfaces irradiated with femtosecond laser pulses polarized parallel to the scan direction for (a) $\upsilon = 3$ mm/s, (b) $\upsilon = 6.5$ mm/s, (c) $\upsilon = 8$ mm/s at $\lambda_L = 515$ nm and fluence (power) $F = 6.36$ J/cm^2 (4 mW), (d) $\upsilon = 3$ mm/s, (e) $\upsilon = 6$ mm/s, and (f) $\upsilon = 8$ mm/s at $\lambda_L = 1030$ nm and fluence (power) $F = 8.36$ J/cm^2 (21 mW).

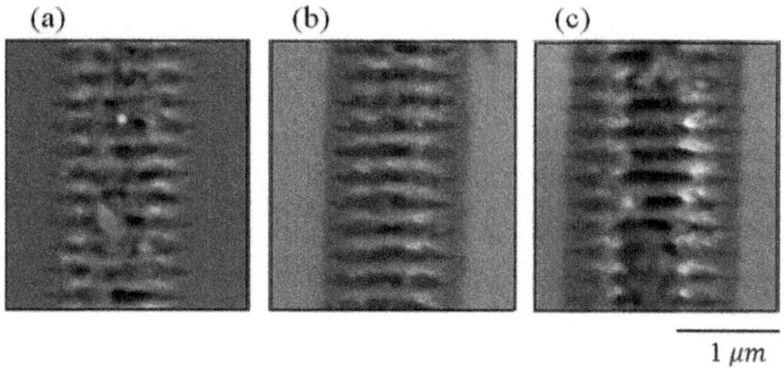

FIG. 3. SEM images of diamond surfaces irradiated with the parallel polarized laser pulses at λ_L=1030 nm with υ = 8 mm/s at (a) F= 7.96 J/cm^2, (b) F= 8.36 J/cm^2 and (c) F= 9.15 J/cm^2 (power 20, 21 and 23 mW respectively).

during the interaction with the femtosecond laser, which is attributed to the incubation of nano-ripples when greater energy density is delivered to the lattice.

Figure 4 shows a recapitulation of the effect of fluence (power) and speed (number of pulses) on the nanostructure period for 515 nm (a) and 1030 nm (b) wavelengths. The period of the nanostructures increases with scan speed for both wavelengths. The minimum period is 110 nm and 190 nm for 515 nm and 1030 nm wavelengths, respectively, for the lowest speed (3 mm/s). From this, we deduce that the formation of nano-ripples has a strong dependence on the wavelength, with the period increasing with wavelength.

One can also observe, according to Fig. 4, that the nanostructures are not obtained for all values of fluence or speeds, due to the large band gap of diamond (5.5 eV), where the electrons can be excited by one photon if it has enough energy, or by a multiphoton absorption if the photon energy is below the gap.[33] To observe nanostructures for powers lower than 4 mW (F = 6.36 J/cm^2) at 515 nm writing wavelength Fig. 4(a), we need to increase the number of pulses by decreasing the scan speed. However at the green wavelength, we found a narrower energy window between ablation and no modification, making it more difficult to observe the formation of nanostructures.

Next we simulate the evolution of the periodicity of nanostructure HSFL type via the generalized plasmonic model proposed above. When the pulse number increases, the excited electron density n_{eh} increases until the condition $\varepsilon_{pm} = -\varepsilon_{air}(m = 0)$ is reached. At this point, the **F**irst **O**rdered **N**anostructure (**FON**) can be observed, with a period of Λ_{FON} = 250 nm at λ_L = 515 nm (Fig. 5(a)), and a period of Λ_{FON} = 510 nm at λ_L = 1030 nm (Fig. 5(b)). This corresponds to the minimum Plasmon Resonance RP_{min}. This RP_{min} is caused when the plasmons tend to coherently couple with the incident laser wave at the pseudo–metal/air interface as if the original material (diamond) did not exist.

The smallest nanostructure or the **L**ast **O**rdered **N**anostructure (**LON**) can be observed when $\varepsilon_{pm} = -\varepsilon_{di}(m = 1)$ with period of Λ_{LON} = 83 nm and 197 nm at wavelength 515 and 1030 nm, respectively (Fig. 5). This corresponds to the maximum Plasmon Resonance RP_{max}. In the data shown in Fig. 5, we are varying the factor of mixed state for 0 to 1 for the purpose of seeing the

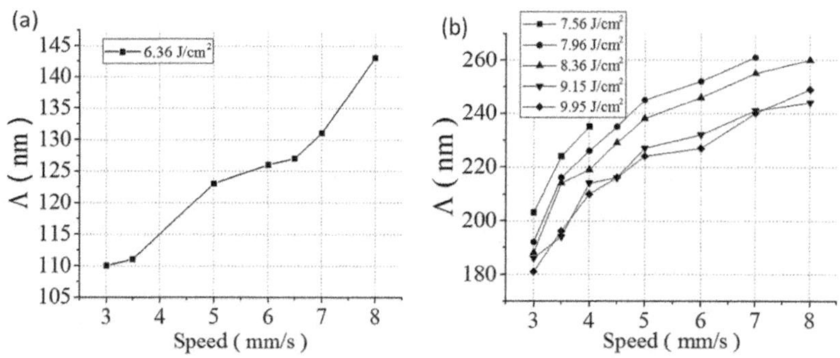

FIG. 4. Evolution of periodicity with writing speed and fluence at wavelength 515 nm (a) and 1030 nm (b).

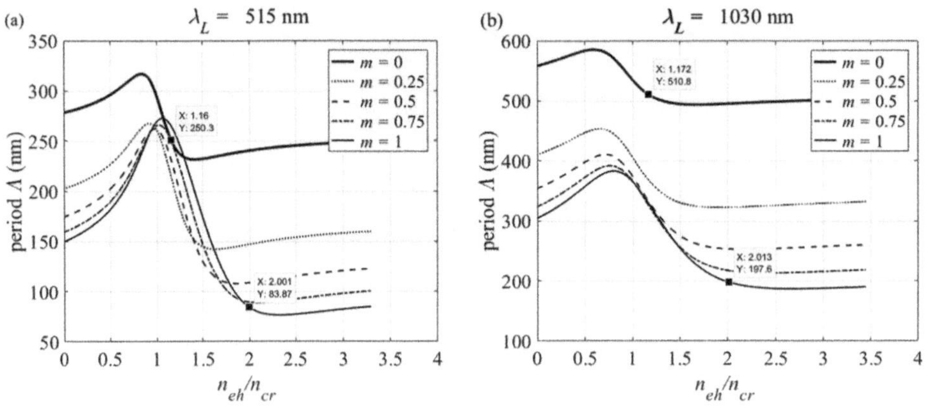

FIG. 5. Calculation of LIPSS period evolution as a function of electron-hole plasma excitation by varying the dielectric surrounding medium, at wavelength 515 nm (a) and 1030 nm (b). X is the fraction n_{eh}/n_{cr} correspond to $\varepsilon_{pm} = -\varepsilon_{air}$ or $-\varepsilon_{di}$ where Y is the period, then (X, Y) represent **FON** or **LON**.

effect of the penetration of plasmonic evanescent wave at both interfaces of the pseudo-metal layer on the periodicity of the nanostructures with the increase of the number of incident laser pulses. Fig. 6 shows the penetration depth of the irradiated wave as a function of electron-hole plasma excitation, $d = \frac{c}{2\omega k}$ (Ref. 18) where all the parameters were previously defined. d is the penetration depth of laser during the irradiation. When the number of pulses increases, the density of excited electrons and the absorption coefficient k increase, then the penetration depth decreases. Based on (Fig. 6), the depth of the pseudo-metal layer is tens of nanometers. Therefore, the pseudo metal layer remains a mix state (from the point of view of the dielectric function) between air and bulk diamond when ε_{pm} decrease from $-\varepsilon_{air}$ to $-\varepsilon_{di}$. Therefore, we can assume that there is a transverse migration of the SPP from the pseudo-metal/air interface to the diamond/pseudo-metal as the number of femtosecond laser pulses increases, where the nanostructure period decreases from Λ_{FON} to Λ_{LON}. So we deduce that the ordered nanostructures (HSFL type) can be observed only inside the range $[RP_{\min}, RP_{\max}]$. It is important to note that this model is valid in a specific carrier density corresponding to the plasmonic excitation. Our experimental results are in agreement with these calculations, where the measured periods (Fig. 4) are included in the plasmonic range calculated for the both wavelengths (Fig. 5). The period Λ_{LON} is also called saturation period as it is the minimum pitch of the nanostructures.[34]

We have highlighted in Fig. 6 the case corresponding to $RP_{\min}$ and $RP_{\max}$. So based on our reasoning above the d_{max} which corresponds to $RP_{\max}$ is the smallest depth possible, taking on values $d_{max} = 61$ nm, 32 nm at $\lambda_L = 1030$ nm, 515 nm respectively, therefore, all the absorbed energy at the pseudo-metal layer with depth d_{max} to be reserved to the SPP excitation, and this is why the period Λ_{LON} can enhance the surface absorption.[4,5] Thus, it is important to find this period if the aim is to

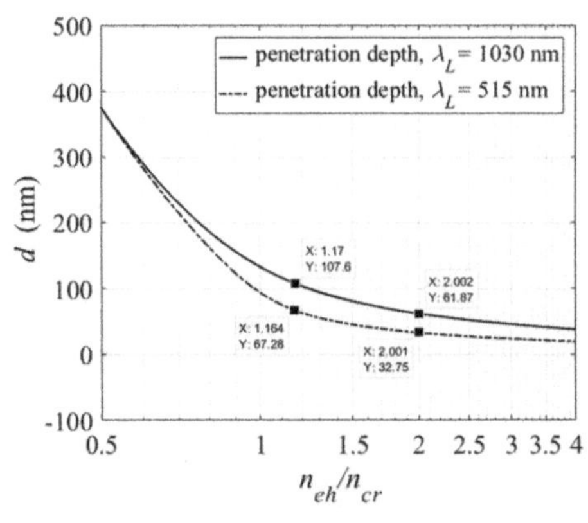

FIG. 6. Penetration depth of the irradiated wave as a function of electron-hole plasma excitation at both wavelengths.

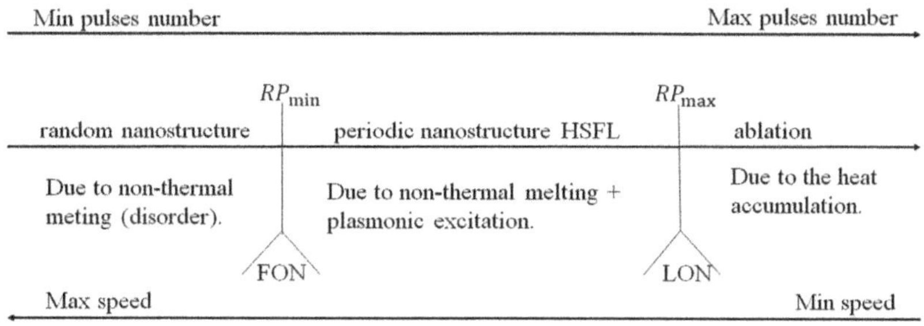

FIG. 7. Evolution of nanostructure formation with the increase in the number of pulses (i.e. reduction of speed).

obtain high efficiency micro–solar cells with femtosecond laser pulses, to avoid energy dissipation inside the material. It has been confirmed experimentally that with only LIPSS the absorption can be enhanced to more than 90% via plasmonic excitation,[2–5] because the plasmons wave oscillateds at a scale of λ_{sp} which is less than λ_L. Therefore, the plasmons are able to efficiently absorb the incident light, i.e. they are able to respond together at the same point of time and space to the incident radiation.

Our model for the formation of HSFL in diamond during the interaction with femtosecond laser pulses can also be applied in general to dielectrics and semiconductors, taking into account the effect of plasmonic excitation in non-thermal melting dynamics. Experimental results show that the period of the nanostructures decreases when the number of laser pulses increases. On the other hand, during the analysis of the LIPSS period versus the electron-hole plasma excitation, we see that the period Λ decreases when the factor of mixed states m decreases until reaching the saturation at the maximum Plasmonic Resonance RP_{max}. This implies that the pseudo–metal layer transitions from crystalline diamond to a graphitic phase occurs progressively when the period decreases from Λ_{FON} to Λ_{LON}. As this transition is due to non-thermal melting (disorder),[19] we can conclude that the ordered nanostructures are formed by a combination of two processes: the random nanoablation due to the near field induced during irradiation, and the plasmonic excitation, that plays a complementary and simultaneous role to order the nanoablation because of its specific wave periodicity (frequency).

Finally, as the random nanostructures are generated by random disorder, it can be called ordinary disorder, while as HSFL are due to disorder forced by the plasmon wave, it can be called plasmonic disorder. The overview of this description is presented in Fig. 7, which also helps to summarize the physical processes involved in the formation of the HSFL during the irradiation with femtosecond laser pulses. It is shown that the origin of random nanostructures is related to one process (disorder), in contrast to order nanostructures of HSFL type that are related to a combination of two processes (disorder + plasmonic excitation).

V. CONCLUSION

In this paper, we have studied the origin of the formation of periodic nanostructures (HSFL) by femtosecond laser pulses on the surface of diamond. The experimental results have shown that the periodicity observed depends on the pulse number for both wavelengths used (1030 nm and 515 nm). We suggest that the non-thermal melting (disorder) and plasmonic excitation have a very important role in the dynamic process. We propose a generalized plasmonic model that can be used directly to simulate the evolution of periodic nanostructure HSFL for dielectric or semiconductor materials at any wavelength. Using this model for the particular case of diamond allows us to calculate the period of nanostructures formed on the surface, and we have confirmed with the experimental results that the period decreases when the number of pulses increases via the variation of both carrier density and the mixed state factor. We conclude that the periodic nanostructures can only be observed in the range of plasmonic excitation bounded by a minimum and maximum plasmon resonance $[RP_{min}, RP_{max}]$, whereas random nanostructures can be formed below RP_{min}. Finally, we propose a physical description that shows the importance of plasmonic excitation in the formation of sub-wavelength nanostructures, where the plasmonic excitation effect for non-thermal melting dynamics (disorder) can be considered as a fundamental process for the periodic nanoablation as origin of HSFL nanostructures.

[1] J. Bonse, S. Höhm, S. V. Kirner, A. Rosenfeld, and J. Krüger, "Laser-induced periodic surface structures—A scientific evergreen," IEEE J. Sel. Topics Quantum Electron **23**, 9000615 (2017).

[2] A. Y. Vorobyev and C. Guo, "Enhanced absorptance of gold following multipulse femtosecond laser ablation," Phys. Rev. B **72**, 195422 (2005).

[3] A. Y. Vorobyev and C. Guo, Solar absorber surfaces treated by femtosecond laser, *Biosciences (BIOSCIENCESWORLD), IEEE International Conference*, 135 (2010).

[4] P. Calvani, A. Bellucci, M. Girolami, S. Orlando, V. Valentini, A. Lettino, and D. M. Trucchi, "Optical properties of femtosecond laser-treated diamond," App. Phys. A **117**, 25 (2014).

[5] P. Calvani, A. Bellucci, M. Girolami, S. Orlando, V. Valentini, R. Polini, and D. M. Trucchi, "Black diamond for solar energy conversion," Carbon **105**, 401 (2016).

[6] E. Rebollar, J. R. V. de Aldana, I. Martín-Fabiani, M. Hernández, D. R. Rueda, T. A. Ezquerra, C. Domingo, P. Moreno, and M. Castillejo, "Assessment of femtosecond laser induced periodic surface structures on polymer films," Physical Chemistry Chemical Physics **15**, 11287 (2013).

[7] M. Huang, F. Zhao, Y. Cheng, N. Xu, and Z. Xu, "Origin of laser-induced near-subwavelength ripples: Interference between surface plasmons and incident laser," ACS Nano **3**, 4062 (2009).

[8] Q. Wu, Y. Ma, R. Fang, Y. Liao, Q. Yu, X. Chen, and K. Wang, "Femtosecond laser-induced periodic surface structure on diamond film," Phys. Rev. Lett **82**, 1703 (2003).

[9] T. Chen, H. Yang, E. Zhao, J. Qian, J. Hao, J. Han, W. Tang, and H. Zhu, "Laser induced nanostructures on the zinc surface," Journal of Russian Laser Research **35**, 561 (2014).

[10] T. J.-Y. Derrien, R. Koter, J. Krüger, S. Höhm, A. Rosenfeld, and J. Bonse, Plasmonic formation mechanism of periodic 100-nm-structures upon femtosecond laser irradiation of silicon in water, J. Appl. Phys **116**, 074902 (2014).

[11] M. Shinoda, R. R. Gattass, and E. Mazur, "Femtosecond laser-induced formation of nanometer width grooves on synthetic single-crystal diamond surfaces," Appl. Phys **105**, 053102 (2009).

[12] D. Dufft, A. Rosenfeld, S. K. Das, R. Grunwald, and J. Bonse, "Femtosecond laser-induced periodic surface structures revisited: A comparative study on ZnO," J. Appl. Phys **105**, 034908 (2009).

[13] R. Le Harzic, F. Stracke, and H. Zimmermann, "Formation mechanism of femtosecond laser-induced high spatial frequency ripples on semiconductors at low fluence and high repetition rate," J. Appl. Phys **113**, 183503 (2013).

[14] G. Miyaji and K. Miyazaki, "Origin of periodicity in nanostructuring on thin film surfaces ablated with femtosecond laser pulses," Opt. Express **16**, 16265 (2008).

[15] K. Miyazaki and G. Miyaji, "Periodic nanostructure formation on silicon irradiated with multiple low-fluence femtosecond laser pulses in water," Physics Procedia **39**, 674 (2012).

[16] G. Miyaji and K. Miyazaki, "Role of multiple shots of femtosecond laser pulses in periodic surface nanoablation," Appl. Phys. Lett **103**, 071910 (2013).

[17] G. Miyaji and K. Miyazaki, "Fabrication of 50-nm period gratings on GaN in air through plasmonic near-field ablation induced by ultraviolet femtosecond laser pulses," Opt. Express **24**, 4648 (2016).

[18] M. Straub, M. Afshar, D. Feili, H. Seidel, and K. König, "Surface plasmon polariton model of high-spatial frequency laser-induced periodic surface structure generation in silicon," J. Appl. Phys. **111**, 124315 (2012).

[19] M. Forster, C. Huber, O. Armbruster, R. Kalish, and W. Kautek, "50-nanometer femtosecond pulse laser induced periodic surface structures on nitrogen-doped diamond," Diamond and Related Materials **74**, 114 (2017).

[20] S. M. Eaton, H. Zhang, M. Li Ng, J. Li, W.-J. Chen, S. Ho, and P. R. Herman, "Transition from thermal diffusion to heat accumulation in high repetition rate femtosecond laser writing of buried optical waveguides," Opt. Express **16**, 9443 (2008).

[21] *Near-field optics and surface plasmon polaritions*, edited by S. Kawata (Springer, Berlin, 2001), p. 19.

[22] *Optical Properties of Diamond*, edited by A. M. Zaitsev (Springer, Bochum, 2001).

[23] K. Sokolowski-Tinten and D. von der Linde, "Generation of dense electron-hole plasmas in silicon," Phys. Rev. B **61**, 2643 (2000).

[24] *Diamond: Electronic Properties And Applications*, edited by L. S. Pan and D. R. Kania (New York, 1995), pp. 251–252.

[25] F. Nava, C. Canali, C. Jacoboni, L. Reggiani, and S. F. Kozlov, "Electron effective masses and lattice scattering in natural diamond," Solid State Communications **33**, 475 (1980).

[26] L. Reggiani, S. Bosi, C. Canali, F. Nava, and S. F. Kozlov, "On the lattice scattering and effective mass of holes in natural diamond," Solid State Communications **30**, 333 (1979).

[27] G. Miyaji, W. Kobayashi, and K. Miyazaki, "Reflectivity change in nanoscale modification of DLC film with femtosecond laser pulses," J. Laser Micro Nanoeng **2**, 146 (2007).

[28] G. Sciaini, M. Harb, S. G. Kruglik, T. Payer, C. T. Hebeisen, F.-J. M. z. Heringdorf, M. Yamaguchi, M. H.-v. Hoegen, R. Ernstorfer, and R. J. D. Miller, "Electronic acceleration of atomic motions and disordering in bismuth," Nature **458**, 56 (2009).

[29] S. K. Sundaram and E. Mazur, "Inducing and probing non-thermal transitions in semiconductors using femtosecond laser pulses," Nature Materials **1**, 217 (2002).

[30] C. V. Shank, R. Yen, and C. Hirlimann, "Time-resolved reflectivity measurements of femtosecond-optical- pulse induced phase transitions in silicon," Phys. Rev. Lett **50** (1983).

[31] T. Zier, E. S. Zijlstra, A. Kalitsov, I. Theodonis, and M. E. Garcia, "Signatures of nonthermal melting," Structural Dynamics **2**, 054101 (2015).

[32] *Handbook of Optical Constants of Solids*, edited by E. D. Palik (Elsevier, Academic Press, USA, 1985), p. 192.

[33] D. Von der Linde, K. Sokolowski-Tinten, and J. Bialkowski, "Laser–solid interaction in the femtosecond time regime," Applied Surface Science **109**, 1 (1997).
[34] S. Richter, M. Heinrich, S. Döring, A. Tnnermann, and S. Nolte, "Formation of femtosecond laser-induced nanogratings at high repetition rates," Appl. Phys. A **104**, 503 (2011).
[35] F. Garrelie, J. P. Colombier, F. Pigeon, S. Tonchev, N. Faure, M. Bounhalli, S. Reynaud, and O. Parriaux, "Evidence of surface plasmon resonance in ultrafast laser-induced ripples," Optics Express **19**, 9035 (2011).

9

Role of phase synchronisation in turbulence

Sara Moradi,[1,a] Bogdan Teaca,[2,b] and Johan Anderson[3,c]
[1]*Laboratory for Plasma Physics - LPP-ERM/KMS, Royal Military Academy, 1000 Brussels, Belgium*
[2]*Applied Mathematics Research Centre, Coventry University, Coventry CV1 5FB, United Kingdom*
[3]*Department of Space, Earth and Enivronment Sciences, Chalmers University of Technology, SE-412 96 Göteborg, Sweden*

The role of the phase dynamics in turbulence is investigated. As a demonstration of the importance of the phase dynamics, a simplified system is used, namely the one-dimensional Burgers equation, which is evolved numerically. The system is forced via a known external force, with two components that are added into the evolution equations of the amplitudes and the phase of the Fourier modes, separately. In this way, we are able to control the impact of the force on the dynamics of the phases. In the absence of the direct forcing in the phase equation, it is observed that the phases are not stochastic as assumed in the Random Phase Approximation (RPA) models, and in contrast, the non-linear couplings result in intermittent locking of the phases to $\pm \pi/2$. The impact of the force, applied purely on the phases, is to increase the occurrence of the phase locking events in which the phases of the modes in a wide k range are now locked to $\pm \pi/2$, leading to a change in the dynamics of both phases and amplitudes, with a significant localization of the real space flow structures.

I. INTRODUCTION

In electrically neutral fluids or in plasma flows the presence of nonlinear interactions can lead to the development of turbulence. In general, turbulence is characterized by energetic couplings between different scales of a flow. However, in the context of turbulence driven transport, such as the case of magnetically confined plasmas[1,2] or the diffusion of cosmic rays,[3] typical flow structures are identified by dominant modes and the global turbulent state is approximated by a superposition of linear contributions (waves in general). These theoretical studies consider the amplitudes of the fluctuating quantities, but disregard the dynamics of the phases by using the so-called random-phase approximation (RPA) for which the existence of a Chirikov-like criterion for the onset of wave stochasticity[4,5] is assumed. In this approximation one assumes that the dynamical amplitudes have a slow variation compared to the rapid change of the phases, which are considered to be distributed uniformly over a 2π interval.[1,2]

The RPA approach is widely used in turbulence transport theory since it has a clear intuitive picture. While it has not been rigorously formulated, there have been attempts in this direction.[6] Its underlying assumption of stochasticity for the phases of Fourier modes in a nonlinearly interacting waves cannot be justified since the phases as well as the amplitudes evolve due to nonlinear interactions that act on the same time scales. Thus, stochastization of phases and amplitudes, if any, occur on the same time scale, see further discussion in Refs. 6,7.

---

[a]smoradi@ulb.ac.be
[b]bogdan.teaca@coventry.ac.uk
[c]anderson.johan@gmail.com

The correlation of phases due to the existence of nonlinear interactions determines how the spectral energy density, information contained in the amplitude of the Fourier modes, is distributed in real space. This is linked to the intermittent nature of turbulence. Ignoring the phase synchronization, or phase locking between scales, neglects the impact made by coherent structures.[8] Phase dynamics play the crucial role in the self-organization and the formation of coherent structures, as was shown in Ref. 9 for plasma turbulence. The self-organization phenomena is not limited to turbulence, albeit the interest of this work, and is common in other dynamical systems, notable examples being: biological clocks, physiological organisms and chemical reactors.[10–19]

In this work, we investigate the role of phase dynamics on turbulent flows. For this analysis, we chose a simple system described by the Burgers equation with an known external force. The prescribed force allows us to control its individual impact on the norm and the phases of the system. In the presence of a force that leaves the phases unaffected, we observe that the phases are not randomized as assumed in the RPA models, and in contrast, the non-linear couplings result in intermittent locking of the phases to $\pm \pi/2$. Employing this force to inject the same power in the system, we can affects the structures developed by the flow through a phase force. Due to the co-evolving and interacting amplitudes and phases, a forcing interacting purely on the phases increases the number of phase locking events, during which the phases are locked to $\pm \pi/2$. In contrast to the stochastically driven oscillator models studied previously[20] the present phase force may be thought of as a strongly coherent force that ultimately change the dynamics of both phases and amplitudes.

In the following we will present the details of our model and the results of the numerical simulations, aiming to elucidate on a series of questions: In which way does a pure phase force affect the structures of the flow? As the dynamics of turbulence is affected and the phase correlations change, how are the spectral energy transfers affected and finally, does a phase force impede or tend to enhance the validity of the RPA approximation?

II. THE SYSTEM OF INTEREST

A. The Burgers equation

The Burgers equation is a simplified mathematical model for the motion of a viscous compressible flow.[21,22] In one dimension (x is the spatial coordinate), it reads

$$\partial_t u(x,t) + u(x,t)\partial_x u(x,t) = \nu \partial_{xx} u(x,t) + \delta f(x,t), \tag{1}$$

where u is the fluctuating velocity field, ν is the kinematic viscosity, and $\delta f(x, t)$ is a prescribed external force.

The Fourier representation ($x \to k$) of the Burgers equation, together with the polar form of the complex fields, i.e.

$$u(k,t) = \psi_k(t) e^{i\theta_k(t)}, \tag{2}$$

$$\delta f(k,t) = f_k(t) e^{i\phi_k(t)}, \tag{3}$$

allows for a decomposition of the dynamics in terms of the norms and the phases as

$$\partial_t[\psi_k(t)e^{i\theta_k(t)}] = -\sum_{k=p+q} ip\,\psi_p(t)\psi_q(t)e^{i(\theta_p(t)+\theta_q(t))} - \nu k^2 \psi_k(t) e^{i\theta_k(t)} + f_k e^{i\phi_k(t)}. \tag{4}$$

Hence, we obtain two coupled equations

$$\partial_t \theta_k = -\sum_{k=p+q} \frac{p\psi_p\psi_q}{\psi_k}\cos(\theta_p + \theta_q - \theta_k) + \frac{f_k}{\psi_k}\sin(\phi_k - \theta_k), \tag{5}$$

$$\partial_t \psi_k = +\sum_{k=p+q} p\psi_p\psi_q \sin(\theta_p + \theta_q - \theta_k) - \nu k^2 \psi_k + f_k\cos(\phi_k - \theta_k). \tag{6}$$

In the absence of an external force, the dynamics of the phases and the norms are linked by the nonlinear terms (first terms on the left hand side of eqs. 5 and 6), while the viscous term acts solely to damp the amplitude of the Fourier modes. Here, the energy for a scale (denoted by the wavenumber k) is defined as $E(k,t) = \frac{1}{2}|u(k,t)|^2 = \frac{1}{2}\psi_k^2$. The subsequent energy balance equation, obtained from multiplying Eq. (6) by ψ_k, shows that the energy transfer to a scale k, i.e.

$$T(k,t) = \sum_{k=p+q} p\,\psi_k\psi_p\psi_q \sin(\theta_p + \theta_q + \theta_k), \tag{7}$$

is maximum for $\sin(\theta_p + \theta_q + \theta_k) = \pm 1$. For maximal energy transfer and in the absence of forcing, the nonlinear contribution to the time variation of the phases (Eq. 5) is zero and all the phases will be locked (i.e. $\partial_t\theta_k = 0$).

B. The prescribed force

Typically, the external force represents a mechanism to control the energy injection in the system and obtain an energetic steady state. This is highly desirable in the study of dissipative turbulent systems. Here, in a small wavenumber corresponding to the large scales in the system, we choose an *amplitude* force proportional to the velocity,

$$\delta f_A(k,t) = \frac{\epsilon_A}{N_A}\frac{u_k}{|u_k u_k^*|} = \frac{\epsilon_A}{N_A}\frac{1}{\psi_k}e^{i\theta_k(t)}, \tag{8}$$

which allows for the phases of the force to match the phases of the flow for each mode, i.e. $\phi_k = \theta_k$. The number of modes in the forcing range is denoted by N_A and the same amount of energy is injected in each mode. In the steady state, the energy injected by the force balances viscous dissipation ($\epsilon_A \approx \sum_k 2\nu k^2 \psi_k^2$). By construction, we see that $\delta f_A(k)$ does not contribute to the evolution of the phases, i.e. the sin function is zero in Eq. (5). This represents a good test-bed to determine the emerging phase correlations solely due to the nonlinear interactions.

To control the phases in a steady state, a general form of the external force can be used,

$$\delta f(k,t) = \delta f_A(k,t) + \delta f_P(k,t), \tag{9}$$

where we take the *phase* force contribution as

$$\delta f_P(k,t) = \frac{\epsilon_P}{N_P}\frac{u_k}{|u_k u_k^*|}e^{i\pi/2} = \frac{\epsilon_P}{N_P}\frac{1}{\psi_k}e^{i[\theta_k(t)+\pi/2]}. \tag{10}$$

Thus, the $\delta f_P(k)$ contribution to the force does not affect the amplitudes of the modes. Here, the δf_P part to the overall force is acting on the same prescribed wavenumber interval as δf_A (i.e. $N_P = N_A$) and ϵ_P represents a control parameter. Note that the reality condition implies that $f(k)^* = f(-k)$. Directly affecting the evolution of the phases, this force allows to gauge the impact made on structure formation by collective synchronization.

C. Numerical set-up

We solve the equations using a Runge-Kutta 4th order scheme (RK4) with an adaptive time stepping length dt while the sampling time step is $\Delta t = 0.001$. The numerical integration is performed in a periodic domain of length $L = 2\pi$ using N modes (i.e. $k \in [-N/2, N/2]$) and the nonlinear term is computed following a pseudo-spectral method with three-half dealiasing method. In our analysis we use $N = 2048$ modes and consider the initial value $u_k(0) = k^{-2}\exp(i\theta_k)$, where θ_k are taken as random with equal probability between $[0, 2\pi]$. In this study a weak level of Burgers' turbulence is considered, thus allowing for the chosen resolution to be sufficient for the lessons drawn in this work.

Throughout the work, we consider that the force acts in a small wavenumber interval ($|k| \in [3,7]$) and we take the same energy injection level, i.e. $\epsilon_A = 4 \times 10^{-3}$. The distinction between the runs analyzed is given by the value of the ϵ_P parameter. In Fig. 1, we plot the energy spectra for these runs. We recover the k^-2 spectral slope in the inertial range, a fact expected for Burgers turbulence. The fact that the system is well resolved is seen by the values of the energy in the highest modes (values that reach machine precision).

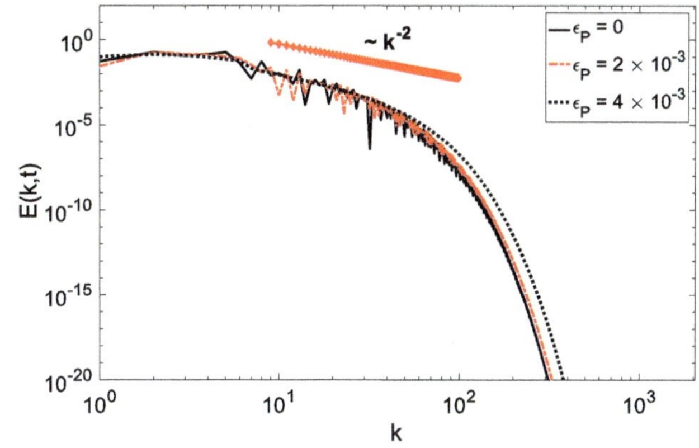

FIG. 1. Typical energy spectra for the steady states. The k^{-2} spectral slope is recovered, a fact expected for Burgers turbulence. Here $\epsilon_A = 4 \times 10^{-3}$.

III. RESULTS

A. Localization of the flow structures in the real space

We know that the phases and their correlations are important in the development of structures in the flow and are ultimately responsible for the emergence of intermittency. However, how is the phase force employed here affecting the structures of our simple flow?

Fig. 2 shows the computed solution of the Burgers equation, i.e. $u(x, t)$. Here, we have increased the strength of the phase force ϵ_P in three steps, with equal time intervals, from 0 to 2×10^{-3} and subsequently to 4×10^{-3} while maintaining $\epsilon_A = 4 \times 10^{-3}$ constant throughout the simulation. During the time interval without the phase force i.e. $t = 0 - 3 \times 10^3$ the turbulent fluctuations are born, and dissipated homogeneously in (x, t) space. By increasing ϵ_P to 2×10^{-3}, the propagating fluctuations are more and more localized in x as well as follow a preferred direction (up-down). This trend is further amplified as the ϵ_P is increased to 4×10^{-3}, where we observe that smaller flow structures rapidly disappear while one structure becomes dominant and propagates as one single traveling wave.

B. Dynamic of the phases

To study the dynamic of the phases we have computed the order parameter shown in Fig. 3 (top). An analytic expression for the order parameter,

$$Z(t) = \sum_{k=1}^{N} \frac{1}{N} e^{i\theta_k}, \tag{11}$$

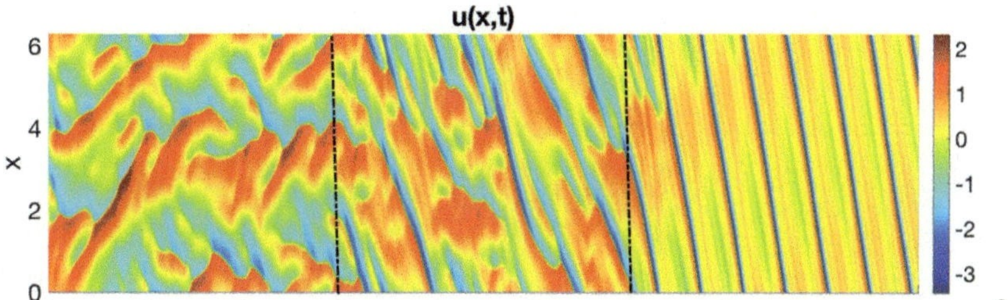

FIG. 2. The solution of Burgers equation, u, as a function of x and t. Here $\epsilon_A = 4 \times 10^{-3}$. The value of ϵ_P is increased in 3 steps denoted by straight lines from 0 to 2×10^{-3} and than to 4×10^{-3}.

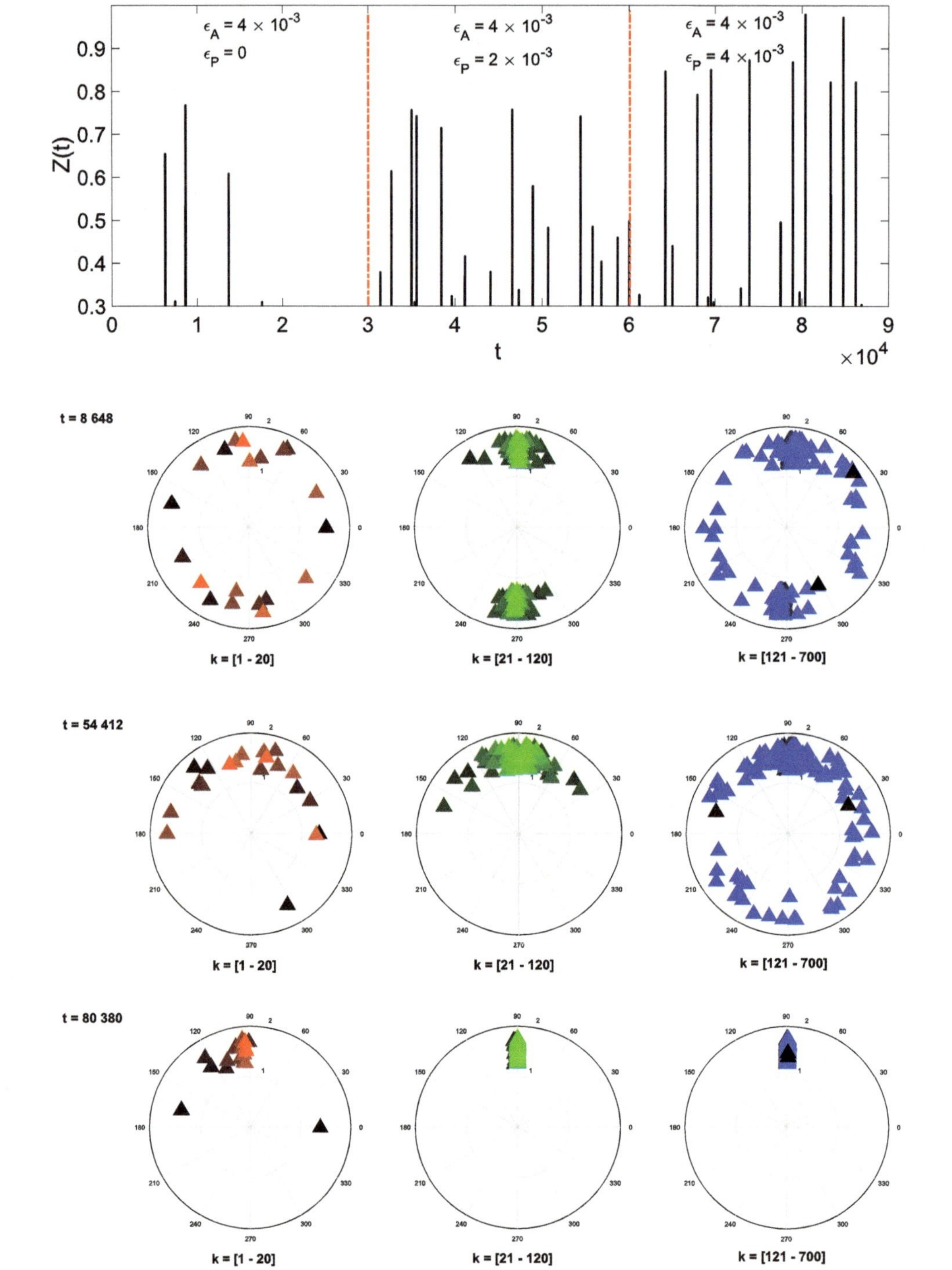

FIG. 3. (top) The order parameter $|Z(t)|$ as function of time, and polar views of the phases at three different time slices corresponding to the highest peaks in $|Z(t)|$ in each case. Here, as the mode number, k, is increased, the colors are varied from darker to lighter shades.

was derived by Kuramoto that describes the quality of the synchronization of the ensemble of phases with $0 \leq |Z| \leq 1$, where $|Z| = 0$ corresponds to a completely asynchronous state, while $|Z| = 1$ corresponds to a totally synchronous state. Here, to eliminate the numerical noise corresponding to the highest modes (modes unexcited from a physical perspective), we have used a low pass filter with $k_c = 700$ and, for a better illustration of the differences, we have offset the y-axis by 0.3.

We observe that in the absence of the phase force, the phases of the fluctuations are asynchronous with the exception of a few occasions where the phases become partially synchronized with peaks in $|Z|$ of ~ 0.8. By applying the phase force, these occurrences become more frequent, and from 3 significant events in the case with no forcing phase, it increases to 11 and 14 significant events in the cases where phase forcing was applied. Also the quality of synchronization has increased strongly reaching peaks of the order of $|Z| \sim 1$ in the last case.

The synchronization of the phases at these occurrences are more clearly observed from the polar view representations of the phases shown in Figs. 3. In the absence of the phase force, at a phase locking event no synchronization between the phases of $|k| \in [1, 20]$ modes is obvious, and the phases of the $|k| \in [21, 121]$ show synchronization to $\pm\,\pi/2$ with equally divided numbers locked to $\pi/2$ and $-\,\pi/2$. Here, we find that the quality of the synchronization in the high-k range of $|k| \in [121, 700]$ is not as good. This is induced by the dissipation since as the dissipation dampens the amplitudes, it results in a reduction of the non-linear coupling strength, i.e. $\psi_p \psi_q / \psi_k$, between the phases. Indeed this was tested and we observed that by increasing ν, the range of the stochastic phases expands further into lower k region, whereas by decreasing ν it will recede into higher k's. Thus, the non-linear interactions result in phase locking in the inertial range rather than stochastization of the phases as is commonly assumed in RPA based models. As seen in the two last panels of the polar views in Fig. 3, a force applied exclusively to the phases facilitates phase synchronization allowing for further locking of modes to $\pi/2$ alone. In Ref 6, it is found that synchronization of the phases in wave turbulence occurs over short time intervals as is seen here.

C. Maximization of the energy transfers due to locking

When synchronized, the phases of all three cases shown in the polar representation in Fig. 3, are $\pm\,\pi/2$ which yield maximum energy transfer for all possible triads, $k + p + q = 0$. Because of the wide locked range at the inertia/dissipative scales i.e. mid-to high-k modes, the locking to $\pi/2$ phase (note that due to the reality condition $u^*(k) = u(-k)$ the phase of the negative wave numbers will be locked to $-\,\pi/2$, not shown in the polar views) should allow for an enhanced removal of the inertia scale energies to the small dissipative scales.

From the definition of the energy transfer, Eq. (7), we see that in general the maximal transfer is obtained for $|\sin(\theta_p + \theta_q + \theta_k)| = 1$ and can be conceptually defined as

$$T^{\max}(k,t) = \sum_{k=p+q} p\,\psi_k \psi_p \psi_q. \tag{12}$$

As we increase the intensity of the phase force, the $\pm\,\pi/2$ locking events become more frequent and $T(k)$ should tend towards $T^{\max}(k, t)$ more often. To capture this tendency on a simple curve we use the definition provided by Ref. 23 for the intensity of the nonlinear transfers and define the maximal ratio as:

$$M(t) = \frac{\sum_k |T(k)|}{\sum_k |T^{\max}(k)|}. \tag{13}$$

Not only that $M(t)$ tends towards unity during synchronization events, it also allows us to gauge the intensity of the transfers compared to the maximal allowed for the reminder of the time. Figure 4 shows the values of M as function of time for the three considered cases (from top to bottom the value of ϵ_P is increased) during a time window around the locking events shown in polar panels in Fig. 3. Here, also the values of the order parameter $|Z(t)|$ are illustrated for convenience. As can be seen here, in the case of $\epsilon_P = 0$ this ratio, the ratio is around $M \sim 0.5$ and at the moment of strong phase locking, as indicated by the peaks in $|Z(t)|$, it increases significantly to $M \sim 0.85$. Similarly, in the case of $\epsilon_P = 2 \times 10^{-3}$ and $\epsilon_P = 4 \times 10^{-3}$, the moments of phase locking coincides with the increase in the ratio of transfers, M. In the case of $\epsilon_P = 4 \times 10^{-3}$, an oscillatory behaviour is observed with an average of $M \sim 0.55$ and peaks reaching $M \sim 0.85$. This increase in the value of M occurs also at lower value of $|Z(t)|$, where a few subpopulations of the phases are locked at different values e.g. $\pm\,\pi/2$ and $\pm\,\pi$, as shown in the polar representations in Fig. 5.

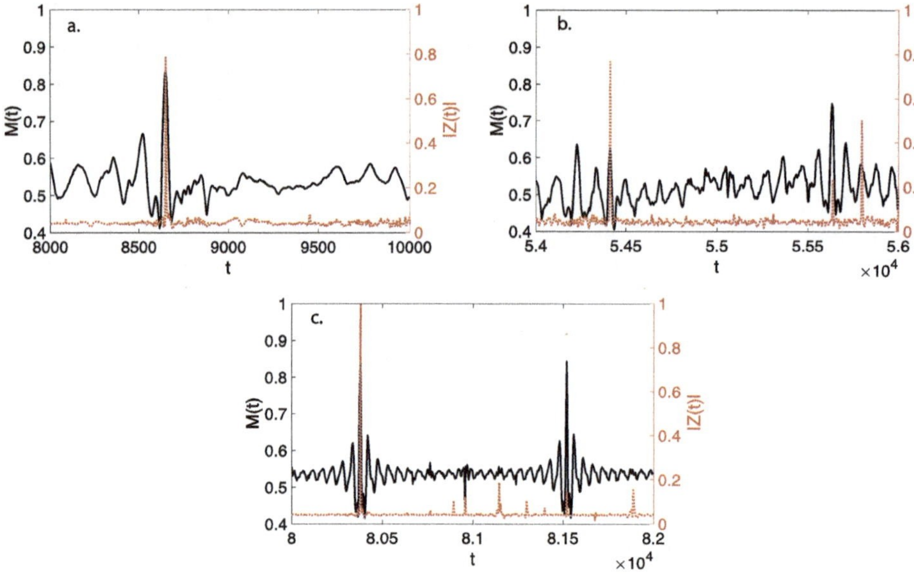

FIG. 4. (On the left y-axes)$M(t)$ as function of time for (a) $\epsilon_P = 0$, (b) $\epsilon_P = 2 \times 10^{-3}$, and (c) $\epsilon_P = 4 \times 10^{-3}$. (On the right y-axes) the order parameter $|Z(t)|$ for the same time window is also shown. Note, the range of the above sums are $|k| \in [21, 700]$.

D. Maximization the energy flux towards small scales due to locking

While the energy transfers are important, the energy fluxes offer a better measurement of the energy passing from large to small scales. The energy flux across a scale k is defined as

$$\Pi(k,t) = \sum_{k'=k}^{\infty} T(k',t), \tag{14}$$

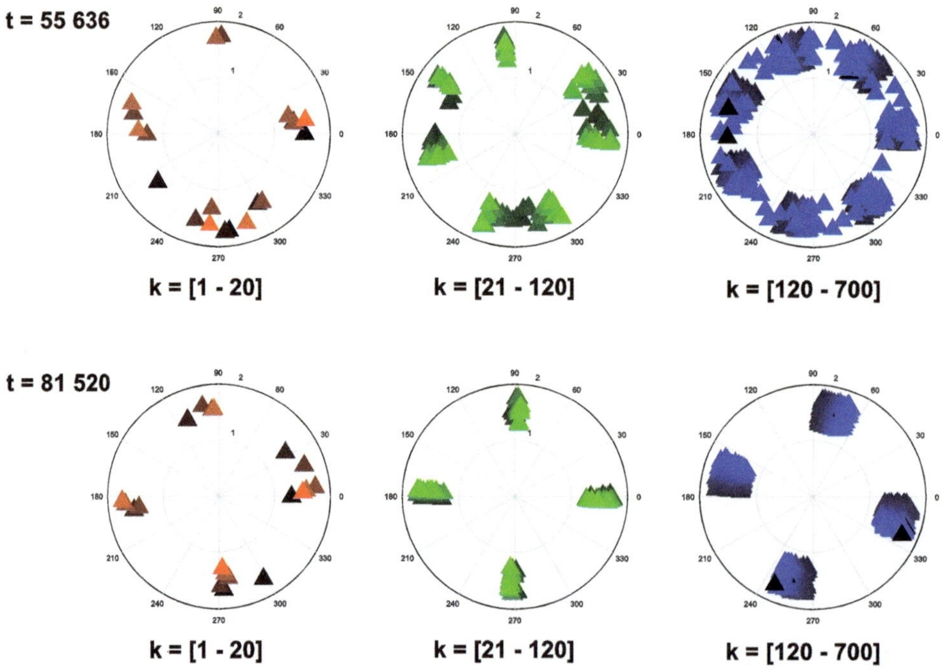

FIG. 5. The polar views of the phases at the time slices corresponding to the second highest peak in $M(t)$ as seen in Fig. 4 (b and c panels respectively). Here, as the mode number, k, is increased, the colors are varied from darker to lighter shades.

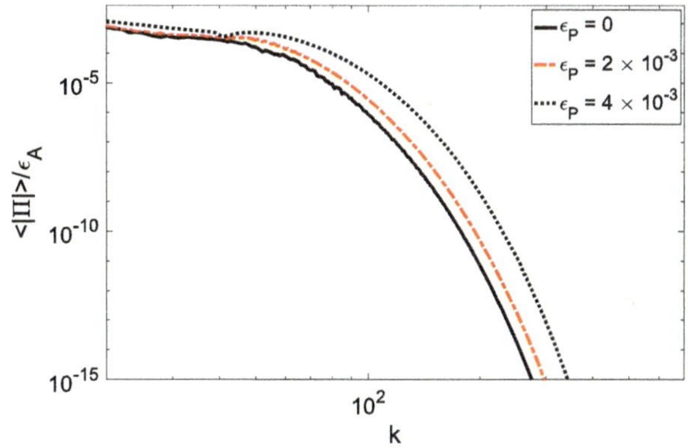

FIG. 6. Time average values of the flux normalized to amplitude forcing strength, $\langle|\Pi|\rangle/\epsilon_A$, as function of mode number k, (black solid line) $\epsilon_P = 0$, (red dashed-dotted line) $\epsilon_P = 2 \times 10^{-3}$, and (blue dotted line) $\epsilon_P = 4 \times 10^{-3}$. The last 2000 time steps of the simulations are used for the computation of the average.

and it is positive for a cascade of energy from large to small scales. Figure 6 shows the time average values of the flux normalized to amplitude forcing strength, $\langle|\Pi|\rangle/\epsilon_A$, as function of mode number k, for the three considered cases. Here, the last 2000 time steps of the simulations, where the steady state conditions are reached, is used for the time averaging. An increase in the values of the averaged flux for the $k = [21 - 120]$ is observed as the strength of the phase force raised. This is due to the phase force with results in the locked events to become more frequent, as well as spread into a wider range of k's, both in lower-k where the forcing is active, and higher-k where dissipation is stronger. Therefore, during the phase locking events the removal of the large scale energies is further increased and as the number of these events increases, the overall energy of the system is reduced. Indeed, as is shown in Fig. 7, the total energy of the system is reduced as the phase forcing is introduced. For the time interval where the ϵ_P is raised to 4×10^{-3}, the energy oscillates as the wave propagates through the system and finally converges to a finite but lower level as compared to the time period without the phase force. Thus, the observed increase in energy flux of Fig. 6 confirms the enhancement in the transfer of the large scales energy to small scales via phase locking.

A maximal flux, $\Pi^{\max}(k,t)$, can be defined as well by simply employing $T^{\max}(k,t)$ in the definition above to obtain the ratio of the flux to its maximal allowed as

$$M_\Pi(t) = \frac{\sum_k |\Pi(k)|}{\sum_k |\Pi^{\max}(k)|}. \tag{15}$$

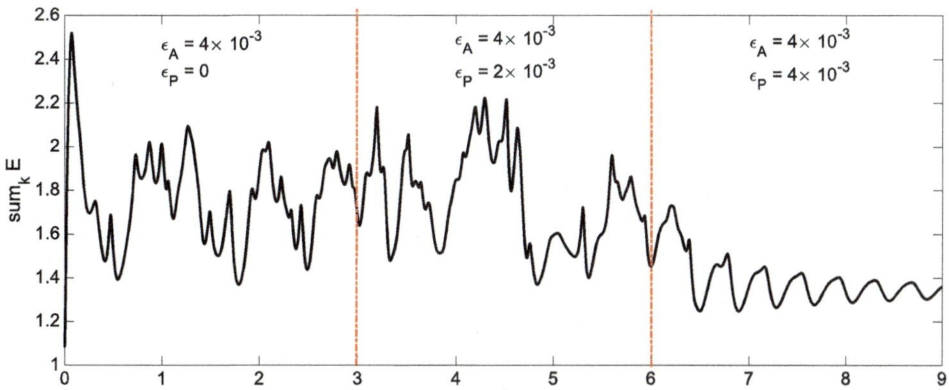

FIG. 7. The total energy ($\sum_k E_k(t)$) as a function of time t with $\epsilon_A = 4 \times 10^{-3}$, and the ϵ_P was increased in 3 steps denoted by straight lines (0, 2×10^{-3}, 4×10^{-3}).

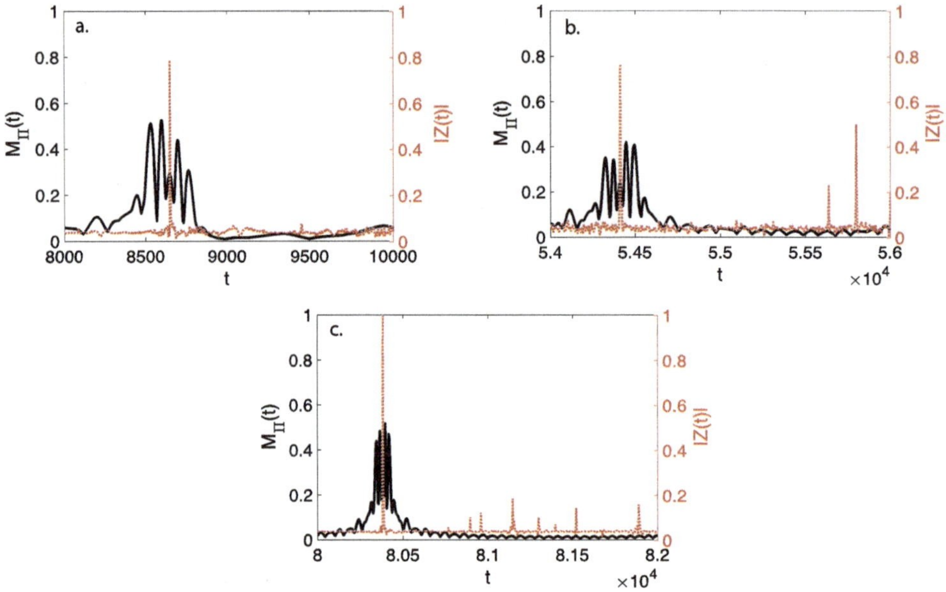

FIG. 8. (On the left y-axes)$M_{\Pi}(t)$ as function of time for (a) $\epsilon_P = 0$, (b) $\epsilon_P = 2 \times 10^{-3}$, and (c) $\epsilon_P = 4 \times 10^{-3}$. (On the right y-axes) the order parameter $|Z(t)|$ for the same time window is also shown. Note, the range of the above sums are $|k| \in [21, 700]$.

Figure 8 shows the values of this ratio i.e. $M_{\Pi}(t)$, for the three considered cases of Fig. 4. In all the cases, we observe similar patterns in the value of M_{Π} around the main locking event where M_{Π} grows from very small values of about $\sim 0.02 - 0.05$ to $\sim 0.3 - 0.6$ with few oscillations around the locking event. This behaviour suggests that there is a build up towards the locking event which also lasts for some time after the actual moment of locking. This build up manifest itself as an increase in the energy flux passed through to the small scales. By increasing the phase force, the time window of the build up becomes shorter, however the pattern remains the same.

IV. CONCLUSION

The phase dynamic in a simple 1D model of Burgers turbulence is examined. External forces which act independently on the amplitudes and phases are introduced. By introducing a special force which only affects the phases, we explored the possibilities of artificially locking the phases.

In real space we observe that in the absence of the phase force the turbulent flows are born, and dissipated homogeneously in (x, t) space. However, by increasing the phase force, the propagating flows are more and more localized in x and eventually the smaller flows disappear and one structure becomes dominant and propagates as one single traveling wave.

The observed phase dynamic shows significant departure from the well known RPA assumptions, with phases locking occasionally (but not in the dissipative high-k range). In the well-known Kuramoto[19] non-linear system with two-body interactions of limit cycle oscillators, it was shown that these types of systems are prone to locking if the coupling strength between the two-bodies has passed a threshold. However, the dynamic of the three-body interactions between the phases in the non-linear Burgers' turbulence differs from this simplified picture, and the phases lock intermittently and only in the low to mid-k range. This is due to the k dependence of the coupling strength in the non-linear term which reduces strongly for high-k range due to the dampening effect of the dissipation which does not allow locking of the phases of the small scales. In a locked event, therefore the evolution of the phases (eq. 5) is determined by the contributions of the small scales to the non-linear sum and the forcing which results in an immediate offsetting of the phases from the locked state and the intermittent property of the observed phase locking.

Our results show an increase in the occurrence of the locking events by inclusion of the phase force, thus, further invalidating the RPA assumptions in the studied system.

As the locking increases through the effect of the phase force, we observe a reduction in the total energy. This is attributed to the fact that during the locking events the phases lock to $\pm \pi/2$ which results in maximisation of the transfers, $T(k, t)$. The computed values of the energy flux $\Pi(k, t)$, confirms the increase in the energy flux passed from large scales through to the small scales, as the phase force is raised to higher levels allowing for more locking events to take place. A build up of increased energy flux prior to the moment of locking is observed which also lasts for some time after the locking.

ACKNOWLEDGMENTS

B. Teaca is partially supported by EPSRC grant No. EP/P02064X/1.

[1] P. H. Diamond, S.-I. Itoh, and K. Itoh, *Modern Plasma Physics: Volume 1, Physical Kinetics of Turbulent Plasmas* (University of Cambridge, 2010).

[2] J. A. Krommes, "Fundamental statistical descriptions of plasma turbulence in magnetic fields," Physics Reports **360**, 1–352 (2002).

[3] D. Gaggero, *Cosmic Ray Diffusion in the Galaxy and Diffuse Gamma Emission*, Springer Theses, (Springer-Verlag Berlin Heidelberg, 2012).

[4] G. M. Zaslavski and R. Z. Sagdeev, "Limits of statistical description of a nonlinear wave field," Zh. Eksp. Teor. Fiz. **52**, 1081 (1967) [Sov. Phys. JETP **25**, 718 (1967)].

[5] V. E. Zakharov, "Statistical descriptions and plasma physics," in *Handbook of Plasma Physics*, edited by A. A. Galeev and R. N. Sudan (North-Holland, Amsterdam, 1984), Vol. 2, Chap. 5.1, p. 3.

[6] Y. Choi, Y. V. Lvov, and S. Nazarenko, Physica D **201**, 121–149 (2005).

[7] S. Nazarenko, *Lecture Notes in Physics 825, Wave turbulence* (Springer, 2011), Volume 825.

[8] H. Branover and Y. Unger, *Progress in turbulence research* (American Institute of Aeronautics and Astronautics, Inc., Washington DC., 1993), Volume 162.

[9] Z. B. Guo and P. H. Diamond, Phys. Rev. Let. **114**, 145002 (2015).

[10] A. T. Winfree, *The Geometry of Biological Time* (Springer, New York, 1980).

[11] H. Daido, Prog. Theor. Phys. **77**, 622 (1987).

[12] H. Daido, Phys. Rev. Lett. **68**, 1073 (1992).

[13] J. D. Crawford, J. Stat. Phys. **74**, 1047 (1994).

[14] H. Daido, Phys. Rev. E **61**, 2145 (2000).

[15] S. H. Strogatz, Physica D **143**, 1–20 (2000).

[16] H. Hong and S. H. Strogatz, Phys. Rev. Lett. **106**, 054102 (2011).

[17] B. Sonnenschein and L. Schimansky-Geier, Phys. Rev. E **88**, 052111 (2013).

[18] E. Kim and P. Diamond, Phys. Plasmas **10**, 1698 (2003).

[19] Y. Kuramoto, *Chemical Oscillations, Waves and Turbulence* (Springer, Berlin, 1984).

[20] S. Moradi and J. Andersson, Phys. Plasmas **23**, 052310 (2016).

[21] G. B. Whitham, *Linear and Nonlinear Waves* (John Wiley Sons, 1974).

[22] Sharath S. Girimaji and Y. Zhou, "Spectrum and Energy transfer in steady Burgers turbulence," NASA Contractor Report 195047 ICASE Report No. 95-13, 1995.

[23] B. Teaca, F. Jenko, and D. Told, New J. Phys. **19**, 045001 (2017).

10

Investigation of phase evolution and control over phase transformation temperature and thermal hysteresis using stoichiometry and co-doping in VO_2 thin films

G. Hassnain Jaffari[a] and Wardah Mahmood
Department of physics, Quaid-i-Azam University, Islamabad 45320, Pakistan

Structural phase transition temperature and its associated hysteresis in VO_2 have been controlled by high valent dopant induced local structural modification which acts as phase nucleation site during phase transformation. Monoclinic phase intensity loops plotted against temperature has been observed to exhibit thermal hysteresis. It is shown that the size, shape and central position of hysteresis loop depend on stoichiometry and doping concentration. Highest reduction in phase transformation temperature and thermal hysteresis width has been observed in case of W^{+6}-Mo^{+6} co-doping and W^{+6} doping, respectively. Hence energy barrier associated with the structural phase transition has been successfully manipulated to vary T_c and hysteresis width. These findings have implications for designing the phase switching devices and smart window applications.

I. INTRODUCTION

Vanadium dioxide (VO_2) is a promising material due to its electrical and optical switching properties for various applications.[1–4] VO_2 exhibits well-known first-order metal–insulator transition (MIT).[5] This transition accompanies increase in electrical conductivity by a factor greater than 10^4,[2] structural phase transition[6] and decrease in infrared transmittance.[2] The dimerization of V^{+4} ions along the *c*-axis direction results in lattice transformation from tetragonal phase to monoclinic phase as temperature decreases.[7] In the monoclinic phase, the neighboring vanadium atoms are paired in groups of two by overlap of their $d_{\parallel}$ bands. The electrons present in the $d_{\parallel}$ band are therefore restricted within this bonding and thus conductivity of material is broken. Above the critical transition temperature, the vanadium atoms re-arrange themselves such that the $d_{\parallel}$ band overlaps the π^* band. The effective band gap diminishes because the π^* band is thermally achievable by the $d_{\parallel}$ band electrons[8] and system becomes metallic.

The metallic rutile VO_2, also referred as T phase, belongs to the tetragonal crystal structure and space group $P4_2/mnm$.[9] While the insulating phase (M1 phase, space group $P2_1/c$) is a distorted version of the tetragonal (T) phase. During phase transition (PT), the unit cell of the T phase increases by a factor of 2, and the vanadium ions start to pair and tilt with respect to the *c*-axis, transforming into M1 phase.[2] This brings pairs of neighboring vanadium ions close enough to form covalent bonds, restricting the mobility of *d*-electrons and so conductivity decreases. The MIT is highly sensitive to chemical and physical local perturbations.[10] The structural PT triggered by strain is shown to be caused by the electronic orbital occupation.[11] Suitable dopant can also affect electronic orbital occupancy; dimerization, etc, hence affects PT temperature.

Hysteresis has been reported in a various physical systems such as magnets, superconductors, ferroelectrics etc. Thermodynamically, metastability in general gives rise to hysteresis.[12] Occurrence of hysteresis is related with nanoscale phase separation during the transition which depends

[a]Corresponding Author, email: hassnain@qau.edu.pk

on nano/micro-scale structural defects and the quenched disorder.[6] These features give rise to complex free energy landscape where different phases coexist.[13] It has been shown that VO_2 thin films exhibit non-conventional metastable resistance states in both insulating and the metallic phase.[14] MIT depends on the stoichiometry, and any external contribution which may drive phase transition, such as strain[15] or doping.[16] Aim of the present study is to observe the role of stoichiometry and doping in structural PT and thermal hysteresis associated with the PT. For this purpose high valent Mo and W ions are selected as dopants and are used to vary PT temperatures and hysteretic response, for application purposes.

II. EXPERIMENTAL SECTION

V_2O_5 (Sigma Aldrich 99.6% purity) powder precursor was loaded in a typical CVD setup. Substrates were cleaned in 2-propanol, acetone and distill water using an ultrasonic bath. Silicon (111) substrate was placed 1 cm away from the precursor, towards the gas outlet end of the tube. The system was initially evacuated down to 8×10^{-2} torr. The furnace was set to final temperature which is 1000°C with temperature ramp 12°C/min and stay time equal to 15 minutes. For oxygen rich sample A, when the temperature reached 300°C, O_2 was supplied into the tube such that the pressure inside the tube was maintained at 9×10^{-1} torr. At 1000°C, O_2 was turned off and Ar was provided to maintain a pressure of 1.5 torr. At the end of stay time, gas supply was disconnected and sample was left to cool in vacuum. For sample B, above parameters was used, however when the system reached 300°C, 95% Ar + 5% H_2 was flowed through the tube. Sample A and B were prepared to vary oxygen stoichiometry and its effect on structural phase transformation was studied. All doped samples were synthesized on route of sample B, except that the precursors, with proper dopant concentration including pure V_2O_5 powder, were calcined before evaporation in CVD setup. For temperature dependent studies metal ceramic heater (part no. HT24S) was used in combination with a resistive temperature detector (part no. TH100PT) manufacture by Thorlabs.

III. RESULTS

A. Morphology

In general, all CVD samples exhibited grains with plate-like and/or rod like morphology with an evidence of presence of facets, as shown in figure 1. SEM micrographs in figures 1(a) and (b)

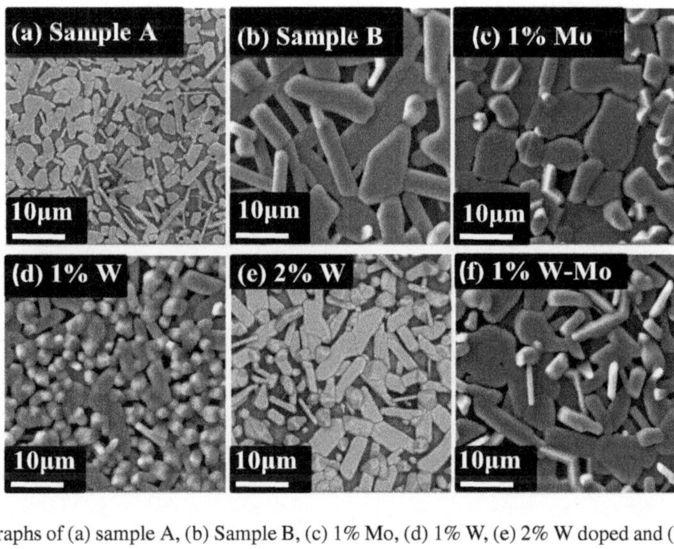

FIG. 1. SEM micrographs of (a) sample A, (b) Sample B, (c) 1% Mo, (d) 1% W, (e) 2% W doped and (f) 1% W-Mo co-doped samples.

are associated with sample A and sample B. Sample B was prepared in Ar+H_2, as explained in experimental section, hence oxygen vacancy density is expected to be higher than sample A. Therefore, presence of oxygen vacancies seems to increase diffusion of the evaporated material on interacting with the substrate, therefore grain size increases in comparison to sample A, as shown in figures 1(a) and (b). All doped samples, were prepared in same conditions as sample B. With addition of Mo, as shown in figure 1(c), in comparison with sample B, grain size was observed to increase and morphology is flat plate-like. While with W addition, reduction in grain size and morphological transformation towards rod like has been observed, as shown in figure 1(d). In this case rods were also aligned perpendicular to the substrate. For higher W concentration, grain morphology became random with formation of nanorods with much smaller aspect ratio, as shown in figure 1(e). Similar behavior has been observed in literature where the orientation of nanorods largely depends on concentration of W doping.[16] In case of co-doped sample, consistently both flat plates and rods have been observed, as shown in figure 1(f). Overall presence of dopant, doping concentration and oxygen vacancies effects diffusivity of the CVD precursors on substrate which further defines final morphology. In the present case, contact-less measurement for PT analysis using XRD for discontinuous films is most suitable characterization technique.

B. Structural properties

1. Role of stoichiometry

Figure 2(a) shows XRD images of pure samples. Films are clearly oriented along [011] and [022] directions. Room temperature (RT) data of Sample A matches with monoclinic phase (JCPDS-01-072-0514). Inset of figure 2(a), shows high resolution scan of (011) reflection of both samples. For sample B peak lies at lower angle. The peak positions of sample A and B is mainly affected by the differences in stoichiometry. Sample B is prepared in Ar+H_2 environment, which results in higher concentration of oxygen vacancies. The presence of oxygen vacancies promotes the formation of V^{+3} ions which is a consequence of charge neutrality.[17] Since V^{+3} (78pm) has greater ionic radius than V^{+4} (72pm), V-O bond length increases.[18] This in turn induces strain in the octahedral environment around vanadium ions, leading to an increase in inter-planar distance which results in a shift in XRD peaks towards lower angles.

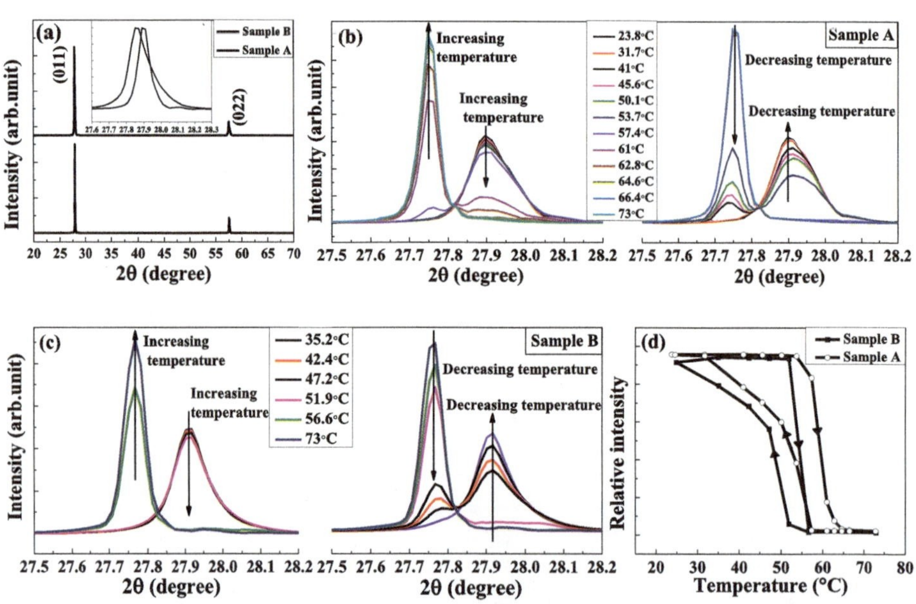

FIG. 2. XRD pattern of (a) sample A and sample B, inset shows zoomed in regions of (011) reflection, (b) XRD patterns of sample A, recorded at various temperatures (c) XRD patterns of sample B, recorded at various temperatures (d) relative intensity plots for sample A and B.

The monoclinic phase transforms to tetragonal phase at higher temperature for both sample A and B, as shown in figures 2(b) and (c), respectively. The relative intensity (I_R) of the monoclinic phase decreases with an increase in temperature and increases on decreasing temperature. These results are consistent with temperature dependent resistivity behavior of VO_2.[19] In order to highlight finer details about the phase transformation temperature, a plot of relative intensity (I_R) of monoclinic phase at each temperature for increasing and decreasing temperature cycle is shown in figure 2 (d). These plots show that phase transition also exhibits hysteresis. In general it is the same case where perturbation strength to induce PT in one way is different compared to case when reversed.[13] For quantitative and comparative analysis, average transition temperature (T_c) is extracted by taking average value of temperatures where relative intensity is half of its highest value. While difference of these temperatures is represented as ΔT, which gives hysteresis width. Both these parameters will be critical for comparison purposes. T_c for Sample A and B lie at 55.5°C and 51°C, respectively. Due to the presence of oxygen vacancies, less energy is required to transform the monoclinic bond arrangement to tetragonal bond arrangements.[20] Thus increase in oxygen vacancy density increases distortion via formation of V^{+3} ions in the monoclinic phase, which leads to a lower PT temperature, as observed in sample B.

2. *Effects of doping*

In all doped samples, shift in the main peak is observed with respect to the pure sample. Tang et al. discussed the substitution of V^{+4} with W^{+6} ions, which results in the breaking of V^{+4}-V^{+4}

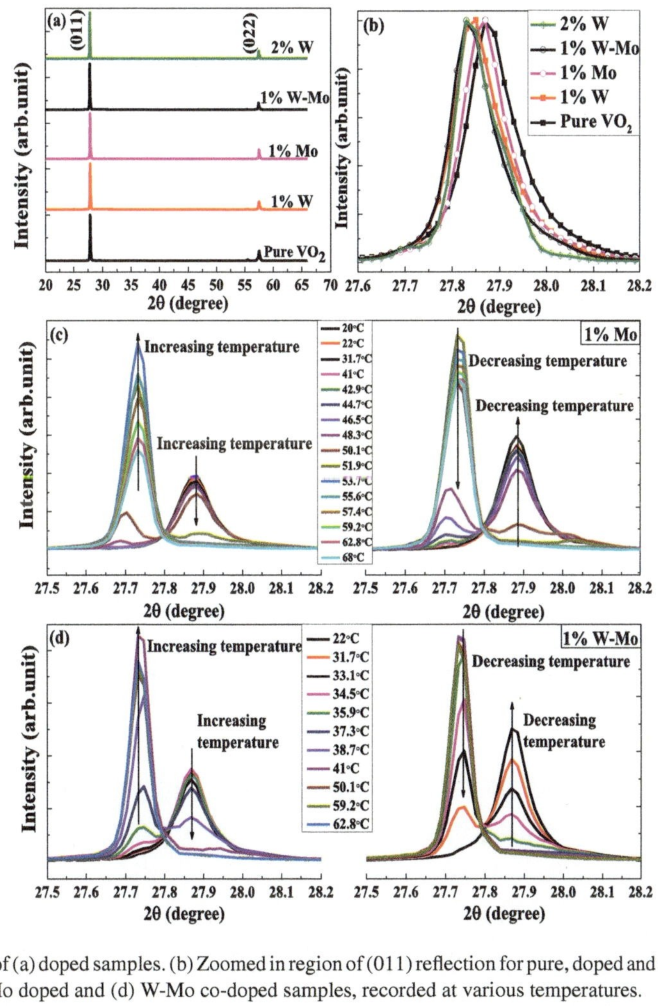

FIG. 3. XRD pattern of (a) doped samples. (b) Zoomed in region of (011) reflection for pure, doped and co-doped compositions. XRD patterns of (c) Mo doped and (d) W-Mo co-doped samples, recorded at various temperatures.

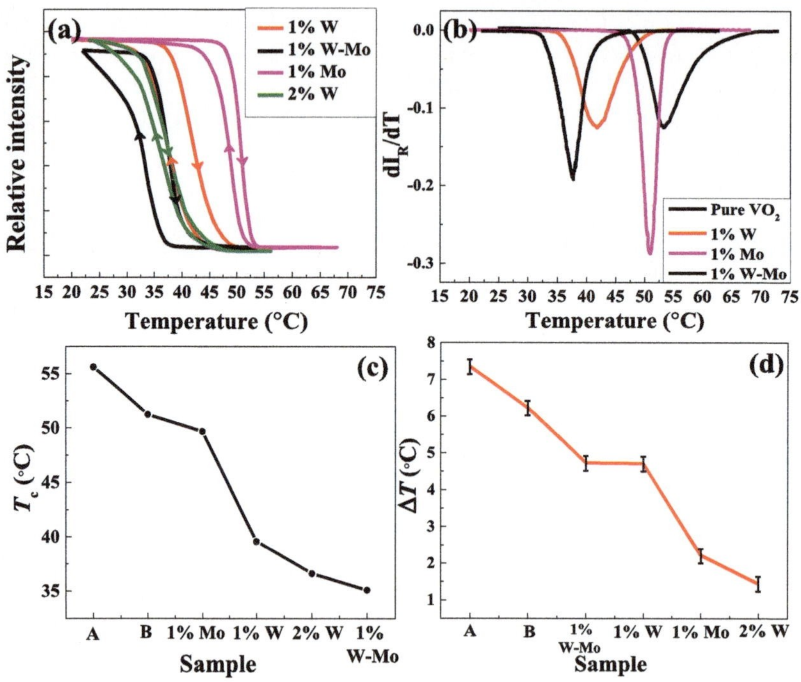

FIG. 4. (a) Relative intensity loops (b) derivative of monoclinic phase intensity with respect to temperature (c) Tc and (d) hysteresis width of various dopants.

bonds and formation of V^{+4}-W^{+6} and V^{+3}-V^{+4},[3] in order to keep charge balance in the system.[21] W^{+6} ions along with newly reduced V^{+3} ions have greater ionic radii than V^{+4} ions, thus corresponding monoclinic peak is observed to be shifted towards lower angles, as shown in figure 3(b). Similarly molybdenum is reported to exist in Mo^{+6} state in VO_2 host lattice.[1] However, Mo^{+6} doped sample exhibited small shift in comparison with W^{+6} doped samples, which is due to slightly smaller ionic radii of Mo^{+6} (73 pm) compared to W^{+6} (74 pm). In doping, charge compensation forms V^{+3} ions. Here dominant effect should come from formation of V^{+3} (78 pm) ions, whose ionic mismatch is highest compared to V^{+4} (72 pm). In case of co-doping, the dopants still prefer to exist in Mo^{+6} and W^{+6} states.[22] Highest shift in T_c has been observed for co-doped samples where doping concentration is highest i.e. 2%.

XRD patterns at various temperatures were recorded for all doped samples. Complete temperature dependent data for 1% Mo doped and Mo-W co-doped VO_2 thin films, is shown in figures 3(c) and 3(d). For 1% Mo doped composition, T_c lies around 48.3°C. In this case the transition is not sharp and phase changes are gradual as evident from the hysteresis plots shown in figure 4(a). The Mo^{+6} ions seem to cause distortion in the V-V bonds, leading to a decrease in phase transformation temperature compared to undoped samples. While in case of co-doped composition, highest decrease in PT temperature has been observed, as shown in figure 3(d). Structural phase transformation occurs around 35.9°C. These results are in general consistent with MIT results based on resistivity measurement where dopants are used to reduce T_c.[23]

IV. PHASE TRANSFORMATION TEMPERATURES AND THERMAL HYSTERESIS

As shown in figures 3(c) and (d), evidence of presence of thermal hysteresis associated with the structural PT is obvious. In order to highlight this feature and to quantify phase transition temperature (T_c), relative intensity loops for all compositions are shown in figure 4(a). First derivative plots of intensity are shown in figure 4(b) which also reflects T_c and hysteresis width. Here as temperature increases, intensity of M1 phase decreases. During cooling, again M1 phase increases but with non-retracing heating curve, hence system exhibits hysteresis. Key features are overall shift in loop towards lower temperature and decrease in degree of hysteresis with doping.

V. DISCUSSION

Sample with higher oxygen vacancies exhibited lower value of T_c. Recently Lanli Chen et al. reported in a DFT based study on oxygen vacancies in VO_2 and showed that the electronic structural modification in VO_2 is a consequence of oxygen vacancies using Bader charge analysis.[18] Oxygen vacancies in rutile phase lead to dimerization of V-V chains in rutile (metallic) phase with a corresponding increase in near-IR absorption. In pure VO_2 monoclinic phase, the V atoms lose electron which is gained by O atoms, while in case when oxygen vacancies were present, V atoms were shown to lose fewer electrons. Hence, if oxygen vacancies are present, there are more electrons to be offered than to be accepted by the anions. In such vicinity, the vanadium ion switches from V^{+4} to V^{+3} state, which results in increase of ionic radius of vanadium ions and the bond length V-O increases, causing lattice distortion due to the strain. The increased strain in the host system affects the MIT temperature.[15] Higher value of T_c in sample A shows that most of vanadium ions are in pristine state. Formation of V^{+3} ions weakens the strong electron correlations and enable the nonvolatile PT,[24] which is the case for sample B.

Tungsten incorporation was able to lower the MIT temperature more than molybdenum because of its slightly higher ionic radii mismatch in comparison with vanadium. Tungsten was able to create more lattice disorder, which made monoclinic phase weaker, resulting in a lower T_c compared to Mo doped and undoped samples, as shown in figures 3 and 4. Mo^{+6} doped sample exhibited smaller shift in XRD, hence generated lower structural distortion in comparison to W doping. Lowest T_c was observed for the co-doped composition. Co-doped samples have shown the most change i.e. higher XRD peak shift which is comparable with 2% W composition. In addition to that, in contrary to co-doping, W substitution affects hysteresis, as ΔT is lowest for 2% W doped case.

It has been shown using micro-Raman based data analysis that T phase is unevenly distributed due to presence of dopant and stress in VO_2 nanowires.[25] Theoretical work with an aid of experiment showed that dopants induces large variation in atomic and electronic structure of neighboring lattice.[26] Local structure around W atom is tetragonal-like and is intrinsically symmetric. Local lattice disorder induced by the dopants such as W ions with larger volume, and the pile-up of itinerant electrons released by higher valent dopant on the neighboring V-sites also suppresses the host dimerized lattice distortion and stabilize metallic phase of VO_2. Furthermore, doping site acts as a nucleus for MIT. At the higher doping concentration, these nuclei with local distortion overlap leads to lowering of thermal energy for PT.[26] Hence this picture is similar to the formation of metallic nanopuddles in VO_2 which facilitates MIT processes.[27] We find that the local structure modification with doping stabilizes tetragonal-like structure, depends on type of dopant, and exhibits a concentration dependent evolution by effecting the growth of nuclei and consequently effects MIT in terms of temperature and hysteresis by lowering the thermal energy barrier for PT.

First order phase transition is associated with nucleation and local symmetry breaking which is characteristic of MIT. In the same context, temperature driven MIT is shown to be associated with nanoscale formation of insulating domains in $NdNiO_3$ epitaxial thin films.[28] In the present study, it has been observed that nucleation and growth/annihilation of T and M1 phases depends on the initial phases, dopant, stoichiometry etc. Growth of phase T (M1) in phase M1 on heating (cooling) proceeds differently compared to case when the process is reversed. Complex energy landscape with generation and modification of metastable regions leads to hysteresis. Finally energy distribution is successfully manipulated where T_c and ΔT, can be tuned for application purposes.

Oxygen vacancies at the grain boundaries are shown to nucleate the PT by lowering the energy barrier associated with the PT.[20] In the same context, our results indicate that the oxygen vacancies were incorporated through stoichiometry change which led to generation of larger ionic mismatch V^{+3} ions. However stronger effect has been observed in case of higher valent ion incorporation where charge compensation enhances V^{+3} formations. Hence we found that incorporation of dopants significantly enhances nucleation and growth kinetics to trigger PT.

VI. CONCLUSION

A detailed study on temperature dependent PT in doped VO_2 is presented. Significant variation in T_c (up to 36.9% decrease) and hysteresis width (up to 80.4% decrease compare to pure composition) is observed which makes it interesting from application view point where thermal cycling is unavoidable in case of smart window applications. Band structure modification affect energy barrier associated with the PT which also effect thermal hysteresis.

[1] S. Xu, H. Ma, S. Dai, and Z. Jiang, "Study on optical and electrical switching properties and phase transition mechanism of Mo^{6+}-doped vanadium dioxide thin films," Journal of Materials Science **39**(2), 489–493 (2004).

[2] M. Sahana, G. Subbanna, and S. Shivashankar, "Phase transformation and semiconductor-metal transition in thin films of VO_2 deposited by low-pressure metalorganic chemical vapor deposition," Journal of Applied Physics **92**(11), 6495–6504 (2002).

[3] B. Rajeswaran and A. M. Umarji, "Effect of W addition on the electrical switching of VO_2 thin films," AIP Advances **6**(3), 035215 (2016).

[4] N. Shukla, A. V. Thathachary, A. Agrawal, H. Paik, A. Aziz, D. G. Schlom, S. K. Gupta, R. Engel-Herbert, and S. Datta, "A steep-slope transistor based on abrupt electronic phase transition," Nature Communications **6**, 7812 (2015).

[5] X. Xue, M. Jiang, G. Li, X. Lin, G. Ma, and P. Jin, "Photoinduced insulator-metal phase transition and the metallic phase propagation in VO_2 films investigated by time-resolved terahertz spectroscopy," Journal of Applied Physics **114**(19), 193506 (2013).

[6] J. Cao and J. Wu, "Strain effects in low-dimensional transition metal oxides," Materials Science and Engineering: R: Reports **71**(2), 35–52 (2011).

[7] B. A. Kruger, A. Joushaghani, and J. K. Poon, "Design of electrically driven hybrid vanadium dioxide (VO_2) plasmonic switches," Optics Express **20**(21), 23598–23609 (2012).

[8] J. B. Goodenough, "The two components of the crystallographic transition in VO_2," Journal of Solid State Chemistry **3**(4), 490–500 (1971).

[9] T. Peterseim, M. Dressel, M. Dietrich, and A. Polity, "Optical properties of VO_2 films at the phase transition: Influence of substrate and electronic correlations," Journal of Applied Physics **120**(7), 075102 (2016).

[10] J. Pouget, H. Launois, J. D'haenens, P. Merenda, and T. Rice, "Electron localization induced by uniaxial stress in pure VO_2," Physical Review Letters **35**(13), 873 (1975).

[11] N. B. Aetukuri, A. X. Gray, M. Drouard, M. Cossale, L. Gao, A. H. Reid, R. Kukreja, H. Ohldag, C. A. Jenkins, and E. Arenholz, "Control of the metal-insulator transition in vanadium dioxide by modifying orbital occupancy," Nature Physics **9**(10), 661 (2013).

[12] D. Knittel, S. Pack, S. Lin, and L. Eyring, "A thermodynamic model of hysteresis in phase transitions and its application to rare earth oxide systems," The Journal of Chemical Physics **67**(1), 134–142 (1977).

[13] Y. Zhou and S. Ramanathan, "Mott memory and neuromorphic devices," Proceedings of the IEEE **103**(8), 1289–1310 (2015).

[14] D. D. Gioacchino, A. Marcelli, A. Puri, C. Zou, L. Fan, U. Zeitler, and A. Bianconi, "Metastability phenomena in VO_2 thin films," Condensed Matter **2**(1), 10 (2017).

[15] L. Fan, S. Chen, Z. Luo, Q. Liu, Y. Wu, L. Song, D. Ji, P. Wang, W. Chu, and C. Gao, "Strain dynamics of ultrathin VO_2 film grown on TiO_2(001) and the associated phase transition modulation," Nano Letters **14**(7), 4036–4043 (2014).

[16] L. Whittaker, T.-L. Wu, C. J. Patridge, G. Sambandamurthy, and S. Banerjee, "Distinctive finite size effects on the phase diagram and metal–insulator transitions of tungsten-doped vanadium (iv) oxide," Journal of Materials Chemistry **21**(15), 5580–5592 (2011).

[17] S. Zhang, I. S. Kim, and L. J. Lauhon, "Stoichiometry engineering of monoclinic to rutile phase transition in suspended single crystalline vanadium dioxide nanobeams," Nano Letters **11**(4), 1443–1447 (2011).

[18] L. Chen, Y. Cui, S. Shi, B. Liu, H. Luo, and Y. Gao, "First-principles study of the effect of oxygen vacancy and strain on the phase transition temperature of VO_2," RSC Advances **6**(90), 86872–86879 (2016).

[19] C. Wu, F. Feng, and Y. Xie, "Design of vanadium oxide structures with controllable electrical properties for energy applications," Chemical Society Reviews **42**(12), 5157–5183 (2013).

[20] L. Fan, S. Chen, Y. Wu, F. Chen, W. Chu, X. Chen, C. Zou, and Z. Wu, "Growth and phase transition characteristics of pure M-phase VO_2 epitaxial film prepared by oxide molecular beam epitaxy," Applied Physics Letters **103**(13), 131914 (2013).

[21] C. Tang, P. Georgopoulos, M. Fine, J. Cohen, M. Nygren, G. Knapp, and A. Aldred, "Local atomic and electronic arrangements in $W_xV_{1-x}O_2$," Physical Review B **31**(2), 1000 (1985).

[22] Y. Jiazhen, Z. Yue, H. Wanxia, and T. Mingjin, "Effect of Mo-W Co-doping on semiconductor-metal phase transition temperature of vanadium dioxide film," Thin Solid Films **516**(23), 8554–8558 (2008).

[23] L. Mai, B. Hu, T. Hu, W. Chen, and E. Gu, "Electrical property of Mo-doped VO_2 nanowire array film by melting−quenching sol−gel method," The Journal of Physical Chemistry B **110**(39), 19083–19086 (2006).

[24] Z. Zhang, F. Zuo, C. Wan, A. Dutta, J. Kim, J. Rensberg, R. Nawrodt, H. H. Park, T. J. Larrabee, and X. Guan, "Evolution of metallicity in vanadium dioxide by creation of oxygen vacancies," Physical Review Applied **7**(3), 034008 (2017).

[25] E. Strelcov, A. Ievlev, A. Belianinov, A. Tselev, A. Kolmakov, and S. V. Kalinin, "Local coexistence of VO_2 phases revealed by deep data analysis," Scientific Reports **6** (2016).

[26] X. Tan, T. Yao, R. Long, Z. Sun, Y. Feng, H. Cheng, X. Yuan, W. Zhang, Q. Liu, and C. Wu, "Unraveling metal-insulator transition mechanism of VO_2 triggered by tungsten doping," Scientific Reports **2** (2012).
[27] M. M. Qazilbash, M. Brehm, B.-G. Chae, P.-C. Ho, G. O. Andreev, B.-J. Kim, S. J. Yun, A. Balatsky, M. Maple, and F. Keilmann, "Mott transition in VO_2 revealed by infrared spectroscopy and nano-imaging," Science **318**(5857), 1750–1753 (2007).
[28] G. Mattoni, P. Zubko, F. Maccherozzi, A. J. van der Torren, D. B. Boltje, M. Hadjimichael, N. Manca, S. Catalano, M. Gibert, and Y. Liu, "Striped nanoscale phase separation at the metal–insulator transition of heteroepitaxial nickelates," Nature Communications **7** (2016).

11

Realization of manipulating acoustic surface waves radiation direction with rectangular-groove structure

Jun Cui,[1] Jiehui Liu,[1] Yiwei Mao,[1] Yifeng Li,[2] and Xiaozhou Liu[1,a]
[1]*Key Laboratory of Modern Acoustics, Institute of Acoustics and School of Physics, Collaborative Innovation Center of Advanced Microstructures, Nanjing University, Nanjing 210093, P. R. China*
[2]*College of Computer Science and Technology, Nanjing Tech University, Nanjing 211800, P. R. China*

Acoustic surface waves (ASWs) can be generated through a one-dimensional array of grooves. Sound can be collimated by ASWs. However, in previous studies, the groove period and grating period have been the same. In this work, we propose a structure where the groove period is different from the grating period, and collimates sound waves with very small side lobes. The structure can alter the acoustic wave radiation direction by manipulating ASWs and the relationship between the radiation direction and the frequency for different groove depths are investigated. Furthermore the incident direction of the sound wave, which can be coupled into ASWs, can be manipulated by changing the period of rectangular grating. We theoretically illustrate the physical mechanism of controlling the ASW radiation direction by wave-number analysis. These theoretical predictions are verified using numerical simulations. Using this proposed structure, we can manipulate the ASW radiation direction, which is very important for practical applications of directional acoustic propagation.

I. INTRODUCTION

Diffraction of sound waves is a significant problem in realizing directional acoustic propagation. Acoustic directivity can save energy and is important in keeping transmitted signals confidential. Previous approaches to realize directivity have mostly relied on modifying acoustic sources.[1,2] However, they are complicated in design and operation. The direction of propagation is also difficult to control, owing to the diffraction of the sound wave at low frequency. After surface plasmons (SPs) were reported to exist on groove structures,[3] application of such structures has become a subject of growing interest. SPs have been shown to result in Raman signal enhancement,[4,5] extraordinary optical transmission,[6,7] light beaming.[8] Extraordinary optical transmission and light collimation stand out as notable discoveries in SP research. Mimicking SPs in acoustics with groove structures was first proposed by Christensen *et al.*,[9] who found that the coupling of acoustic surface waves (ASWs) can be demonstrated to realize collimation of sound. Since then, groove structures have been extensively studied in acoustics. For example, acoustic rainbow trapping has been demonstrated by utilizing periodic grooves with a depth gradient,[10] a rigid surface has been decorated with an array of grooves of graded widths in order to slow down ASWs,[11] and groove structures have been used in acoustic imaging and focusing.[12,13] Propagation of ASWs has also been studied,[14–17] and dipole-like radiation patterns through the adjustment of acoustic boundary impedance based on cascaded Helmholtz resonators have been reported.[18,19] However, it remains difficult to control the radiation direction. In this work, we propose a structure that can control ASW radiation direction, which is important for

[a]Author to whom correspondence should be addressed. Electronic mail: xzliu@nju.edu.cn

applications requiring directional propagation. We theoretically illustrate the physical mechanism of controlling the ASW radiation direction by wave-number analysis. These theoretical predictions are verified using numerical simulations.

II. DESCRIPTION OF THE PROPOSED STRUCTURE AND PRINCIPLE

Figure 1(a) shows the schematic configuration of the proposed structure, which consists of a thin metal plate with a subwavelength center slit. The grooves surround the center slit where the groove period is d, the groove width is a, and the groove depth is h. When a plane sound wave encounters this structure, an ASW is generated. Depending on the wavelength of the impinging sound wave, the sound wave travels along the groove structure or is radiated away from the structure. Strong acoustic wave localization can be observed.[16] When the ASWs reach the end of the structure, this surface wave is scattered and yields strong sound radiation at the structure tip. However, the direction of these radiated ASWs cannot be controlled. In the following part, we will illustrate how to manipulate the ASW radiation direction.

Before discussing the directional radiation of sound, let us review the basic features of ASWs. The periodic grooves support the propagation of ASWs along the surface of the structure. In the lossless case, the dispersion relationship of ASWs can be given by[20]

$$k_{asw} = k_0\sqrt{1 + (a/d)^2 \tan^2 k_0 (h - a\log 2/\pi)} \tag{1}$$

where $k_0 = \omega/c_0$, ω is angular frequency and c_0 is the sound velocity in air.

Now, we demonstrate collimation of sound by manipulating ASWs. First, we consider the mechanism of collimation of sound. Sound radiation can be characterized by wave-number k_0 which is composed of the in-plane component k_{0P} and the out-of-plane component $k_{0\perp}$. In the proposed structure, k_{0P} is determined by surface wave diffraction theory and the grating equation, which can be written as

$$k_{0P} = -k_{asw} + Nk_g = k_0 \sin\varphi \tag{2}$$

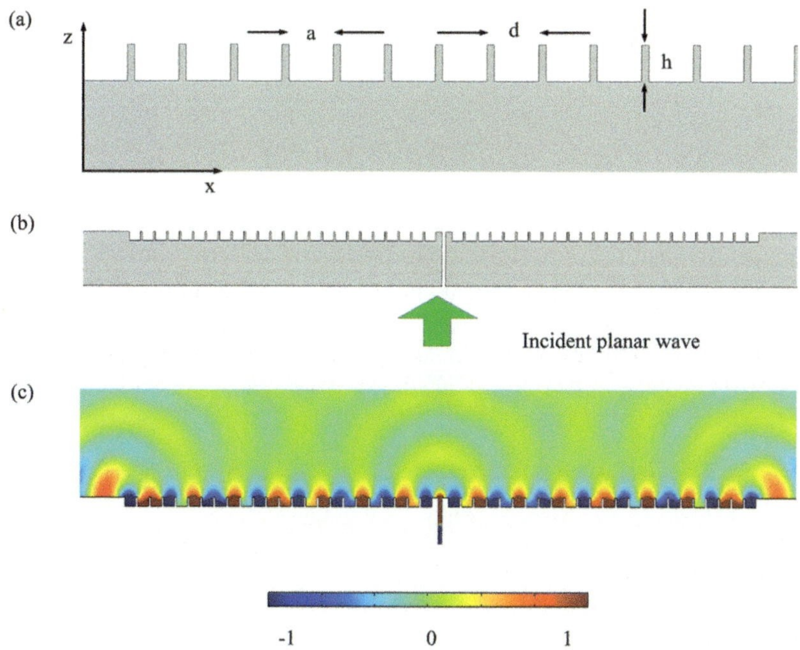

FIG. 1. (a) Structural parameters of the periodic grooves. (b) Metal plate consists with a single slit surrounded by finite periodic grooves. (c) Acoustic pressure distribution in the output side of the plate; the frequency of the incident sound wave is 10750 Hz.

where k_g is wave number of the rectangular grating, N is an integer, φ is the angle between the z-axis and the direction of radiation beaming.

III. SIMULATION RESULTS AND DISCUSSIONS

The simulations are carried out using the finite element method (COMSOL Multiphysics). The structural parameters used in the simulation are set as $a = 6$ mm, $h = 5$ mm, $d = 7$ mm and the width of center slit is 2 mm. The thickness of the structure is 28.6 mm. An acoustic groove structure with a narrow slit is shown in Fig. 1(b). The material parameters are mass density $\rho_0 = 1.21\text{kg/m}^3$ and sound speed $c_0 = 340\text{m/s}$ for air, and mass density $\rho = 8500\text{kg/m}^3$, Young's modulus E=104GPa, and Poisson's ratio $\sigma = 0.37$ for brass. Brass is chosen as the solid material due to its high impedance relative to air. The other solid materials, such as aluminum and steel were also used for the calculation. The results show that the influence of material properties on power transmission is negligible. A normal incident planar sound wave propagates in air and impinges at the single slit. After traveling through the slit, the incident acoustic wave strongly couples into the collective surface oscillations and travels along the surface of the plate on both sides, as shown in Fig. 1(c).

We can obtain the wave-number k_{asw} from Eq. (1) at 10750 Hz. For collimation of sound, φ=0 must be satisfied. Therefore we need to add the momentum of the rectangular gratings to satisfy the Eq. (2). The groove parameters are the same as those in Fig. 1(a). When $N = 1$, we get $k_g = 245$ and $\lambda_g = 26$ mm (λ_g is the grating period), according to Eq. (2). The structure is shown Fig. 2(a), where

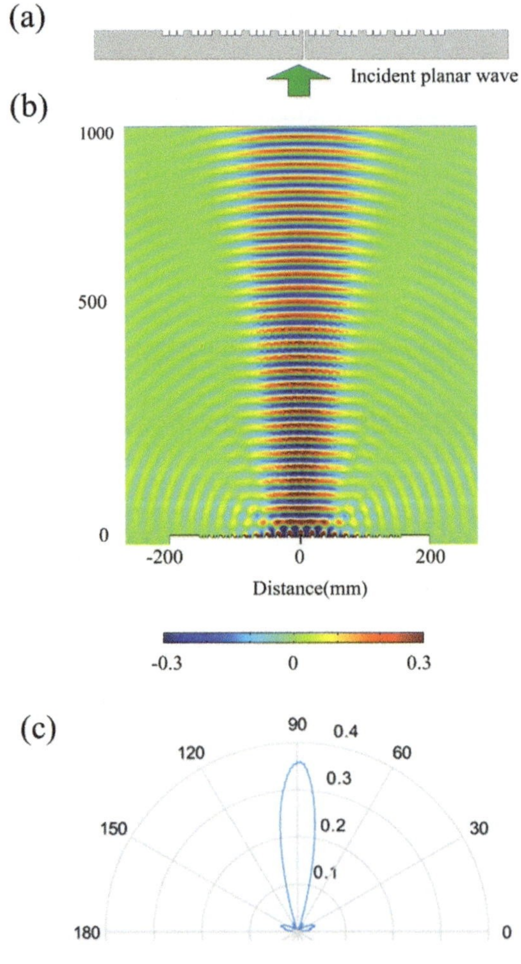

FIG. 2. (a) Grating structure at the surface of the plate. (b) Sound intensity distribution when collimation beam occurs. (c) Acoustic intensity pattern at 50 cm from the sound source at 10750 Hz.

we add the rectangular gratings that act as antennae to couple the collective surface oscillations to a radiation wave at a specific wavelength. This results in the collimation effect shown in Fig. 2(b). In contrast to previous studies,[9,21] we find that the collimated beam obtained is nearly without side lobes. When ASWs are generated through the structure shown in Fig. 1(a), strong acoustic waves are confined in the grooves, and are not radiated out at this frequency [Fig. 1(c)]. When we add rectangular gratings in this structure, the groove parameters do not change. Furthermore, the grooves are equivalent to a series of acoustic capacitance and acoustic mass, which will resonate when the frequency is 10750 Hz and h = 5mm, and the acoustic wave will be absorbed. Therefore, when the condition is matched, a part of the ASWs near the slit radiate out to achieve collimation of acoustics. The remaining acoustic waves are confined in the grooves. It can be demonstrated more clearly using the radiation pattern, as shown in Fig. 2(c). The small side lobe results from the scattering of sound by the edge of the structure. As in the previous structure, the groove period is same as the grating period, but in the present study we chose the period of the rectangular gratings to achieve collimation of acoustic waves at a certain frequency.

Furthermore, this structure can alter the acoustic wave radiation direction by manipulating ASWs. The details of the decorated plate and the rectangular gratings are shown in Fig. 3(a). The parameters of the grooves are the same as those in Fig. 1(a) and the parameters of rectangular gratings are the same as those in Fig. 2(a). When the planar wave is incident from the left side of the structure, the sound wave is well confined on the surface of the plate.[16] Here the incident frequency is 10750 Hz. If there are no rectangular gratings, the ASWs will propagate along the surface of the structure, as shown in Fig. 3(b). If there are rectangular gratings, as shown in Fig. 3(c), the ASWs will radiate away from the surface of this structure. We can manipulate the radiation angle by modifying the grating period. The match condition[22] is $k_{out}\sin\theta + Nk_g = k_{asw}$. When N =1, the grating period

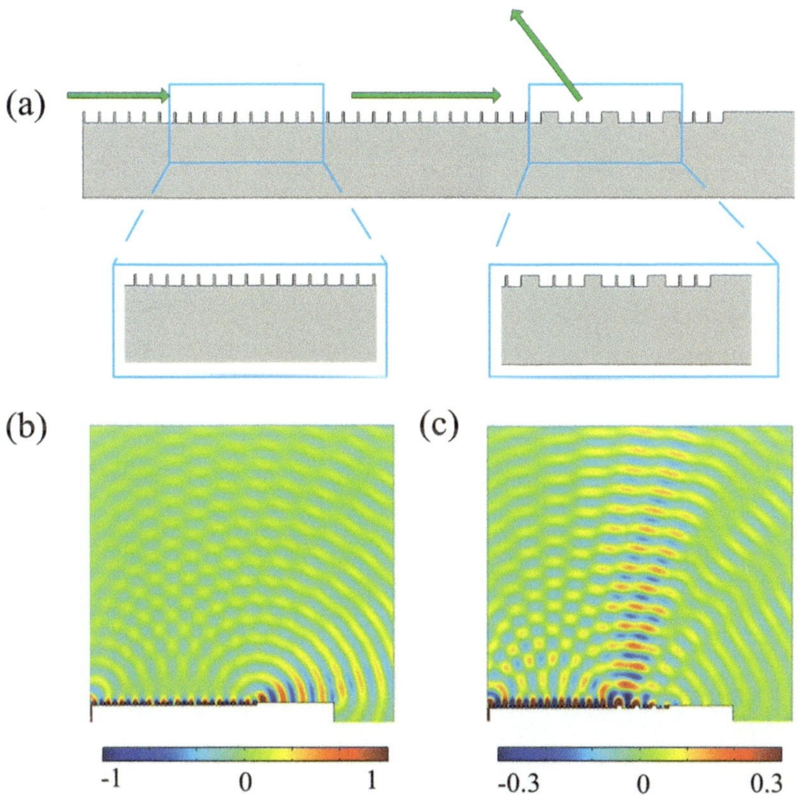

FIG. 3. (a) Schematic diagram of the structure. The structure is a rigid plate decorated with periodic grooves on the surface and several rectangular gratings on the right side. (b) Acoustic pressure distribution with no rectangular gratings. Acoustic surface waves propagate along the surface of the plate. (c) Acoustic pressure distribution when there are rectangular gratings. The radiation angle can be manipulated, and here is almost vertical radiation.

is 26 mm, and the radiation direction is vertical radiation because the match condition is satisfied. Fig. 3(c) presents the simulation results showing that the radiation angle is almost vertical, which correlates well with the theoretical prediction. Through this example, we confirm that this structure can be used to control the acoustic wave radiation direction using the grating period of the structure.

Furthermore, we investigated the relationship between the radiation direction and the frequency for different groove depths, as shown in Fig. 4(a). We used a frequency range from 5500 Hz to 9000 Hz and groove depths from 5 mm to 10 mm. In this frequency range, the incident acoustic wave strongly couples into the ASWs, and the ASW radiation direction is from 81° to 18°. The numerical simulation is shown in Fig. 4(b-e), where h = 8 mm correlates well with the theoretical

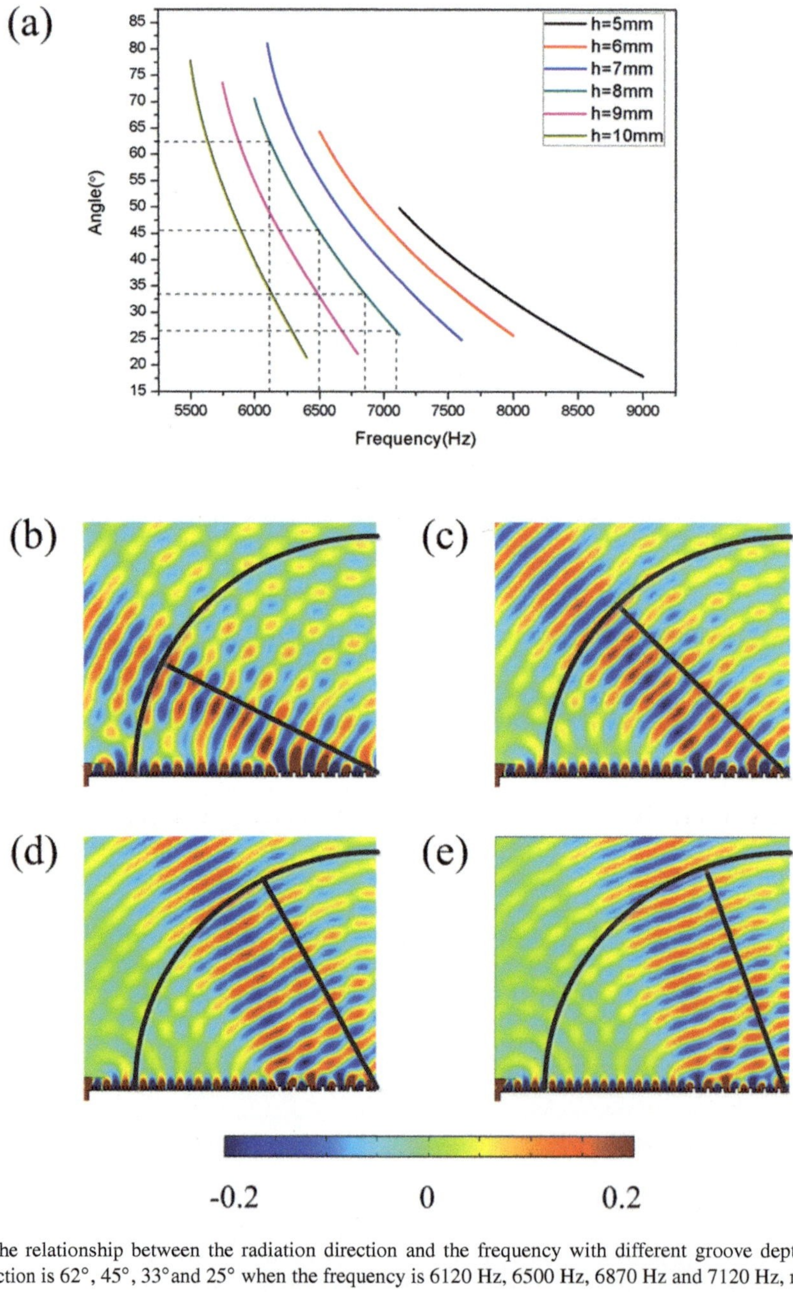

FIG. 4. (a) The relationship between the radiation direction and the frequency with different groove depths. (b)-(e) The radiation direction is 62°, 45°, 33°and 25° when the frequency is 6120 Hz, 6500 Hz, 6870 Hz and 7120 Hz, respectively, for h=8 mm.

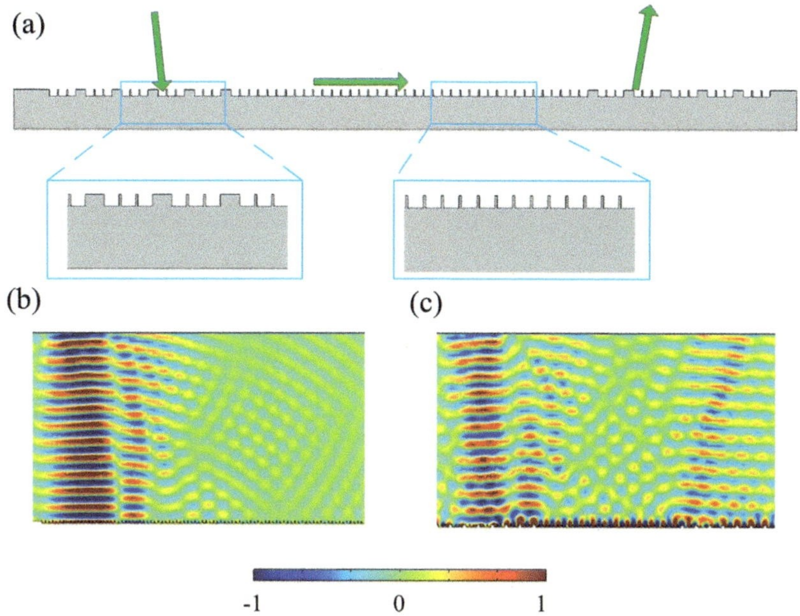

FIG. 5. (a) Schematic diagram of the structure to excite and manipulate ASWs. The structure is a rigid plate decorated with periodic grooves on its surface and several rectangular gratings on both sides. (b) Acoustic pressure distribution when match condition is not satisfied. (c) Acoustic pressure distribution when match condition is satisfied. Here the frequency is 11025 Hz.

prediction. Therefore the radiation direction can be controlled by adjusting the frequency of the incident wave.

Moreover, the incident direction of the sound wave, which can be coupled into ASWs, can be manipulated by changing the period of rectangular grating. The structural parameters are unchanged but rectangular gratings are added on the left side; the period of rectangular gratings at both ends remain the same. The details of the decorated plate and the rectangular gratings are shown in Fig. 5(a). The introduction of the rectangular gratings adds momentum to the incident wave. For an incident wave with wave number k_{in} and incident angle φ, if the match condition, $k_{in}\sin\varphi + Nk_g = k_{asw}$, is satisfied then the incident wave can be greatly coupled into surface modes and the collective surface oscillation is stimulated. The green arrows indicate the path of sound propagation. When $N = 1$ and the frequency of the incident acoustic wave is 11025 Hz, we can get $\varphi = 4°$ if the match condition is satisfied, which is close to vertical incidence. Fig. 5(b) presents the simulation results when the match condition is not satisfied because there are no rectangular gratings. But when the match condition is satisfied, there is a strong collective surface oscillation, as shown in Fig. 5(c), and the ASWs propagate to the output end of the plate. Finally, the ASW radiation angle is close to vertical. This angle is the same as the incident angle when the match condition, $k_{out}\sin\theta + Nk_g = k_{asw}$, is satisfied at the output end of the plate. Therefore, we can manipulate the radiation and incidence directions of the sound wave, which can be coupled into ASWs by changing the rectangular grating period or the frequency of the incident acoustic wave.

IV. CONCLUSIONS

In conclusion, the manipulation of ASWs has been demonstrated by adding rectangular gratings to a groove structure. The physical mechanism of the manipulation is described, and has been used to collimate sound waves and to change the sound wave radiation direction. Our results show that it is possible to steer ASWs by adding gratings of an appropriate period. This results of this study are important for acoustic device miniaturization, noise control, medical ultrasound, and other applications involving directional propagation of sound waves.

ACKNOWLEDGMENTS

We acknowledge financial support of the National key R & D program (No.2017YFA0303700), National Natural Science Foundation of China (No.11474160, No.61571222), Fundamental research funds for the Central Universities (No.020414380001), State Key Laboratory of Acoustics, Chinese Academy of Science (No.SKLA201609).

[1] M. Yoneyama, J. Fujimoto, Y. Kawamo, and S. Sasabe, J. Acoust. Soc. Am. **73**, 1532 (1983).
[2] J. Durnin, J. J. Miceli, Jr., and J. H. Eberly, Phys. Rev. Lett. **58**, 1499 (1987).
[3] H. J. Lezec, A. Degiron, E. Devaux, R. A. Linke, L. MartinMoreno, F. J. Garcia-Vidal, and T. W. Ebbesen, Science **297**, 820 (2002).
[4] K. Kneipp, H. Kneipp, I. Itzkan, R. R. Dasari, and M. S. Feld, J. Phys.: Condens. Matter **14**, R597 (2002).
[5] J. Bravo-Abad, F. J. Garci-Vidal, and L. Martin-Moreno, Phys. Rev. Lett. **93**, 227401 (2005).
[6] H. F. Ghaemi, T. Thio, D. E. Grupp, T. W. Ebbesen, and H. J. Lezec, Phys. Rev. B **58**, 6779 (1998).
[7] T. W. Ebbesen, H. J. Lezec, H. F. Ghaemi, T. Thio, and P. A. Wolff, Nature **391**, 667 (1998).
[8] H. J. Lezec, A. Degiron, E. Devaux, R. A. Linke, L. Martin-Moreno, F. J. Garcia-Vidal, and T. W. Ebbesen, Science **297**, 820 (2002).
[9] J. Christensen, A. I. Fernandez-Dominguez, F. de Leon-Perez, L. Martin-Moreno, and F. J. Garcia-Vidal, Nat. Phys. **3**, 851 (2007).
[10] J. Zhu, Y. Y. Chen, X. F. Zhu, F. J. Garcia-Vidal, X. Y. Yin, W. L. Zhang, and X. Zhang, Sci. Rep. **3**, 1728 (2013).
[11] H. Jia, M. Lu, X. Ni, M. Bao, and X. Li, J. Appl. Phys. **116**, 124504 (2014).
[12] H. Jia, M. Lu, Q. Wang, M. Bao, and X. Li, Appl. Phys. Lett. **103**, 103505 (2013).
[13] Y. Ye, M. Ke, Y. Li, T. Wang, and Z. Liu, J. Appl. Phys. **114**, 154504 (2013).
[14] C. W. Horton and J. D. Jones, J. Acoust. Soc. Am. **52**, 1456 (1972).
[15] O. R. Asfar and M. A. Hawwa, J. Acoust. Soc. Am. **93**, 2468 (1993).
[16] J. Christensen, P. A. Huidobro, L. Martín-Moreno, and F. J. García-Vidal, Appl. Phys. Lett. **93**, 083502 (2008).
[17] L. Quan, F. Qian, X. Z. Liu, X. F. Gong, and P. Johnson, Phys. Rev. B **92**, 104105 (2015).
[18] L. Quan, X. Zhong, X. Z. Liu, X. F. Gong, and P. Johnson, Nat. Commun. **5**, 3188 (2014).
[19] E. L. Ding, Y. W. Mao, and X. Z. Liu, J. Acoust. Soc. Am. **141**, 1936 (2017).
[20] L. Kelders, J. F. Allard, and W. Lauriks, J. Acoust. Soc. Am. **103**, 2730 (1998).
[21] Y. Zhou, M. Lu, L. Feng, X. Ni, Y. Chen, Y. Zhu, S. Zhu, and N. Ming, Phys. Rev. Lett. **104**, 164301 (2010).
[22] H. Kim, J. Park, and B. Lee, Opt. Lett. **34**, 2569 (2009).

12

Modeling energy band gap of doped TiO_2 semiconductor using homogeneously hybridized support vector regression with gravitational search algorithm hyper-parameter optimization

Taoreed O. Owolabi,[1,2,a] Kabiru O. Akande,[3] Sunday O. Olatunji,[4]
Nahier Aldhafferi,[5] and Abdullah Alqahtani[5]
[1]*Physics Department, King Fahd University of Petroleum and Minerals, Dhahran Postal code 34464, Kingdom of Saudi Arabia*
[2]*Physics and Electronics Department, Adekunle Ajasin University, Akungba Akoko Postal code 342111, Ondo State Nigeria*
[3]*Institute for Digital Communications, School of Engineering, University of Edinburgh, Edinburgh Postal code EH8 9AB, United Kingdom*
[4]*Computer Science Department, College of Computer Science and Information Technology, Imam Abdulrahman Bin Faisal University, Dammam Postal code 31433, Saudi Arabia*
[5]*Computer Information System Department, College of Computer Science and Information Technology, Imam Abdulrahman Bin Faisal University, Dammam Postal code 31433, Saudi Arabia*

Titanium dioxide (TiO_2) semiconductor is characterized with a wide band gap and attracts a significant attention for several applications that include solar cell carrier transportation and photo-catalysis. The tunable band gap of this semiconductor coupled with low cost, chemical stability and non-toxicity make it indispensable for these applications. Structural distortion always accompany TiO_2 band gap tuning through doping and this present work utilizes the resulting structural lattice distortion to estimate band gap of doped TiO_2 using support vector regression (SVR) coupled with novel gravitational search algorithm (GSA) for hyper-parameters optimization. In order to fully capture the non-linear relationship between lattice distortion and band gap, two SVR models were homogeneously hybridized and were subsequently optimized using GSA. GSA-HSVR (hybridized SVR) performs better than GSA-SVR model with performance improvement of 57.2% on the basis of root means square error reduction of the testing dataset. Effect of Co doping and Nitrogen-Iodine co-doping on band gap of TiO_2 semiconductor was modeled and simulated. The obtained band gap estimates show excellent agreement with the values reported from the experiment. By implementing the models, band gap of doped TiO_2 can be estimated with high level of precision and absorption ability of the semiconductor can be extended to visible region of the spectrum for improved properties and efficiency.

I. INTRODUCTION

Fabrication of titanium dioxide (TiO_2) semiconductor of desired band gap still remains a challenge in the field of material sciences for photo-catalysis application.[1] Desired semiconductor for photo-catalysis should possess band gap that enhances absorption of light in the visible spectrum that constitutes 45% of the whole electromagnetic spectrum.[2] The choice of TiO_2 for this application

[a]Correspondence: Taoreed O. Owolabi, Physics Department, King Fahd University of Petroleum and Minerals, Dhahran, Saudi Arabia. owolabi@kfupm.edu.sa, owolabitaoreedolakunle@gmail.com

is due to its unique features which include, among others, tunable band gap, low cost and chemical stability. The tunable band gap of TiO_2 coupled with other properties such as refractive index strengthens its application in lithium insertion-based devices, dye-sensitized solar cells and integrated circuits to mention but few.[3] Absorption of photon across the energy band gap and generation of charge carriers are required for photocatalytic activity. Many recent researches on photo-catalysis application of TiO_2 as well as other applications aim at controlling the band gap of the semiconductor to suit their applications through doping with other foreign atoms or compounds called dopants.[1,2,4–12] Dopants substitution in TiO_2 activates electron-hole separation and promotes surface charge carries to the surface of semiconductor. The need for modeling technique that can quickly and accurately determine the band gap of doped TiO_2 becomes necessary so as to relieve the experimental stress therein and concurrently enhance the light absorbing ability of the semiconductor.

Major crystal structures of titanium dioxides include rutile (tetragonal), brookite (orthorhombic) and tetragonal anatase. However, only anatase and rutile are responsible for most of applications of titanium dioxide. In these structures, each oxygen atom is coordinated with three titanium atoms while each titanium atom is coordinated with six oxygen in slightly distorted octahedral configuration.[13] TiO_2 crystal structure is built up from octahedral and results into anatase phase when vertices are connected, rutile when edges are connected and brookite phase when both vertices and edges are connected. Due to the ionic size of the dopant, creation of oxygen ion vacancy as well as crystal defect becomes necessary for neutralizing charge imbalance attributed with titanium oxide structure. In band gap engineering, dopants distorts the lattice structures of TiO_2 and results into a structural elongation or contraction along *a–axis* and *c–axis* depending on the nature of the dopants. In this work, crystal lattice distortion is used as descriptor for modeling band gap of doped TiO_2.

Support vector regression (SVR) is a machine learning intelligent algorithm which has been reported to have excellent generalization and predictive ability for both regression and classification problems.[14–18] It employs kernel trick in transforming non-linear function to high dimensional space in which construction of linear regression is carried out. Application of SVR extends to a wide varieties of condensed matter physical problems such as superconductivity,[19] surface sciences,[18,20] perovskites and many others.[16,21–24] The purpose of employing SVR in these fields centered on the unique characteristics of SVR which revolve around generalization of error bond even in some cases where descriptive features are few.[25] The generalization ability of this algorithm is promising even when developed using few data-points and this has rendered it a good candidate of choice in this present work. Due to non-linear nature of the relationship between the descriptors (crystal structural lattice parameters) and the target (band gap), homogeneously hybridized SVR (HSVR) was developed. Since the accuracy of SVR-based model is affected by the proper selection of its hyper-parameters, these hyper-parameters were optimized using a novel gravitational search optimization algorithm (GSA) that is developed based on gravitational law and mass interaction.[26] The developed model, GSA- HSVR, has two stages of implementation; the first stage uses the crystal structural lattice parameters as the descriptors to SVR algorithm (resulting in GSA-SVR) while the estimated band gap of the first stage serves as the descriptor to the second SVR algorithm (leading to GSA-HSVR). By this technique, a robust and more precise model (GAS-HSVR) was developed and its result outperforms ordinary SVR model.

Evaluation of the generalization capacity of the developed GSA-HSVR model to unseen data shows that the model can effectively estimate the band gap of doped TiO_2 with mean absolute error as low as 0.07eV when validated using unseen dataset. This precision demonstrated by the proposed model is of high significance in band gap engineering and can effectively improve photo-catalytic activity of TiO_2 semiconductor for better performance and efficiency.

II. FORMULATION OF THE PROPOSED MODEL

A. Mathematical description of support vector regression

Support vector regression is regarded as a computational intelligence learning tool which constructs optimal hyper-plane and minimizes the sum of the distances of training samples.[27,28]

The superiority of SVR over other computational intelligence tools includes utilization of structural risk minimization principle for its formulation as opposed to other tools that are based on empirical risk minimization principle. In SVR mathematical formulation, the generalized error is bounded by the sum of a term (which is controlled by the Vapnik-Chervonenkis dimension) and the training error set. Meanwhile, the objective of SVR algorithm is to formulate input-output relationship on a set of observations contained in the training dataset. With training data points $(x_1,y_1),\ldots\ldots\ldots,(x_m,y_m)$ of m number of samples and n-dimensional vector input $x_i \in R^n$, a regression function is presented in equation (1) where $\langle w \bullet x \rangle$ represents dot product in R^n space.

$$F(x, w) = \langle w \bullet x \rangle + b \tag{1}$$

where $w \in R^n$ and b represents a bias term that can be neglected for zero mean preprocessed data.[29]

This work implements $\varepsilon - SVR$ in which the maximum allowed deviation of the predicted band gaps from the experimental results does not exceed the specified threshold called epsilon (ε) for all training data points.[27,30] This necessitates the use of Vapnik's linear loss function defined in equation (2) and gives zero loss function when the difference between the estimated and experimental band gap is less than epsilon.

$$L(e_j) = \begin{cases} 0, \text{ if } |e_j| \leq \varepsilon \\ |e_j| - \varepsilon, \text{ otherwise} \end{cases} \tag{2}$$

where

$$e_j = y_j - f(x)$$

In order to find a vector **w** that simultaneously ensures flatness of the function and minimizes the error so as to avoid over-fitness, the Euclidean norm presented in equation (3) needs to be minimized and subjected to the condition contained in equation (4)

$$\text{minimize } \frac{1}{2}\|w\|^2 \tag{3}$$

$$|e_j| \leq \varepsilon \tag{4}$$

Equation (3) is modified and presented in equation (5) in order to include a penalty term that regularizes large residual and ensures sparse solution.

$$\text{Minimize } \frac{1}{2}\|w\|^2 + C\sum_{j=1}^{m} L(e_j) \tag{5}$$

The parameter C in equation (5) is called penalty or regularization factor. It measures the trade-off between the flatness of equation (1) and the amount to which the deviations higher than epsilon is allowed.

Actually, it is assumed that there exists a function that approximates all pairs of the data set with accuracy of epsilon. However, non-negative variables called slack variables (ξ and ξ^*) are introduced so as to ensure the existence of the function in a case of infeasible constraints. With this inclusion, the new optimization problem is presented in equation (6).

$$\text{minimise } \frac{1}{2}\|w\|^2 + C\sum_{j=1}^{m} \left(\xi_j + \xi^*{}_j\right)$$

$$\text{subject to} \begin{cases} y_j - \langle w \bullet x_j \rangle - b \leq \varepsilon + \xi_j \\ \langle w \bullet x_j \rangle + b - y_j \leq \varepsilon + \xi^*{}_j \\ \xi_j \text{ and } \xi^*{}_j \geq 0 \quad \text{for all } j = 1, 2, \ldots\ldots, k \end{cases} \tag{6}$$

By invoking Lagrange multipliers (η_j,$\eta_j{}^*$) for dual space transformation as described in Refs. 20 and 31, the regression hyperplane is presented in equation (7)

$$F(x,\eta) = \sum_{j=1}^{m} \left(\eta_j^* - \eta_j\right)\varphi\left\langle x_i, x_j\right\rangle + b \tag{7}$$

Where $\varphi(\overrightarrow{x_i}, \overrightarrow{x_j})$ represents Laplacian radial basis kernel function presented in equation (8)

$$\varphi(\overrightarrow{x_i}, \overrightarrow{x_j}) = \exp\left(-\frac{\left\|\overrightarrow{x_i} - \overrightarrow{x_j}\right\|}{\sigma}\right) \tag{8}$$

The essence of the non-linear kernel mapping function is to transform the input data to high dimensional space where linear regression is constructed. Therefore, SVR successfully caters for non-linear problems and maintain its excellent generalization and predictive strength in several non-linear real life problems.

B. The proposed homogeneously hybridized support vector regression (HSVR)

For the essence of effectively capturing the non-linear relationship between energy band gap for doped TiO_2 and the crystal distortion, two SVR algorithms is hybridized and further optimized using GSA algorithm. Among the advantages of the proposed hybridization is that it aggregates intrinsic features of two different SVR algorithms and allow the utilization of different kernels for data transformation.[32,33] The first stage of the proposed GSA-HSVR model is considered as a stage where all the intricacies and information needed to make accurate final estimation is extracted. The proposed hybrid system provides multiple steps through which generalization bound error could be minimized and thereby allows enhancement of the generalization and predictive capacity of the proposed model. The regression functions of the initial and final stage of the proposed HSVR model are presented in equation (9) and (10) respectively assuming $f(x) = y$.

$$f(x) = \sum_{i=1}^{N_{\text{support vectors}}} \left(\eta_i - \eta_i^*\right)\exp\left(-\frac{\left\|\overrightarrow{x_i} - \overrightarrow{x}\right\|}{\sigma}\right), 0 \le \eta_i^* < C, 0 \le \eta_i \le C \tag{9}$$

$$Z(y) = \sum_{j=1}^{\textit{Number of}\ \text{support vectors}} \left(\lambda_j - \lambda_j^*\right)\exp\left(-\frac{\left\|\overrightarrow{x_j} - \overrightarrow{x_i}\right\|}{\sigma}\right), 0 \le \lambda_j^* < C, 0 \le \lambda_j \le C \tag{10}$$

C. Description of the gravitational search optimization algorithm

Gravitational search optimization algorithm (GSA) is a newly proposed population based optimization methodology that is governed by the Newton law of gravitation as well as mass interactions.[26] It generates iterative mechanisms and approaches through which masses are allowed to interact and move under the force of gravity through a search space of multiple dimensions.[34–36] In attaining the global minimum, masses associated with each of the agents determine its performance while lighter agents migrate towards massive ones (global movement).[37–39] Consider a system of N number of agents in which the position of *nth* agent in d-dimension is defined by equation (11).

$$X_n = (x_n^1, x_n^d \ldots\ldots, x_n^d),\ \textit{where}\ \mathrm{n} = 1, 2, \ldots, N \tag{11}$$

Equation (12) gives the value of the force experienced by an agent of mass $M_k(t)$ due to its interaction with an agent of mass $M_j(t)$ under the influence of gravity with a gravitational constant of $G(t)$ which is time dependent.

$$\mathrm{F}_{kj}^d(t) = G(t)\frac{M_k(t)M_j(t)}{R_{kj}(t) + \varepsilon}\left(x_j^d(t) - x_k^d(t)\right) \tag{12}$$

where $R_{kj}(t)$ and ε respectively represent Euclidian distance and a constant.

The total force experienced by *ith* agent in *dth* direction is depicted in equation (13) where random number $rand_j$ spans through [0,1] interval and enforces randomness in the search procedure.

$$F_k^d(t) = \sum_{j=1, j \neq k}^{N} rand_j F_{kj}^d(t) \tag{13}$$

The acceleration, velocity and position of the *kth* agent are represented by equation (14), (15) and (16) respectively.

$$a_k^d(t) = \frac{F_k^d(t)}{M_{ik}(t)} \tag{14}$$

$$v_k^d(t+1) = rand_k v_k^d(t) + a_k^d(t) \tag{15}$$

$$x_k^d(t+1) = x_k^d(t) + v_k^d(t+1) \tag{16}$$

$$M_{ik} = \text{inertial mass}$$

With the equality assumption of inertial and gravitational masses, fitness function plays a significant role in masses evaluation and the agent of heavier mass moves slowly because of its large attractive force. With mass equality assumption, the masses are updated each time in accordance with equation (17) and (18).

$$m_k(t) = \frac{fit_k(t) - worst(t)}{best(t) - worst(t)} \tag{17}$$

$$M_k(t) = \frac{m_k(t)}{\sum_{j=1}^{N} m_j(t)} \tag{18}$$

where $fit_k(t)$ = fitness of *kth* agent at t time step.

D. Evaluation of the generalization and predictive ability of the model

Root mean square error (RMSE), mean absolute error (MAE) and correlation coefficient (CC) between the experimentally reported band gaps and the estimated values using the proposed models are the parameters through which the predictive strength of the proposed models was evaluated. These parameters were formulated using the relations presented in equation (19), (20) and (21) respectively.

$$RMSE = \sqrt{\frac{1}{n}\sum_{i=1}^{n} e_i^2} \tag{19}$$

$$MAE = \frac{1}{n}\sum_{i=1}^{n} |e_i| \tag{20}$$

$$CC = \frac{\sum_{i=1}^{n}\left(E_{g_i(\exp)} - E'_{g(\exp)}\right)\left(E_{g_i(est)} - E'_{g(est)}\right)}{\sqrt{\sum_{i=1}^{n}\left(E_{g_i(\exp)} - E'_{g(\exp)}\right)^2 \sum_{i=1}^{n}\left(E_{g_i(est)} - E'_{g(est)}\right)^2}} \tag{21}$$

Where e_i and n respectively represent error and number of data points. $E_{gi(\exp)}$ and $E_{gi(est)}$ denote experimental and estimated band gap of *ith* data sample while $E'_{g(\exp)}$ and $E'_{g(est)}$ respectively stand for their mean value.

TABLE I. Results of the statistical analysis.

	a (Å)	c (Å)	Band gap (eV)
Mean(eV)	3.783	9.498928	2.94327
Maximum(eV	3.807	9.667	3.458
Minimum(eV)	3.763	9.363	1.75
Standard deviation	0.008	0.042441	0.356839
Correlation coefficient (%)	1.42	−42.80	

III. MODEL DEVELOPMENT

A. Dataset acquisition and analysis

Sixty-three experimental data-points consisting of crystal lattice parameters and the corresponding band gaps were used in training and validating the algorithms through which the proposed GSA-HSVR and GSA-SVR models were developed. The data-points are the experimentally reported values obtained from published research works.[1,3,40–43] The outcomes of the statistical analysis are shown in table I. Useful hints about the dataset as revealed from the statistical results include the average of the content of the dataset as well as its range.

The results of standard deviation show degree of consistency in the dataset. The value of correlation coefficient shows that the degree of linear dependency of band gap on crystal lattice parameter *a* is very weak while lattice parameter *c* shows negative correlation with relatively weak linear dependency also. Correlation cross-plots between the descriptors (lattice parameter a and c) and band gaps further confirms the aforementioned weak linear dependency as presented in Fig.1 and Fig.2. Hence, due to this non-linear dependency, linear modeling tool will perform poorly when deployed in this scenario which makes the SVR an appropriate choice.

B. Computational implementation of the proposed GSA-SVR and GSA-HSVR model

The proposed GSA-SVR and GSA-HSVR models were developed and implemented within MATLAB computing environment using SVR and GSA algorithms coupled with experimental crystal lattice constants and the band gaps corresponding to the distortion. The dataset was randomized and systematically divided into training and testing dataset in the ratio of eight to two respectively. By this, fifty one data-points were assigned for training the algorithm while twelve data points were utilized while validating the model. RMSE of the testing set of data was used as the fitness function through which the accuracy of the model was evaluated. In the development of GSA-SVR model,

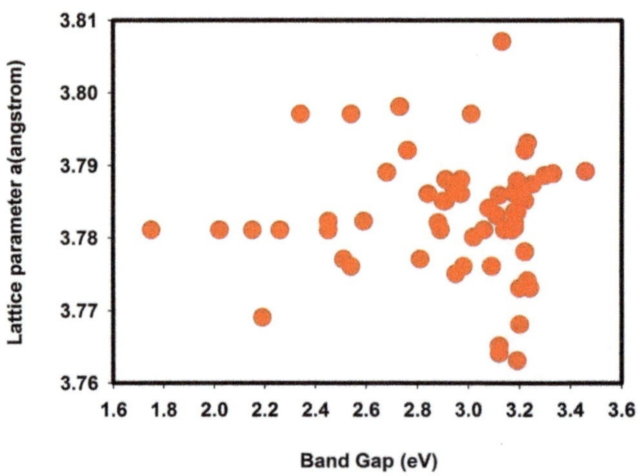

FIG. 1. Correlation cross-plot between band gap and lattice parameter (a).

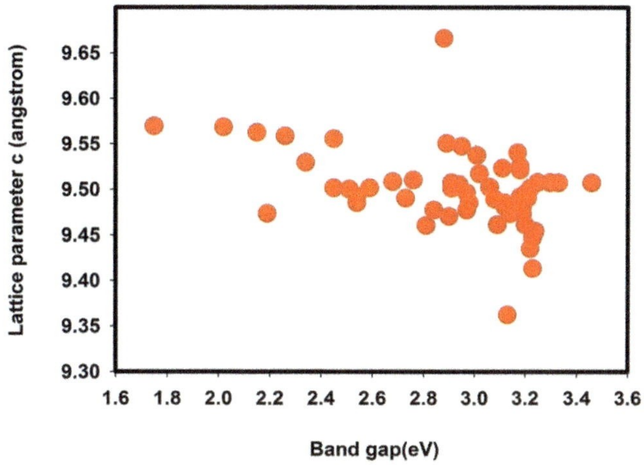

FIG. 2. Correlation cross-plot between band gap and lattice parameter (c).

SVR hyper-parameters which include kernel option (σ), penalty factor (C) and epsilon (ε) were populated using GSA algorithm, each of the population was used to train SVR algorithm using training dataset and the fitness of each of the agent in the population was evaluated with the aid of testing set of data. Through these procedures, SVR hyper-parameters were optimized using GSA algorithm and GSA-SVR model was thereafter developed. The computational description of GSA-SVR model is detailed as follow:

Step 1: Dataset (sixty-three experimental crystal lattice constants and band gaps corresponding to the distortion) was randomized and systematical partitioned in the ratio of eight to two for training and testing respectively.

Step 2: Generation of initial population of agents (these represent potential SVR hyper-parameters) using GSA algorithm

Step 3: Training of SVR algorithm using agents generated in **Step: 2** coupled with training set of data and evaluation of fitness function by employing testing dataset.

Step 4: Update the best agents of the population and calculation of agents' acceleration, velocity and position until SVR hyper-parameters are optimized

Step 5: optimized SVR hyper-parameters and the kernel function corresponding to lowest RMSE of the testing dataset are the optimum SVR hyper-parameters**.**

Effect of the number of agents on model performance was investigated and the result is presented in Fig.3. RMSE of the testing dataset converges to lowest value when the number of agents was ten.

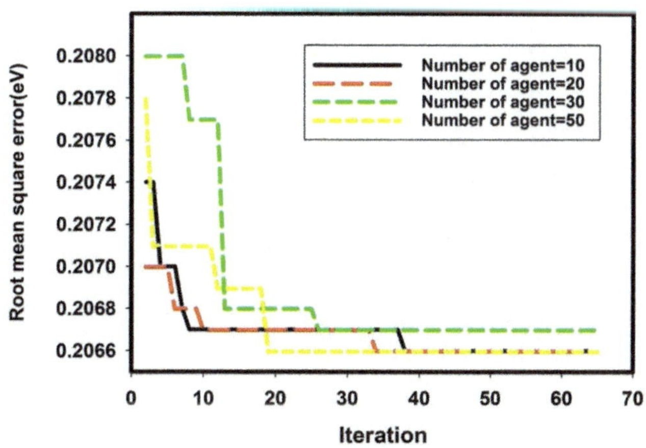

FIG. 3. Sensitivity of the proposed GSA-SVR model to the number of agent.

TABLE II. Optimum SVR hyper-parameters obtained using GSA.

Hyper-parameter	GSA-SVR	GSA-HSVR
Regularization factor	114.0928	475.8765
Epsilon	0.0044	0.0032
Kernel option	0.0065	0.05
Hyper-parameter lambda	E-1	E-1
Kennel function	Gaussian function	Gaussian function
Number of agent	10	10

Although, 10, 20 and 50 number of initial population of agents show similar convergence but the optimum number of agents for our experimental was taken as 10 because it has least computational time complexity. Table II shows the values of optimum hyper-parameters.

Development of GSA-HSVR model was realized by using the output of GSA-SVR model to further train SVR algorithm. The complete computational description of the model is illustrated in fig.4. The optimum values of SVR hyper-parameters for GSA-HSVR are also presented in table II. Similarly, the result of the sensitivity of GSA-HSVR model to the number of agent is also depicted in Fig.5. Minimum convergence was obtained when the initial population of agent was ten.

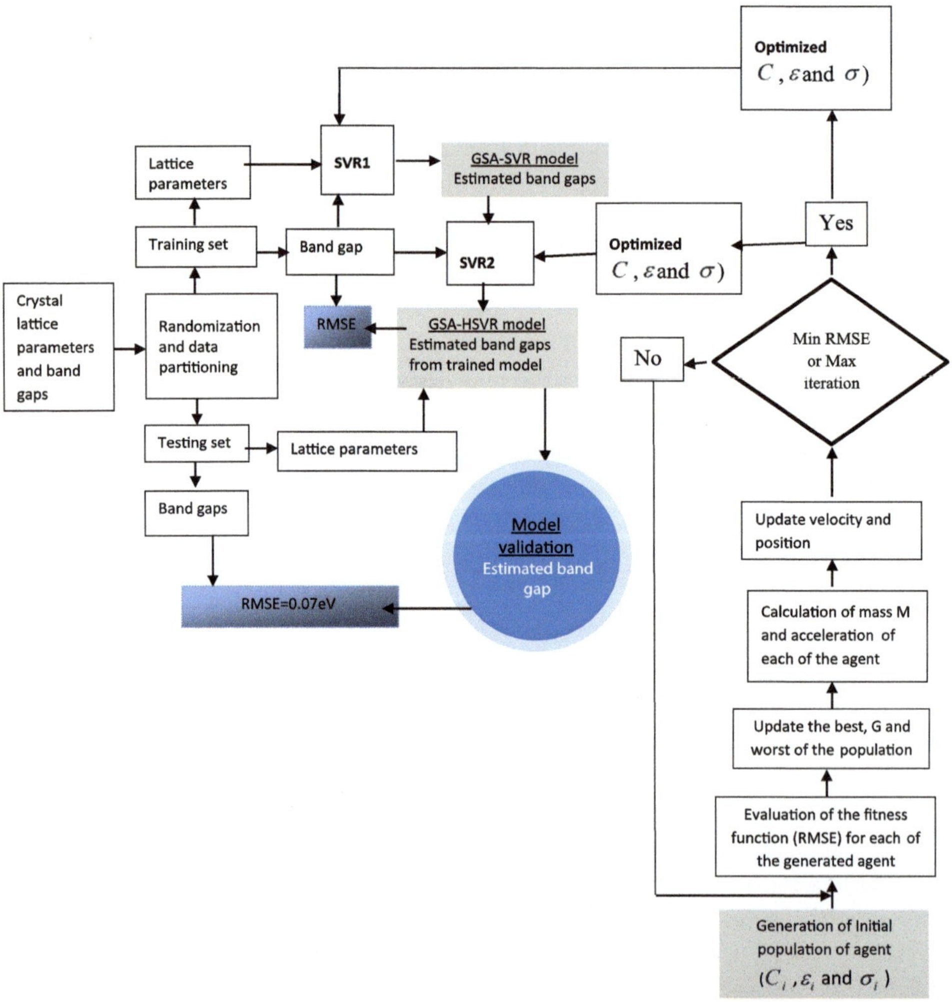

FIG. 4. Computational flow chart for the developed GSA-HSVR model.

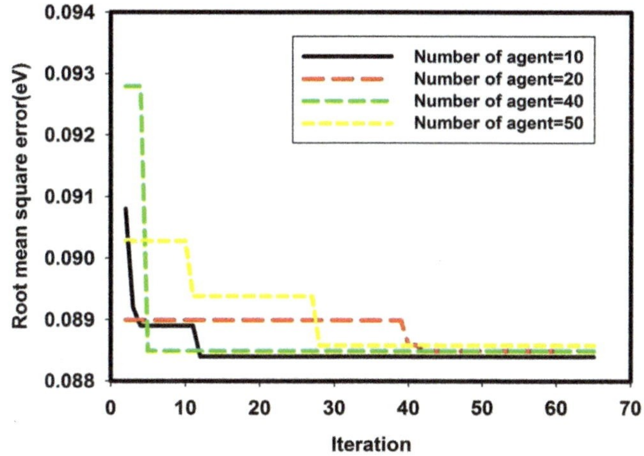

FIG. 5. Sensitivity of the proposed GSA-HSVR model to the number of agent.

IV. RESULTS AND DISCUSSION

A. Model evaluation

Correlation cross-plots between the estimated band gaps using GSA-SVR and GSA-HSVR model and experimental values for testing set of data is presented in fig.6. Data-points representing the estimated band gaps using GSA-SVR model show high degree of disparity with correlation coefficient of 79.08%. Whereas, the data-points for GSA-HSVR model are well aligned with correlation coefficient of 97.7%. The estimation accuracy of the models can be deduced from the alignment of its data-points. GSA-HSVR model outperforms GSA-SVR model with performance improvement of 23.55% on the basis of correlation coefficient. Parameters that measure the generalization strength of the developed model and their corresponding values are presented in table III.

Comparison of the errors (RMSE and MAE) between the estimated band gap and the experimental values for the GSA-SVR and GSA-HSVR model are illustrated in fig.7.

RMSE of 0.2066 eV was obtained for GSA-SVR model while GSA-HSVR model gives 0.0884 eV RMSE. On this basis, GSA-HSVR model demonstrates 57.21% performance improvement as illustrated in fig.7. Similarly, on the basis of MAE, the developed GSA-HSVR enjoys 55.56%

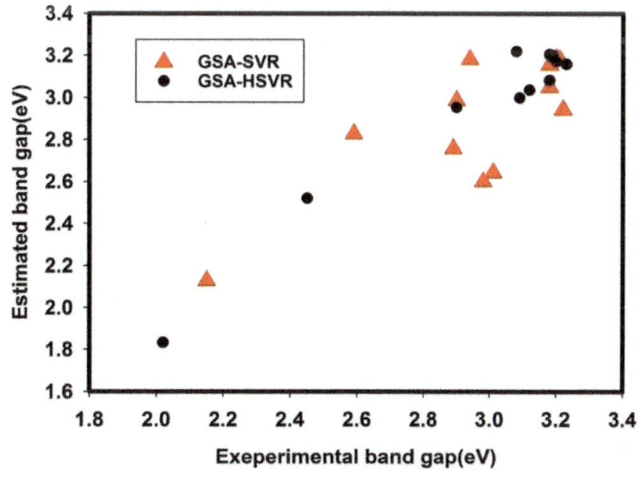

FIG. 6. Correlation cross-plot between experimental and estimated band gap.

TABLE III. Measure of predictive and generalization ability of the developed models.

	GSA-SVR		GSA-HSVR	
	training	testing	training	testing
CC(%)	93.48	79.08	93.12	97.7
RMSE(eV)	0.1328	0.2066	0.1294	0.0884
MAE(eV)	0.095	0.1611	0.0998	0.0716

performance improvement. The improvement in the performance of GSA-HSVR model can be attributed to the combined intrinsic features of two SVR algorithms embedded in GSA-HSVR model where the model has potential of minimizing generalized error bond more than once and effective search for global minimum is thereby enhanced.

B. Further implementation of the developed GSA-SVR and GSA-HSVR model

The predictive capacity and the ability of the model to generalize to unseen data was further investigated by using the model to estimate band gaps of some doped TiO_2 semiconductors that were not included during model development. In this case, the models were supplied with the crystal structural parameters and the developed models utilized the acquired support vectors during the training phase of the model development to estimate the band gap. GSA-SVR estimates were afterwards compared with the experimentally reported values. It is worth mentioning that the developed models perform well and the estimated band gaps are very close to the experimentally reported data. Definitely, this represents the real life application of computational intelligence to condensed matter physical problems and material properties.

1. Effect of cobalt (Co) doping on TiO_2 semiconductor band gap

The impact of Co dopants to TiO_2 semiconductor band gap tuning was investigated and shown in fig.8. Co dopant slightly increases the band gap of the semiconductor from 3.25eV to 3.33eV from the reported experimental results up to the concentration of 0.05 cobalt.[3] Similar trend of band gap increment due to the substituted dopant is captured by the developed models with little or no disparity as depicted in the figure. It can be inferred from the figure that the developed GSA-HSVR performs better than GSA-SVR model as revealed from the closeness of GSA-HSVR estimated band gap to the experimental results in comparison to the results of GSA-SVR model.

2. Effect of Nitrogen-iodine (N/I) ratio on TiO_2 semiconductor band gap

The effect of N/I on TiO_2 semiconductor band gap was obtained using GSA-SVR model and fig.9 compares the experimental values with GSA-SVR estimates. N/I suppress the band gap of the

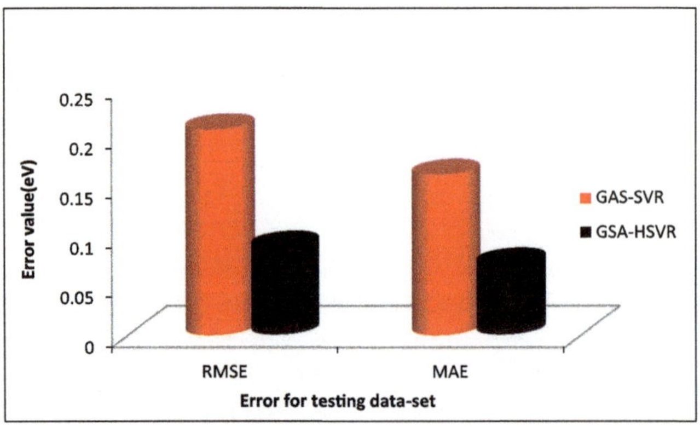

FIG. 7. Measure of generalization and predictive strength of the proposed model.

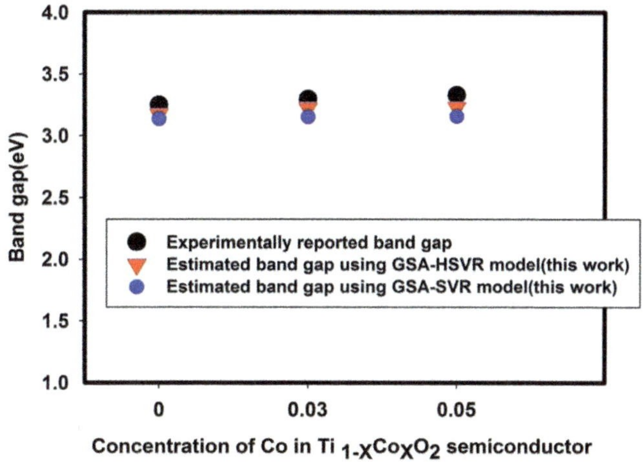

FIG. 8. Sensitivity of the band gap of titanium dioxide to Co dopant.

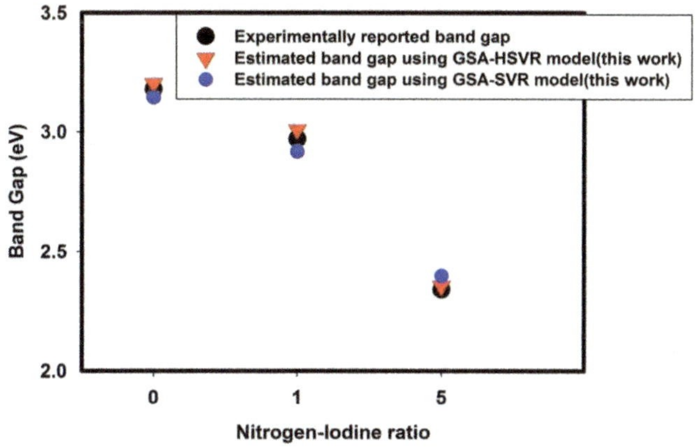

FIG. 9. Effect of Nitrogen-Iodine ratio on the band gap of TiO2.

semiconductor as revealed from the presented figure. The band gaps obtained from GSA-SVR model show excellent agreement with the experimentally reported values.[44] Similarly, GSA-HSVR model outperforms GSA-SVR model in its estimates.

V. CONCLUSION

Crystal lattice distortion that accompanies dopant substitution in TiO_2 lattice structure was used to develop computational intelligence based models for band gap estimation. GSA-SVR model was developed using SVR algorithm optimized with the aid of GSA. The non-linear relationship between the lattice distortion and band gap was well captured by hybridization of two SVR algorithms coupled with GSA to develop GSA-HSVR model. The developed GSA-HSVR model performs better than GSA-SVR model on the basis of CC (23.55% performance improvement), RMSE (57.2% performance improvement) and MAE (55.56% performance improvement). The first stage of GSA-HSVR model is a feature extraction stage in which the needed pattern and information that leads to accurate and precise estimation is extracted. The developed models were further utilized to estimate effect of cobalt doping and nitrogen-iodine co-doping on the band gap of TiO_2 semiconductor and the estimated band gaps agree excellently with the experimentally reported data. Implementation of the developed model would definitely relieve the experimental stress and widen the absorption ability of the semiconductor and thereby leads to improved properties and efficiency.

ACKNOWLEDGMENTS

The support received from University of Dammam is acknowledged

[1] A. Siddiqa, D. Masih, D. Anjum, and M. Siddiq, "Cobalt and sulfur co-doped nano-size TiO2 for photodegradation of various dyes and phenol," J. Environ. Sci. (China) **37**, 100–109 (2015).

[2] M. Mollavali, C. Falamaki, and S. Rohani, "Preparation of multiple-doped TiO2 nanotube arrays with nitrogen, carbon and nickel with enhanced visible light photoelectrochemical activity via single-step anodization," Int. J. Hydrogen Energy **40**(36), 12239–12252 (2015).

[3] W. Khan, S. Ahmad, M. M. Hassan, and A. H. Naqvi, "Structural phase analysis, band gap tuning and fluorescence properties of Co doped TiO 2 nanoparticles," Opt. Mater. (Amst). **38**, 278–285 (2014).

[4] Y. Cheng, M. Zhang, G. Yao, L. Yang, J. Tao, Z. Gong, G. He, and Z. Sun, "Band gap manipulation of cerium doping TiO2 nanopowders by hydrothermal method," J. Alloys Compd. **662** 179–184 (2016).

[5] M. Kumar, A. K. Gupta, and D. Kumar, "Mg-doped TiO2 thin films deposited by low cost technique for CO gas monitoring," Ceram. Int. **42**(1), 405–410 (2015).

[6] D. Yu, W. Zhou, Y. Liu, and P. Wu, "Magnetic and optical properties of Al-doped anatase TiO2(101) surface from density functional theory," J. Magn. Magn. Mater. **404**, 7–13 (2016).

[7] C. McManamon, J. O'Connell, P. Delaney, S. Rasappa, J. D. Holmes, and M. A. Morris, "A facile route to synthesis of S-doped TiO2 nanoparticles for photocatalytic activity," J. Mol. Catal. A Chem. **406**, 51–57 (2015).

[8] S. A. Ahmed, "Annealing effects on structure and magnetic properties of Mn-doped TiO2," J. Magn. Magn. Mater. **402**, 178–183 (2016).

[9] L. Kernazhitsky, V. Shymanovska, T. Gavrilko, V. Naumov, L. Fedorenko, V. Kshnyakin, and J. Baran, "Photoluminescence of Cr-doped TiO2 induced by intense UV laser excitation," J. Lumin. **166**, 253–258 (2015).

[10] A. Arunachalam, S. Dhanapandian, C. Manoharan, and G. Sivakumar, "Physical properties of Zn doped TiO2 thin films with spray pyrolysis technique and its effects in antibacterial activity," Spectrochim. Acta - Part A Mol. Biomol. Spectrosc. **138**, 105–112 (2015).

[11] T. Potlog, P. Dumitriu, M. Dobromir, A. Manole, and D. Luca, "Nb-doped TiO2 thin films for photovoltaic applications," Mater. Des. **85**, 558–563 (2015).

[12] A. Sobczyk-Guzenda, S. Owczarek, H. Szymanowski, L. Volesky, B. Walkowiak, S. Miszczak, and M. Gazicki-Lipman, "Iron doped thin TiO2 films synthesized with the RF PECVD method," Ceram. Int. **41**(6), 7496–7500 (2015).

[13] U. Diebold, "The surface science of titanium dioxide," Surf. Sci. Rep. **48**(5–8), 53–229 (2003).

[14] D. Basak, S. P. And, and D. C. Partababis, "Support vector regression," Neural Inf. Process. (2007).

[15] I. A. Naguib and H. W. Darwish, "Support vector regression and artificial neural network models for stability indicating analysis of mebeverine hydrochloride and sulpiride mixtures in pharmaceutical preparation: A comparative study," Spectrochim. Acta - Part A Mol. Biomol. Spectrosc. **86**, 515–526 (2012).

[16] A. Majid, A. Khan, G. Javed, and A. M. Mirza, "Lattice constant prediction of cubic and monoclinic perovskites using neural networks and support vector regression," Comput. Mater. Sci. **50**(2), 363–372 (2010).

[17] T. O. Owolabi, K. O. Akande, and S. O. Olatunji, "Computational intelligence method of estimating solid- liquid interfacial energy of materials at their melting tem- peratures," J. Intell. fuzzy Syst. (2016).

[18] T. O. Owolabi, K. O. Akande, and S. O. Olatunji, "Estimation of average surface energies of transition metal nitrides using computational intelligence technique," Soft Comput. (2016).

[19] T. O. Owolabi, K. O. Akande, and S. O. Olatunji, "Application of computational intelligence technique for estimating superconducting transition temperature of YBCO superconductors," Appl. Soft Comput. **43**, 143–149, 2016.

[20] T. O. Owolabi, K. O. Akande, and S. O. Olatunji, "Development and validation of surface energies estimator (SEE) using computational intelligence technique," Comput. Mater. Sci. **101**, 143–151 (2015).

[21] Y. Kaneda and H. Mineno, "Sliding window-based support vector regression for predicting micrometeorological data," Expert Syst. Appl. **59**, 217–225 (2016).

[22] M. Ghorbani, G. Zargar, and H. Jazayeri-Rad, "Prediction of asphaltene precipitation using support vector regression tuned with genetic algorithms," Petroleum 1–6 (2016).

[23] G. Zhiqiang, W. Huaiqing, and L. Quan, "Financial time series forecasting using LPP and SVM optimized by PSO," Soft Comput. **17**(5), 05–818 (2013).

[24] L. Zhou, K. K. Lai, and L. Yu, "Credit scoring using support vector machines with direct search for parameters selection," Soft Comput. **13**(2), 149–155 (2009).

[25] T. O. Owolabi, K. O. Akande, and S. O. Olatunji, "Estimation of superconducting transition temperature T C for superconductors of the doped MgB2 system from the crystal lattice parameters using support vector regression," J. Supercond. Nov. Magn. (2014).

[26] E. Rashedi, H. Nezamabadi-pour, and S. Saryazdi, "GSA: A gravitational search algorithm," Inf. Sci. (Ny). **179**(13), 2232–2248 (2009).

[27] R. D. A. Timoteo, L. N. Silva, D. C. Cunha, and G. D. C. Cavalcanti, "An approach using support vector regression for mobile location in cellular networks," Comput. Networks **95**, 51–61 (2016).

[28] K. O. Akande, T. O. Owolabi, and S. O. Olatunji, "Journal of natural gas science and engineering investigating the effect of correlation-based feature selection on the performance of neural network in reservoir characterization," J. Nat. Gas Sci. Eng. **27**, 98–108 (1997).

[29] T. O. Owolabi, M. Faiz, S. O. Olatunji, and I. K. Popoola, "Computational intelligence method of determining the energy band gap of doped ZnO semiconductor," Mater. Des. **101**, 277–284 (2016).

[30] J. B. Gao, S. R. Gunn, C. J. Harris, and M. Brown, "A robabilistic framepwork for SVM regression and error bar estimation," Mach. Learn. **46**(1–3), 71–89 (2002).

[31] T. O. Owolabi, K. O. Akande, and S. O. Olatunji, "Estimation of surface energies of hexagonal close packed metals using computational intelligence technique," Appl. Soft Comput. **31**, 360–368 (2015).
[32] T. O. Owolabi, Y. F. Zakariya, K. O. Akande, and S. O. Olatunji, "Estimation of melting points of fatty acids using homogeneously 3 hybridized support vector regression," Neural Comput. Appl. (2016).
[33] K. O. Akande, T. O. Owolabi, S. O. Olatunji, and A. Abdulraheem, "A novel homogenous hybridization scheme for performance improvement of support vector machines regression in reservoir characterization," Appl. Comput. Intell. Soft Comput. **2016**, 1–10.
[34] J. Xiang, X. Han, F. Duan, Y. Qiang, X. Xiong, Y. Lan, and H. Chai, "A novel hybrid system for feature selection based on an improved gravitational search algorithm and k-NN method," Appl. Soft Comput. **31**, 293–307 (2015).
[35] S. D. Beigvand, H. Abdi, and M. La Scala, "Combined heat and power economic dispatch problem using gravitational search algorithm," Electr. Power Syst. Res. **133**, 160–172 (2016).
[36] X. Han, X. Chang, L. Quan, X. Xiong, J. Li, Z. Zhang, and Y. Liu, "Feature subset selection by gravitational search algorithm optimization," Inf. Sci. (Ny). **281**, 128–146 (2014).
[37] Y. Mohamed Shuaib, M. Surya Kalavathi, and C. Christober Asir Rajan, "Optimal capacitor placement in radial distribution system using Gravitational Search Algorithm," Int. J. Electr. Power Energy Syst. **64**, 384–397 (2015).
[38] D. Goswami and S. Chakraborty, "Parametric optimization of ultrasonic machining process using gravitational search and fireworks algorithms," Ain Shams Eng. J. **6**(1), 315–331 (2015).
[39] M. Rezaei and H. Nezamabadi-pour, "Using gravitational search algorithm in prototype generation for nearest neighbor classification," Neurocomputing **157**, 256–263 (2015).
[40] E. O. Oseghe, P. G. Ndungu, and S. B. Jonnalagadda, "Photocatalytic degradation of 4-chloro-2-methylphenoxyacetic acid using W-doped TiO2," J. Photochem. Photobiol. A Chem. **312**, 96–106 (2015).
[41] M. Tahir and N. S. Amin, "Photocatalytic CO2 reduction with H2 as reductant over copper and indium co-doped TiO2 nanocatalysts in a monolith photoreactor," Appl. Catal. A Gen. **493**, 90–102 (2015).
[42] S. Larumbe, M. Monge, and C. Gomez-Polo, "Comparative study of (N, Fe) doped TiO2 photocatalysts," Appl. Surf. Sci. **327**, 490–497 (2015).
[43] A. E. Giannakas, M. Antonopoulou, Y. Deligiannakis, and I. Konstantinou, "Preparation, characterization of N – I co-doped TiO 2 and catalytic performance toward simultaneous Cr (VI) reduction and benzoic acid oxidation," Applied Catalysis B ;: Environmental **141**, 636–645 (2013).
[44] A. E. Giannakas, M. Antonopoulou, Y. Deligiannakis, and I. Konstantinou, "Preparation, characterization of N-I co-doped TiO2 and catalytic performance toward simultaneous Cr(VI) reduction and benzoic acid oxidation," Appl. Catal. B Environ. **140–141**, 636–645 (2013).

13

X-ray source on the basis of the piroelectric crystal $Sr_{0.61}Ba_{0.39}Nb_2O_6$

V. A. Andrianov,[1,a] A. L. Erzinkian,[1] L. I. Ivleva,[2] and P. A. Lykov[2]
[1]*Skobeltsyn Institute of Nuclear Physics, Lomonosov Moscow State University, Leninskie gory 1(2), GSP-1, Moscow 119991, Russian Federation*
[2]*Prokhorov General Physics Institute RAS, Vavilov Str., 38, Moscow 119991, Russian Federation*

The properties of X-ray source based on $Sr_{0.61}Ba_{0.39}Nb_2O_6$ crystals (SBN-61) having a large pyroelectric coefficient $\gamma = 85$ nC/(cm^{2}K) was studied. When the crystal was heated to 60 ° C, X-rays from a tungsten target with energy of 8.4 keV were detected. The maximum energy of the electron beam was 52 keV. The visualization of the electron beam with a grid electrode and a fluorescent screen showed that the electron beam was inhomogeneous in the polar plane of the crystal and varied during heating.The radiation intensity was unstable in time. The work of the X-ray source was limited to electrical breakdowns between the polar faces of the crystal. When the crystal was cooled, there was no X-ray emission, which could be due to the depolarization of the crystal by the total electric field or as a result of electrical breakdowns. The crystals SBN-61 and $LiNbO_3$ are compared.

I. INTRODUCTION

Pyroelectric crystals can be used to create portable sources of X-ray and neutron radiation.[1–4] In these devices, strong electric fields with strengths up to 100 kV/cm are produced due to uncompensated charges, which are formed on the surface of a pyroelectric crystal under temperature change by 50-100 ° C. In X-ray sources, these fields accelerate electrons, which after collision with a target or a pyroelectric crystal induce the generation of characteristic and bremsstrahlung X-rays. In neutron sources, these electric fields cause ionization of deuterium molecules and their subsequent acceleration to a deuterium target, where the reaction of fusion D+D = He + n occurs.[3,4]

The electric charge Q on the polar face of pyroelectric crystal is the sum of the bound charge $Q_{\rm bond}(T)$ and the compensating surface charge $Q_{\rm comp}$: $Q = Q_{\rm bond}(T) + Q_{\rm comp}$. The bound charge $Q_{\rm bond}(T)$ is due to the macroscopic polarization of the crystal and depends on the temperature T. The charge $Q_{\rm bond}(T)$ is sufficiently large,[5,6] and for the lithium niobate and lithium tantalate crystals it is about 50 μC/cm^2. At atmospheric pressure, the bound charge $Q_{\rm bond}(T)$ is completely compensated by the captured charges $Q_{\rm comp}$, and the total charge Q is zero.

When the temperature changes by ΔT, the bound charge $Q_{\rm bond}$ changes to the next value:

$$\Delta Q_{bond} = -\gamma(T) \bullet \Delta T \bullet S, \quad (1)$$

where $\gamma(T)$ is the pyroelectric coefficient, S is the area of the polar face of the crystal.

If the crystal is placed in a vacuum chamber and its temperature is changed by ΔT(t), then the change in the compensating charge $Q_{\rm comp}(t)$ is delayed compared to $Q_{\rm bond}(t)$. As a result, an uncompensated total charge $Q(t) \approx \Delta Q_{\rm bond}$ appears. This charge can create an electric field with strengths of up to 350 kV/cm.[7]

[a]E-mail: andrva22@mail.ru

Traditionally, pyroelectric sources use lithium niobate crystals or lithium tantalate crystals with large pyroelectric coefficients: $\gamma(LiNbO_3) = 8.3$ nC/(cm$^2\times$ K) and $\gamma(LiTaO_3) = 17.6$ nC/(cm$^2 \times$ K).[5,6] In the present work, we studied an X-ray source based on $Sr_{0.61}Ba_{0.39}Nb_2O_6$ crystals (SBN-61) having a significantly larger coefficient γ(SBN-61) = 85 nC/(cm^{2}K).

SBN crystals are relaxor ferroelectrics with a diffuse phase transition.[8] The temperature of the ferroelectric transition for SBN-61 is small and amounts to T_c = 80° C. This limits the heating of crystals with a temperature of about 60° *C*. Another feature of SBN crystals is a small coercive electric field, which for SBN-61 is E_c = 2 - 2.5 kV/cm. This indicates the possible instability of the domain structure, and, in particular, the possibility of the destruction of the single-domain state during electrical breakdowns.

II. EXPERIMENT

We used SBN-61 crystals grown by the modified Stepanov technique[9] in Institute of General Physics of the Russian Academy of Sciences (Moscow, Russia). The investigated samples had the form of parallelepipeds cut along the crystallographic axes *a*, *b* and *c*. The dimension along the polar axis *c* for all crystals was 13 mm. The dimensions along the *a* and *b* axes were about 10 mm. After cutting and polishing the faces, the crystals were polarized along the *c* axis in an electric field of 6 kV/cm at room temperature. Electrical contacts were made by conductive glue based on colloidal carbon. Before the experiment all surfaces of the crystal were cleaned with pure alcohol and acetone.

The experiments were carried out in a vacuum chamber, the construction of which is shown in Figure 1. The crystal (1) was attached onto the copper plate (2), mounted on the thermoelectric Peltier module (3). The upper face of the crystal had the positive polarity (c+). The Peltier module provided heating and cooling of the crystal. The thermal contact between the copper plate and the Peltier module was made by heat-conducting glue. The temperature was measured with the chromel-alumel thermocouple (4), which was fixed at the plate (2) next to the crystal. The lower face of the crystal together with the plate (2) was grounded to the metal housing of the chamber (5). Above the crystal, at a distance of about 7 mm, there was a target (6), which in our case was a tungsten grid with a pitch of 2 × 2.5 mm. The grid (6) was also grounded on the camera housing. The fluorescent screen (7) was located above the grid (6) at the distance of about 7 mm. The glow of a fluorescent

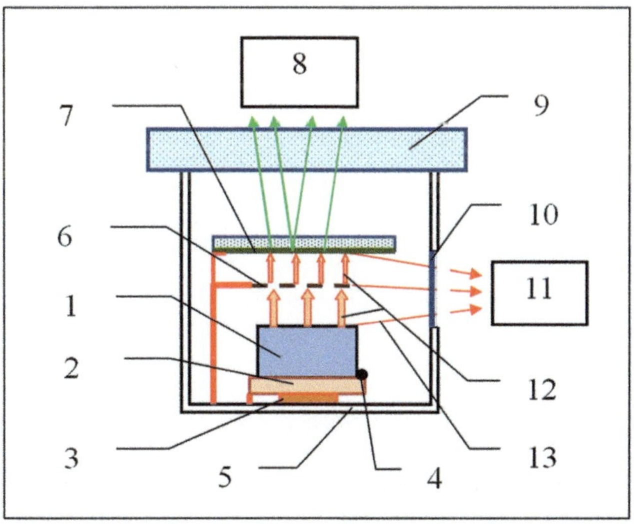

FIG. 1. Scheme of the vacuum chamber. Here, 1 is the pyroelectric crystal; 2 is the copper plate; 3 is the heater (Peltier module); 4 is the thermometer (thermocouple); 5 is the housing of the chamber; 6 is the grid target; 7 is the fluorescent screen; 8 is the photo/video equipment; 9 is the glass window; 10 is the beryllium window; 11 is the photoelectric multiplier; 12 designates the electron flux; 13 designates X-ray radiation.

screen caused by the electron beam (12) coming from the crystal was recorded with digital photo- and video equipment (8) through a glass window (9). The Be-window (10) was located on the side of the chamber. It provided the output of X-ray radiation (13) from the chamber. X-ray radiation was detected by the photomultiplier (11) with a thin NaI(Tl) crystal and the energy resolution of 3.2 keV for 5.9-keV line.

The experiment was carried out as follows. The chamber (5) was evacuated to a pressure better than 10^{-3}Torr. In the initial state, the bound crystal charge on the upper (c+) face was compensated by the surface charge, $Q_{bond}(t = 0) = - Q_{comp}(t = 0)$. During heating the bound crystal charge $Q_{bond}(t)$ decreased in accordance with expression (1). The compensating charge $Q_{comp}(t)$ was delayed compared to $Q_{bond}(t)$. As a result, the total negative charge $Q\ (t)$ was formed on the upper face of the crystal, and this face acquired the negative potential with respect to the grid (6). The electron flux caused by ionization of residual gas and by the injection of electrons from the surface was directed from the crystal to the grid-target (6). The electron flux in the collision with the grid-target caused the generation of bremsstrahlung and characteristic X-ray radiation. The intensity $I(t)$ and the pulse height spectrum S(E) of X-ray radiation were recorded during heating and subsequent cooling of the crystal. The spatial structure of the electron beam on the fluorescent screen (7) was fixed by photo and video equipment. The crystal was heated by the Peltier module to a temperature of 60°C at a rate of approximately 0.05 K/s, then it was held at maximum temperature for approximately 10 minutes to ensure the transition to an equilibrium state. The cooling of the crystal was carried out at the rate 0.05 K/s to a temperature of 10° C.

It should be noted that during heating the direction of the resulting electric field between the polar faces of the crystal coincides with the direction of polarization and can provide additional polarization of the crystal. When cooling, the direction of the resulting field changes to the opposite one and in this case one cannot exclude the partial depolarization of the crystal. In view of this, the cooling of the crystal was carried out in two ways. The main mode was the cooling of samples at elevated pressure, about 2×10^{-1} Torr, when the compensation of surface charges occurred rather quickly and the total electric field was close to zero. When the temperature reached 10 °C, the sample was held for about 5 minutes at this temperature, after which the chamber was again evacuated to a pressure of $P \leq 10^{-3}$ Torr. Then the next heating-cooling cycle began. In a number of experiments, cooling was carried out at a reduced pressure $P<10^{-3}$ Torr. As a rule, this regime was carried out at the end of thermal cycling.

III. RESULTS AND DISCUSSION

Figure 2a shows a typical plot of the radiation intensity I versus the heating temperature ΔT. Radiation occurred when the temperature reached ΔT_0, at which the voltage between the crystal and the target exceeded the binding energy of the electron in the target atoms. In our case, this is the binding energy of the L-electron in the tungsten $E_L(W) \approx 10.2$ keV. With a further increase in temperature, the intensity of the radiation increased sharply, but then an electrical breakdown of the crystal occurred, and the X-ray emission disappeared. With a further heating, it was possible to restore X-ray radiation, which was also limited by electrical breakdown (see Figure 2a).

Figure 2b shows the pulse height spectra obtained during heating in 3 consecutive heating/cooling cycles. The spectra contain the peak of the X-ray L-lines of Tungsten L(W) ≈ 8.4 keV and the continuous spectrum of the bremsstrahlung. The maximum energy of the bremsstrahlung spectrum corresponds to the maximum electron energy, E_{max}. It can be seen from the graph that there is a training effect, that is, in each subsequent cycle an increase in the radiation intensity and the maximum energy E_{max} is observed. In the third cycle, the electron energy reaches the value $E_{max} = 52$ keV. However, in the subsequent 3 cycles, the opposite effect was observed: the intensity and the maximum energy of electrons E_{max} decreased. In the 6th heating cycle, X-ray radiation was practically absent. We note that in this series of experiments cooling was always carried out at elevated pressure $P \approx 2*10^{-1}$ Torr, in the absence of any X-ray radiation.

An analysis of the data shows that the main factor limiting the stable operation of the X-ray source is the electrical breakdown of the crystal. The electrical breakdowns limit the real temperature range, since the discharge between the polar faces resets the voltage between them to zero. Figure 3 shows

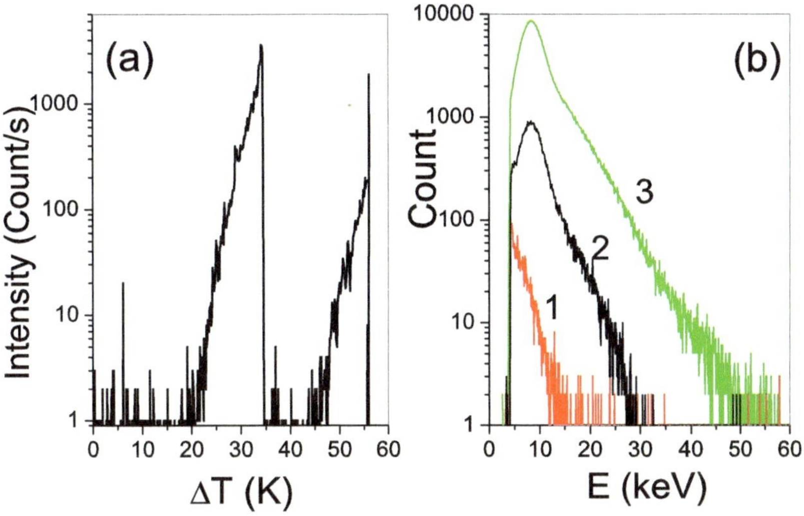

FIG. 2. a) Dependences of the X-ray intensity $I(\Delta T)$ on the heating temperature ΔT. b) Pulse height spectra of X-ray radiation in the logarithmic scale. The digits at the curves correspond to the number of the heating/cooling cycle.

the dependence of the maximum energy of the bremsstrahlung spectrum $E_{\max}$ on the temperature interval between electrical breakdown (ΔT_{br}) and the X-ray emission temperature ΔT_0. The effect of training in the first 3 cycles is associated with an increase in the operating temperature range (ΔT_{br}-ΔT_0). The attenuation of X-rays and the decrease of $E_{\max}$ in subsequent cycles are associated with the reverse process.

Video and photo shooting of the crystal showed that the electrical breakdown between the polar faces occurs mainly along the lateral faces. Heating the crystal to 60 °C in a vacuum can be considered as the additional cleaning of the surfaces that improves the conditions in subsequent heating cycles. However, when heated to higher temperatures and, correspondingly, to higher voltages, more powerful electrical breakdowns occur, accompanied by the ejection of matter and the formation of deep vertical scratches-caverns (see Figure S1 and S2 of the supplementary material).

The observed effect of training can be explained by two factors. The first is cleaning the surface when the crystal is heated in a vacuum. The second one is the additional polarization of the crystal in the field of surface charges. We recall that the electric fields arising during heating reached values of 30-50 kV, which is much higher than the polarizing fields at the preparation of the crystals.

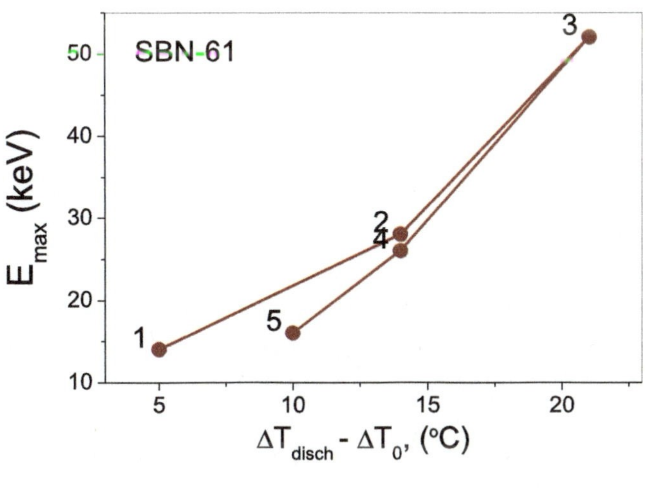

FIG. 3. Dependence of the maximum electron energy $E_{\max}$ on the temperature interval between the beginning of X-ray radiation and the discharge (ΔT_{disch} - ΔT_0). The digits at the experimental points indicate the numbers of the thermal cycles.

Electrical breakdowns, as a rule, occurred at the end of each heating. The negative role of electrical breakdowns consisted not only in the discharge of the crystal and, therefore, in the limitation of the operating temperature range. Electrical breakdowns, like the rapid removal of electrical voltage, could lead to the destruction of the single-domain structure of the crystal, since the coercive force of SBN crystals is not large.[10] In addition, the scratches-caverns that appeared during the discharge prevented the polarization of the crystal, since they were regions of small electrical resistance.

The fluorescent screen made it possible to observe the dynamics of the electron beam. The glow of the screen appeared almost simultaneously with the X-ray radiation and had the form of one or two luminous points (see Figure 4a). As the temperature increased, the luminous region increased in size, and its luminescence became brighter, but remained inhomogeneous. Further, against the background of the luminous region, a luminous grid appeared with dimensions corresponding to the grid-electrode (see Figure 4b). The glowing region was inhomogeneous and consisted of several bright regions, which could change their location during the heating process. The inhomogeneity of the electron beam indicates that the electrical potential of the upper face of the crystal is not uniform, and that the distribution of the potential changes during heating. Similar data were obtained for $LiNbO_3$ crystals.[11] In this work, we also observed an inhomogeneous structure of the electron beam against the background of a luminous grid. In a simple model, the grid-target had to lead to a shadow pattern on the fluorescent screen. The luminous grid means that the grid-target had focusing properties in the geometry of our experiment.

In a series of works of J. D. Brownridge and S. M. Shafroth,[12,13] the effect of self-focusing of the electron beam was considered. In our work, this effect was not observed, since the experiment had a different geometry. The polar face of our crystal had a larger area of about 1.2 cm^2. At a small distance above the crystal (7 mm), there was a grid-target. The resulting electric field actually projected the electron beam onto the fluorescent screen above the grid (see Figure 1). The observed inhomogeneous structure of the electron beam, as well as changing of the beam with time, had nothing in common with the focusing effects. It can be assumed that the observed effects are due to the spatial fluctuations of the leakage currents, which reflects the internal imperfection of the investigated crystals.

In cooling modes at low pressure $P \leq 10^{-3}$Torr, X-ray radiation was not detected. With subsequent heating, X-rays were also absent. The measurements were carried out at the end of the standard heating/cooling cycles, as well as in separate experiments, when heating was carried out at an elevated pressure $P \approx 2 \times 10^{-1}$ Torr, and cooling was carried out at a low pressure of 10^{-3} Torr. The experimental data indicate that upon cooling the crystal is completely depolarized. A possible cause is the depolarization of the crystal by the total electric field directed against the polarization of the crystal. This field destroys the single-domain state of SBN crystal by forming the surface microdomains with an opposite polarization orientation.[14]

The main characteristic of pyroelectric sources is the voltage U_{gap}, which arises between free surface of the crystal and the target as the temperature changes by ΔT. This voltage determines the energy of the electron beam and, accordingly, the maximum energy of the X-ray radiation. U_{gap} is

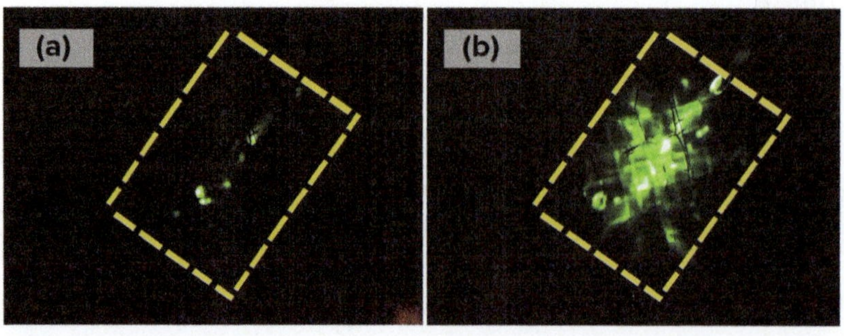

FIG. 4. Photographs of the electron flux on the luminescent screen: a) at the beginning of the X-ray emission; b) at the maximum intensity of X-ray radiation. Dotted rectangles are the positions of the SBN crystal in the pictures.

TABLE I. Main characteristics of pyroelectric crystals SBN-61 and $LiNbO_3$.[5,6]

Characteristic	SBN-61	$LiNbO_3$
Pyroelectric coefficient $\gamma(T)$, $(C \cdot cm^{-2} K^{-1})$	85	8.3
Permittivity ε_{cr}	800	30
Quality parameter for detectivity, $F_I^* = \gamma/\varepsilon_{cr}$, $(C \cdot cm^{-2} K^{-1})$	0.11	0.28
Curie temperature, T_c, $(°C)$	81	1210
Coercive field, E_c, (kV/cm)	2.5	200
Electrical resistivity, ρ, $(\Omega \times m)$	$0.2\ 10^{10}$	10^{13}-10^{14}
Thermal diffusivity, χ, (cm/s)	$2.9*10^{-3}$	$15*10^{-3}$

given by the following approximate expression:[15]

$$U_{\text{gap}} = \frac{\gamma(T)\Delta T}{\varepsilon_0} \frac{d_{\text{gap}}}{1 + \frac{d_{\text{gap}}}{d_{\text{cr}}} \varepsilon_{\text{cr}}} \approx \frac{\gamma(T)\Delta T}{\varepsilon_0 \varepsilon_{\text{cr}}} d_{\text{cr}}. \tag{2}$$

Where ε_0 is the free space permittivity, ε_{cr} is the permittivity of the crystal, d_{gap} is the distance between the target and the crystal, and d_{cr} is the crystal thickness.

It follows from formula (2) that when comparing different crystals, the main parameter is the ratio $F_I = \gamma(T)/\varepsilon_{cr}$. The Table I shows the characteristics of the pyroelectrics SBN-61 and lithium niobate $LiNbO_3$. It follows from the Table that for SBN-61 the parameter $\gamma(T)/\varepsilon_{cr}$. = 0.11 nC/($cm^2$×K), which is 2.5 times smaller than for $LiNbO_3$. The lower value F_I for SBN-61 crystals is determined by the permittivity, which is anomalously large: $\varepsilon_{cr} = 800$. Note that large values of ε_{cr} are related to the fact that SBN-61 is not a classical pyroelectric, but refers to ferroelectric relaxors.[10] The estimate of the ratio $\gamma(T)/\varepsilon_{cr}$ from our experimental data gives a smaller value for the parameter $F_I = (\gamma(T)/\varepsilon_{cr})_{exp} \approx 0.04$ nC/(cm^2×K).

From Table I it follows that SBN-61 has a low Curie temperature, $T_c = 81°$ C, which limits the temperature range in the X-ray source. SBN-61 has also small coercive electric field $E_c = 2.5$ kV/cm, which means an instability of the domain structure and, in particular, the possibility of destruction of the single-domain state as a result of electric breakdowns[14,16] or due to the electric field of the opposite direction. The electrical resistance of SBN-61 is 3 orders of magnitude lower than the resistance of lithium niobate. If we consider a pyroelectric crystal as an R-C chain, then the resistivity will determine the proper discharge time of the crystal $\tau = \varepsilon_0 * \rho$. For the SBN-61 crystal this time appears much shorter than for $LiNbO_3$. Thermal diffusivity χ for SBN is also 5 times worse than for $LiNbO_3$. This means that at the same heating rate of the crystal, in the case of SBN-61, the temperature gradient and mechanical stresses will be appreciably larger.

Thus, analysis of known characteristics and the experimental data obtained show that despite the anomalously high values of the pyroelectric coefficient, the set of properties of SBN crystals substantially limits the possibilities of using this material as an X-ray source.

IV. CONCLUSION

The X-ray pyroelectric source with an SBN-61 crystal was first realized. In the heating mode up to 60 °C, X-ray radiation from a tungsten grid-target with an energy of 8.4 keV was detected. The maximum energy of the electron beam was 52 keV. It is shown that the electron flux is inhomogeneous in the plane of the polar face of the crystal and varies during the heating process. The radiation intensity is unstable in time and is limited by electrical breakdowns between the polar faces of the crystal. Electrical breakdowns cause the formation of the deep scratches-caverns on the side faces of the crystal, which prevent its polarization. In the cooling mode, X-ray radiation was not detected. A possible cause is the depolarization of the crystal by the total electric field or as a result of electrical breakdowns. Comparison of the characteristics of SBN-61 and $LiNbO_3$ crystals and data of the experiments did not reveal the advantages of SBN-61 when creating X-ray or neutron pyroelectric sources.

ACKNOWLEDGMENTS

This work was carried out without any support from Russian scientific foundations, RFBR and RSF. The authors would like to thank Dr. I.V. Andrianov for the possibility of using his digital camera for the visualization of the electronic flux in the experiment.

[1] J. D. Brownridge, Nature **358**, 287 (1992).
[2] J. D. Brownridge and S. Raboy, J. Appl. Phys. **86**, 640 (1999).
[3] D. J. Gillich, A. Kovanen, and Y. Danon, J. Nucl. Mater. **405**, 181 (2010).
[4] B. Naranjo, S. Putterman, and T. Venhaus, Nucl. Instr. Meth. Phys. Res. A. **632**, 43 (2011).
[5] A. Bush, *Pyroelectric Effect and its Applications* (Moscow, MIREA, 2005), ISBN: 5-7339-0538-7, (https://www.researchgate.net/profile/Alexandr_Bush/publications?pubType=book).
[6] A. K. Batra and M. D. Aggarwal, *Pyroelectric Materials: Infrared Detectors, Particle Accelerators and Energy Harvesters* (SPIE Press, Bellingham, Washington USA, 2013), ISBN: 978-0-8194-9331-6.
[7] W. Tornow, S. M. Lyman, and S. H. Shafroth, J. Appl. Phys. **107**, 063302 (2010).
[8] W. H. Huang, D. Viehland, and R. R. Neurgaonkar, J. Appl. Phys. **76**, 490 (1994).
[9] L. I. Ivleva, Bull. Russ. Acad. Sci. Phys. **73**, 1338 (2009).
[10] L. E. Cross, Ferroelectrics **76**, 241 (1987).
[11] V. A. Andrianov, A. A. Bush, A. L. Erzinkyan, and K. E. Kamentsev, J. Surf. Invest: X-ray, Synchr. Neutr. Techn. **11**, 704 (2017).
[12] J. D. Brownridge and S. M. Shafroth, Appl. Phys. Lett. **79**, 3364 (2001).
[13] J. D. Brownridge and S. M. Shafroth, *Trends in Lasers and Electro-Optics Research*, edited by W. Arkin (Nova Science Publisher, 2006), Ch. 2, P. 59.
[14] V. Y. Shur and P. S. Zelenovskiy, J. Appl. Phys. **116**, 066802 (2014).
[15] T. Z. Fullem and Y. Danon, J. Appl. Phys. **106**, 074101 (2009).
[16] E. A. Kolchina, M. M. Neradovskiy, V. A. Shikhova, D. V. Pelegov, V. Y. Shur, L. I. Ivleva, and J. Dec, Ferroelectrics **496**, 149 (2016).

14

Ka-band relativistic diffraction generator with a tapered coaxial Bragg reflector

Feng Lan,[1,2,a] Ziqiang Yang,[1] Pinaki Mazumder,[2] Zongjun Shi,[1] Lin Meng,[1] Dagang Liu,[1] and Wenjie Fu[1]

[1]*School of Physical Electronics, University of Electronic Science and Technology of China, 610054 Chengdu, Sichuan, P. R. China*

[2]*Department of Electrical Engineering and Computer Science, University of Michigan, Ann Arbor, MI 48109-2121, USA*

A coaxial line with periodically tapered inner rode corrugated ripples is introduced to act as a distributed Bragg reflector to increase output microwave power of relativistic diffraction generator (RDG). The RDG device consists of an overmoded cross-section, two-sectional disk-loaded cylindrical slow-wave structure (SWS), and coaxial reflector is placed upstream at the first SWS section. The RDG device with tapered Bragg reflector is investigated by full three-dimensional Particle-in-Cell (PIC) simulations and experiments in this paper. The structural parameters of the reflector are optimized for favorable characteristic of mode selection and frequency response. It is found that oscillation modes varied at different magnetic field magnitude. The measured radiation power at 29~31 GHz is over 300 MW for TM_{01} mode and over 700 MW for TM_{11} mode. Results show that the reflector has a great impact on performance of the device. Operating at proper conditions, the radiation power is remarkably enhanced.

I. INTRODUCTION

Backward-wave Cerenkov radiation sources including backward-wave oscillators (BWOs),[1] multiwave Cerenkov generators (MWCGs)[2] and relativistic diffraction generators (RDG)[3] have been studied widely for several decades. The RDG devices are the promising high-power electromagnetic oscillation sources of effective conversion of axial electron beam energy into millimeter wave radiation.[4] Since 1980s, efforts have been made to achieve sufficient interaction of electron beam with wave in relativistic Cerenkov devices. Typical methods of increasing radiation power include adopting large transverse diameter ($D/\lambda >> 1$) and two-sectional slow-wave structure separated by a drift tube.[5–7] Preventing microwave radiation from infiltrating into the beam diode region is the key factor to the beam generation and output power increment.[8] The cut-off neck was successfully used as a reflector at the beam input section.[9] However, the cut-off neck reflector is detrimental to beam operation efficiency by reducing beam radius.

Recently, the high power microwave generators with Bragg reflector have been under extensive theoretical and experimental researches.[10–13] The Bragg corrugation scatters the incident wave coherently into reflected waves which is along the direction of electron beam, so the radiation power leaking out from the tube will be reduced obviously, and the electron gun will be electrically isolated. Ogura studied a K-band weakly relativistic oversized BWO with cylindrical Bragg reflector at the input end of SWS, which demonstrated considerable interest for practical use.[14] Elfrgani and his co-workers theoretically designed a two-spiral corrugated Bragg reflector placed downstream of the cathode in an X-band relativistic BWO to convert the backward TM_{01}-wave to a forward

[a]correspondence email: lanfeng@uestc.edu.cn

TE_{11}-mode.[15,16] Due to the change of SWS end conditions, the cylindrical Bragg reflector can cause unfavorable modulation of the electron beam for the operating TM_{01}-mode in the BWO. Xiao and et al. experimentally demonstrated that a weak beam-wave modulation occurs along the coaxial Bragg section provides reflection and suppresses the cross excitation instability efficiently in a coaxial relativistic Cerenkov generator.[17] However, an elaborate design of a coaxial Bragg reflector composed of open periodic structures with more favorable beam modulation, better mode selectivity, and higher frequency feedback for an oversized RDG still remains to be intensively studied.

In this paper, we design a coaxial Bragg reflector with tapered inner rode corrugated ripples and phase matching section to substitute the smooth coaxial reflector which is introduced in our previous work.[18–20] With a reduced guiding magnetic field from 2.6 Tesla to 1.8 Tesla, an efficiency of about 35.5% according to a 3D electromagnetic PIC simulation has been obtained, and an efficiency of 22.9% has been measured in preliminary experiments. Compared with the previous design, the novel reflector prevents backward radiation leak more effectively and offers the beneficial electron premodulation, leading to an over twofold output power.

The organization of this paper is as follow. In Section II, the physical model of the system and the theory of the Bragg reflector are presented. Simulations and optimization are given in section III. Experimental researches are shown in section IV. The final conclusions are contained In Section V.

II. PHYSICAL MODEL

A. Brief review of RDG physics

The sketch diagram of the Ka-band RDG with a tapered coaxial Bragg reflector is illustrated in Fig.1 (a). A completely overmoded ($2R/\lambda \approx 4$) cylindrical waveguide periodically loaded by circular disks act as the SWS. The period *d*, disk width *b*, disk depth *h* and outer radius *R* of the cylindrical waveguide are 4 mm, 2 mm, 1.2 mm, and 19.8 mm, respectively. Fig. 1 (b) gives dispersion curves of the disk loaded SWS. Note that the lowest operating point with the beam line for 550-kV electrons is TM_{01} which is around π mode and with a frequency of 32.9 GHz. The electron beam line of 550-kV crosses the dispersion curves of TM_{11} mode at 35.3 GHz, which is near the selected operating point of TM_{01} mode. Because the electron beam interacts with the minus first spatial harmonic of TM mode along the opposite direction, the reflector is located upstream at the SWS to prevent the radiation power from leaking out of the system.

B. Design and numerical analysis of reflector

As shown in Fig. 2(a), the reflector is electrically contacted with the outer wall of the device by a molybdenum support ring, and the front graphite cap is used to absorb the reverse electrons. It is consisted of a periodic coaxial line with tapered inner rode corrugated ripples and phase matching section, as shown in Fig. 2(b). The geometric parameters of the reflector are: the inner rod mean radius a_o=14.4 mm, ripple period p_b=4.6 mm, corrugated cosine ripples with length L=46 mm (10 periods), mean ripple depth h_0=0.6 mm, initial ripple phases φ_{in}=0, descending taper angle θ=-1.25°, and phase matching section length l_p=1 mm.

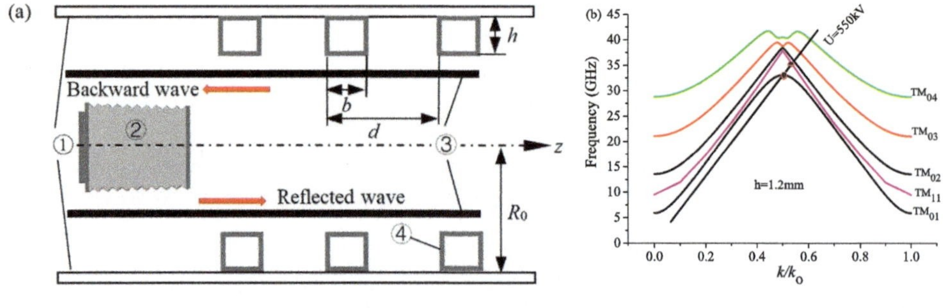

FIG. 1. (a) Schematic diagram of the high frequency system, (b) dispersion curve. ① Circular waveguide,② reflector, ③ electron beam,④ SWS.

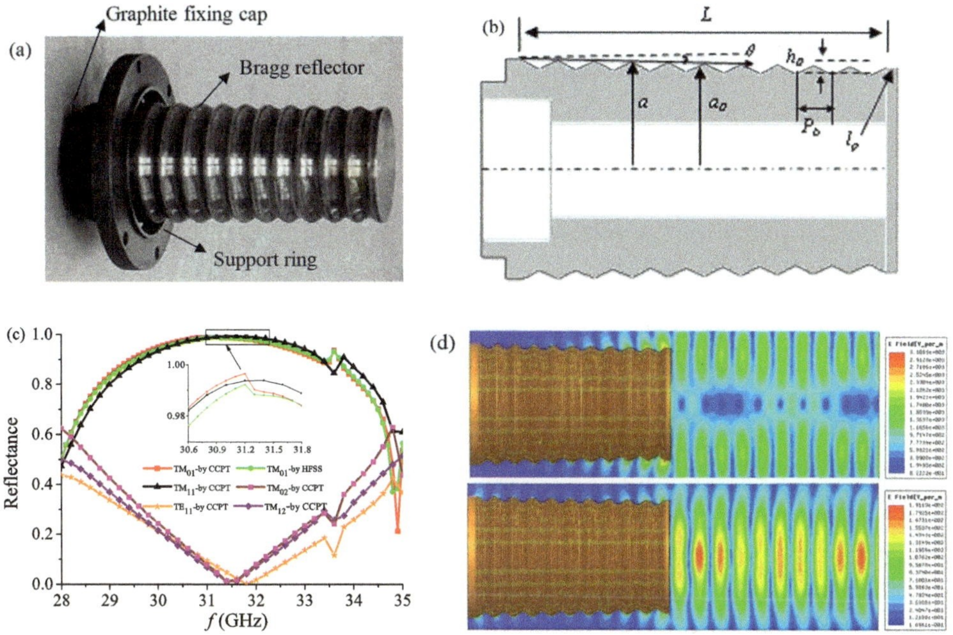

FIG. 2. Coaxial Bragg reflector (a) picture, (b) diagrammatic axial profile, (c) spectral reflectance of different modes, (d) electric field distributions at f_0=31.2 GHz: TM_{01} incidence (up) and TM_{11} incidence (down).

The perturbation of the inner-wall radius on the longitudinal position z can be expressed by:

$$a_{in} = a_0 + h_0 \cos(2\pi z/P_b + \phi_{in}) + z\tan(\theta) \quad (1)$$

Obviously, large cross-section of the reflector is conductive to amplify reflection coefficients. Due to the mode competitions caused by increase of the descending taper angle, the reflection characteristic and frequency response characteristic of the reflector become significantly worse. Too close to the electron beam will cause unfavorable modulation and increase its coupling impedance by the reflector.[17] In the cases of no taper and ascending taper, the electron beam channel through the reflector will be narrower, leading to over modulation by the reflector thus being harmful to the beam-wave interaction. In order to reduce the beam disturbance by the Bragg structure and ensure high reflection concurrently, θ=-1.25° becomes the best trade-off option. In order to make the directly reflected TM_{01} mode and the TM_{01} mode converted from TEM mode in phase, the phase matching section places at the front of the reflector. In addition, as a near cut-off mode in the corrugated coaxial part, the possible working mode TM_{11} can be effectively reflected by the reflector. A generalized scattering matrix method based on the conservation of complex power technique (CCPT) is employed to give full-wave analysis for optimal design of the reflector.[21] Fig. 2 (c) shows that the reflection coefficients for TM_{01} mode and TM_{11} mode are both near to 1 centered at 31.2 GHz. Observed from the inset specific, the CCPT and HFSS results are in an excellent accordance. As shown in Fig.2 (d), the mode transformation inside the Bragg structure indicates that most of the backward TM waves are reflected back into the circular waveguide, while remaining in pure mode patterns.

III. PIC SIMULATIONS

In order to study both symmetrical and asymmetrical operation modes, a full 3D PIC employing a parallel algorithm was carried out by a CHIPIC code.[22] The PIC geometry is shown in Fig.3. The system mainly consists of a cathode, a tapered coaxial Bragg reflector and two sections of SWS separated by a smooth drift tube. In our simulation, the support ring rods are modeled as virtual inductive rods to filter the DC power, which makes the simulation results more accurate. Each support rod is formed as a sector with a thickness of 2 mm and 2 angular grids, or 5 degrees, hence its arc

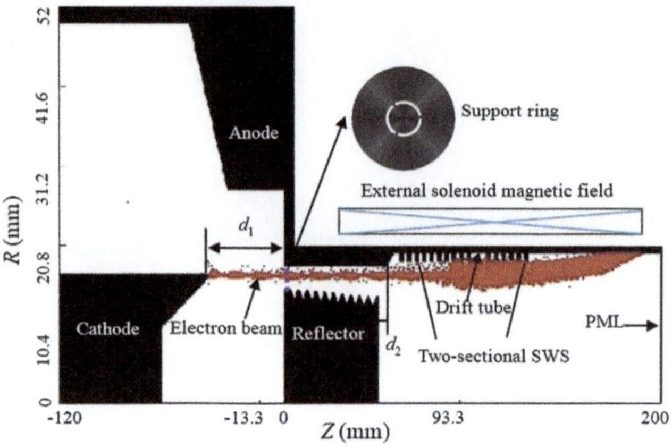

FIG. 3. PIC simulation geometry.

lengths of inner and outer rings are 1.57 mm and 1.31 mm, respectively. In experiment, the support ring is consisted of an inner ring and an outer ring, connected by 3 rods with a width of 1 mm and a thickness of 2 mm in intervals of 120°. As well known to all, it is preferable to remove any obstacle from the beam path for efficient beam-wave interaction. The intense electron beam caused more obvious damage on the rods for the obstruction than the inner and outer rings. Due to the minimized obstacle of the rods in the electron beam path and electron absorption by graphite, there is less than 5 % decrement of the space charge in the small region. However, compared to the cut-off neck reflector and cylindrical Bragg reflector near to the SWS structure, the tapered coaxial Bragg structure with the support ring and rods have enough radial space and axial distance to keep the beam remaining in integrally annular shape and proper distance to the SWS surface. The main parameters for the simulations are summarized in Table I.

An annular electron gun is specifically designed to support surface waves and to provide strong beam-wave coupling at 550 kV. The electron beam is generated from an explosive emission face located at the left end and confined by a uniform external magnetic field to run along the z-axis. The fitting formula for simulating the magnetic field is as below:

$$B_z = B_0(1 + \exp[60(-z - 0.05)] + \exp[100(z - 0.19)])^{-1} \tag{2}$$

As calculated by equation (2), 0-200 mm region is a uniform magnetic field distribution. A tapered transition section placed in front of the SWS is favorable for electron pre-bunching. The first section of the SWS has 10 periods, the second stage has 6 periods, the smooth drift length is 8 mm, the distance of the cathode to the anode is d_1, and the distance of the reflector to the tapered transition section is d_2. The simulation results are given as follows.

Optimum phase matching in the high frequency cavity can be tuned by changing d_2, significantly influencing the coupling output efficiency of the interaction cavity. While d_2 is an integer multiple of

TABLE I. Main parameters for simulation.

Beam voltage	U= 550 kV
Current	I=6 kA
Cathode end thickness	δ=0.4 mm
Cathode position	d_1=40 mm
Outer radius of cathode	r_1=17 mm
Anode radius	r_2=28 mm
SWS period	d=4 mm
Outer radius of WG	R=19.8 mm
Reflector position	d_2=5mm
Drift length	l= 8 mm
External magnetic field	B_0=1.6-2.6 T

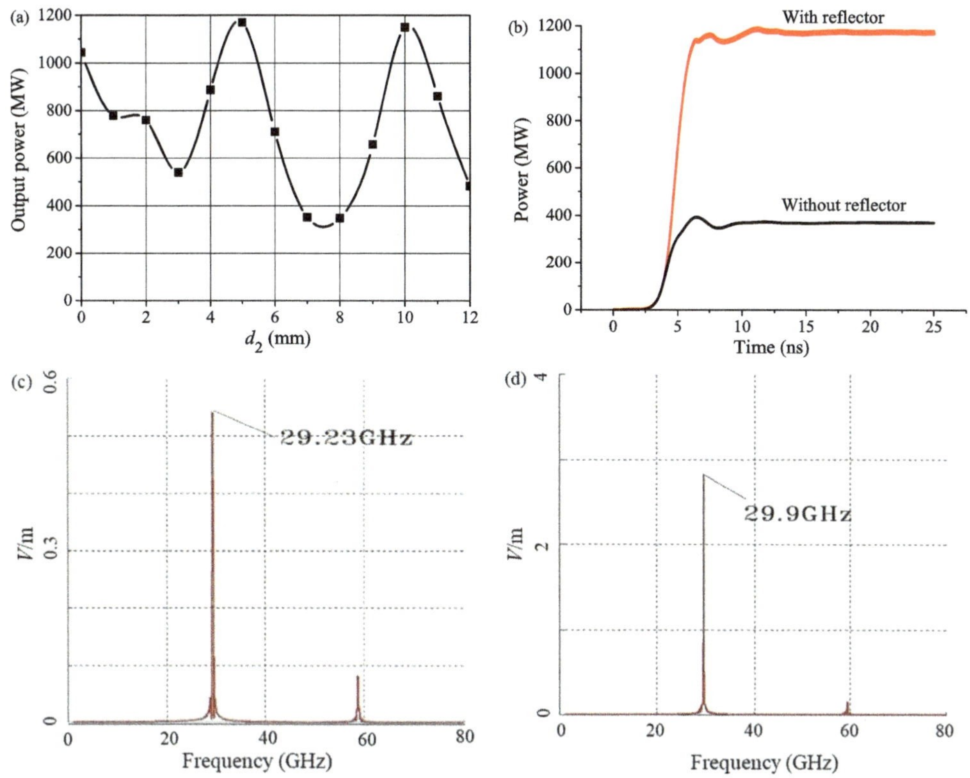

FIG. 4. PIC results (a) output power versus d_2, (b) output power versus time, (c) FFT spectrum at B_0=2.4 Tesla, (d) FFT spectrum at B_0=1.6 Tesla.

half operation wavelength, the output power peaks by the in phase coupling, as shown in Fig.4 (a). Fig.4 (b) displays the output power of the device with reflector at d_2=5 mm and without one.

For comparison sake, the output powers with reflector and without reflector, both simulated under identical conditions. However, the operation modes in the two cases are different. The operation modes in the device without reflector and with reflector are TM_{01} mode and TM_{11} mode, respectively. There is a dramatic increase from 369 MW to 1170 MW for the device maximum strength of the upstream field between the reflector and the anode is about 14.3 kV/cm for 1170 MW microwave power, which does not reach the breakdown values in gases. Due to the introduction of the Bragg reflector, the microwave propagates towards the cathode has been significantly decreased, expectantly no anode plasma of microwave-heated ionization will cause breakdown at that point.

Fig.4 (c) and (d) shows the corresponding FFT spectrum at B_0=2.4 Tesla and B_0=1.6 Tesla. Remarkably, different guiding magnetic field intensities can make electron beam relatively further or closer to the SWS surface, providing favorable oscillation condition for TM_{01} or TM_{11} mode, respectively. The output frequencies of 29.23 GHz and 29.9 GHz are corresponding to TM_{01} and TM_{11} main mode, respectively, which are lower than the ones in Fig.1 (a). Introduction of reflector and drift section into the high frequency cavity is the cause of frequency redshift. Obviously, the output power at lower magnetic field is much higher than that of higher magnetic field.

As shown in Fig.5, there are more particles in the deceleration field with the reflector than without the reflector, indicating that more electron beam energy is converted into electromagnetic wave for the former case. Furthermore, the electron beam premodulation by the reflector will benefit the following electron bunching in the smooth drift tube, thus creating suitable conditions to obtain the maximum output power. By the mode conversion introduced by the reflector, higher order TM modes could be observed. Fig. 6 (a) and (b) show the TM_{11}-like pattern at B_0=1.6 Tesla and TM_{02}-like pattern at B_0=2.4 Tesla, whereby depicting the axial components of the electric field at the output end cross section.

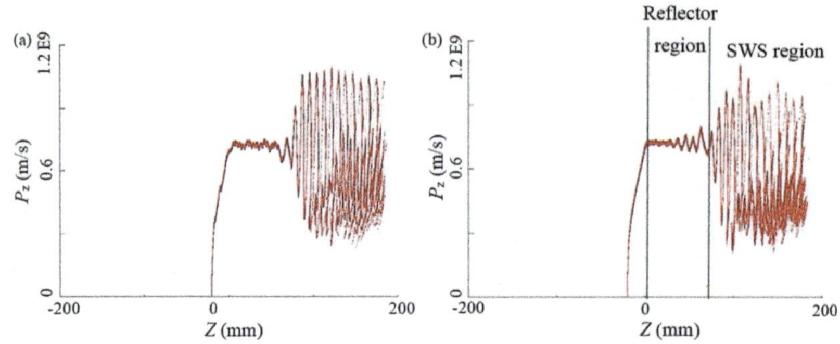

FIG. 5. Electron kinetic energy evolution: (a) without, (b) with reflector.

Lower guiding magnetic field and reflector premodulation at specific position may make electron beam closer to the SWS surface, providing favorable conditions for TM_{11} oscillation.

The output power of the device varies with the confinement of the electron beam under different axial magnetic field strengths. In our simulations, it is found that the beam-wave interaction efficiency and the output power is relatively higher at 1.0 Tesla<B_0<2.0 Tesla than B_0>2.0 Tesla. Not surprisingly, when B_0<1.0Tesla, both electron beam divergence and backflow phenomenon occur through the starting end of the reflector, leading to a poor oscillation condition. Fig. 6 (c) plots the output power as a function of different values of B_0. It is found that the output power of the device in the lower magnetic field is relatively higher than that in the higher magnetic field, peaking at the optimal value of B_0=1.6 T.

In summary, the excellent beam premodulation, backward reflection and mode selection characteristics of the reflector will hopefully benefit the device.

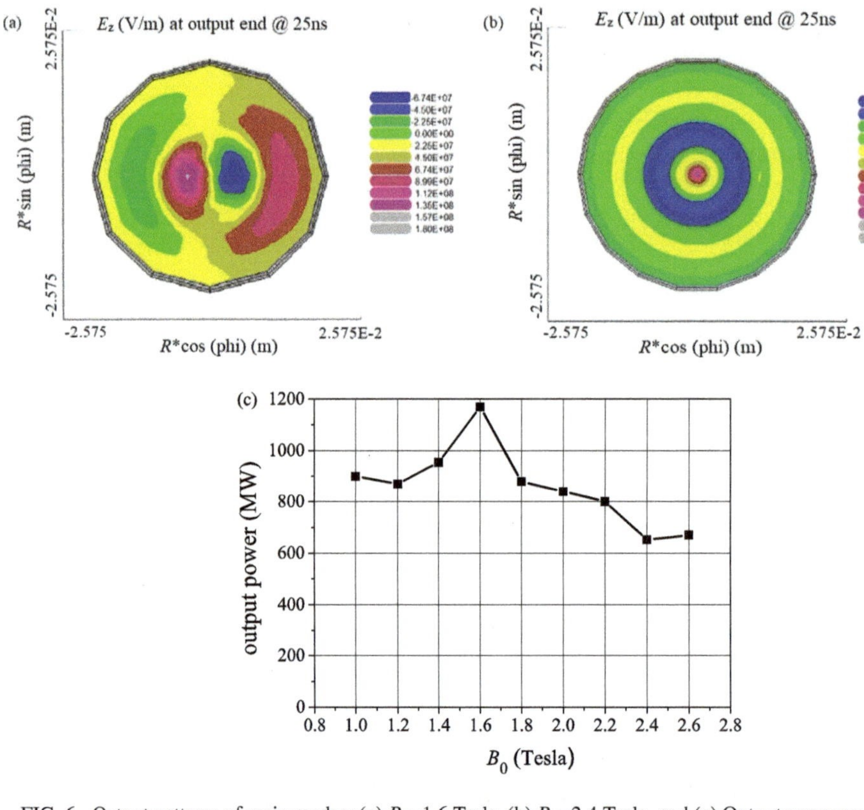

FIG. 6. Output patterns of main modes: (a) B_0=1.6 Tesla, (b) B_0=2.4 Tesla, and (c) Output power versus B_0.

IV. EXPERIMENTAL SETUP AND RESULTS

The schematic diagram of the experimental device is shown in Fig.7 (a), consisted of the accelerator, electron optical system, high frequency system, guiding magnetic field system and signal detection system. The measurement setup is based on the PIC simulations. An adjustable internal reflector is located between the electron gun and the SWS. Two WR-28 open-ended waveguides with attenuators and detectors are acting as vertical and horizontal receivers, moving at a semicircular arc 2 meters away from the output window. Total radiated power is measured by the method of space integral.[23] A neon bead-array screen is placed at 800 mm away from the output horn to display the radiation patterns.

Fig.7 (b) displayed the picture of the experimental device on the particle accelerator of LIA-2006. The SWS consisted of 10 and 6 periodic circular disks, connected by a smooth drift tube. A solenoid magnetic field is to constrain the electron beam. Typical waveform of the injected pulse, radiated microwave signals are shown in Fig.8. The measured diode voltage is 460 kV~550 kV, about 70 ns duration and the beam current ranges from 5.8 kA to 8.8 kA. The microwave pulse length is about 21 ns. For existence of microwave breakdown in SWS slots,[24] the pulse shortening phenomenon is

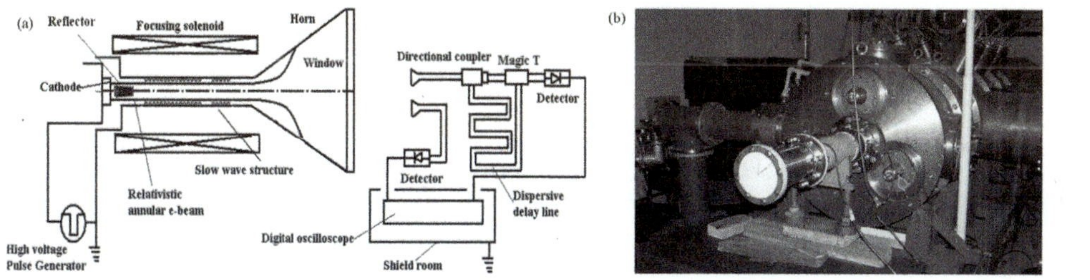

FIG. 7. RDG measurement setup (a) schematic diagram, and (b) photograph.

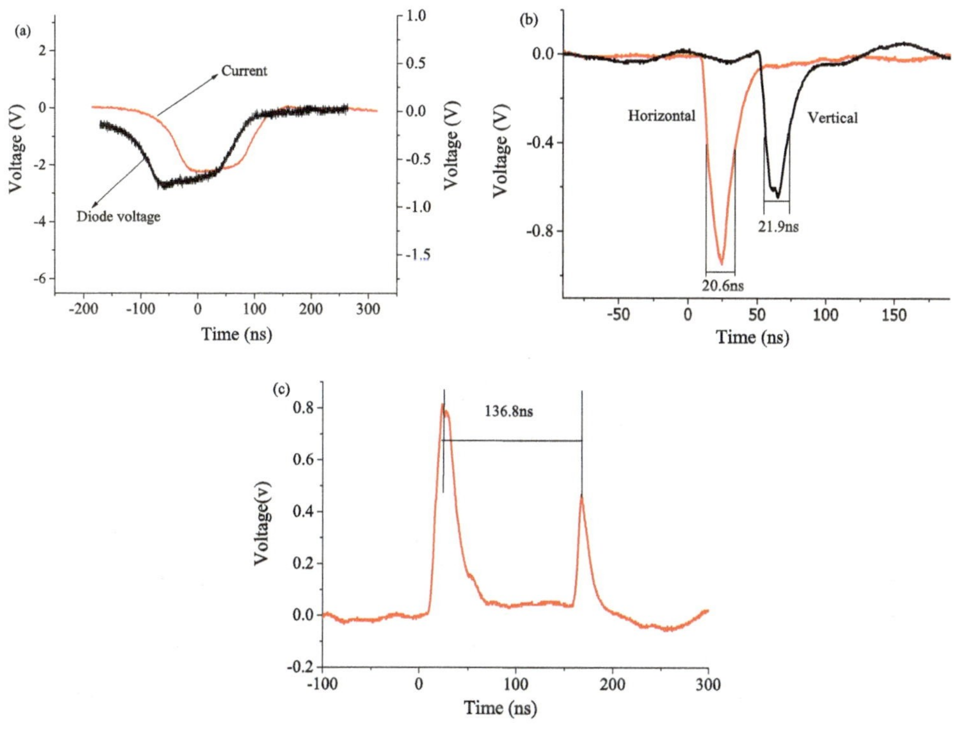

FIG. 8. Typical diagnostic waveforms (a) diode voltage and current, (b) voltage response of the receiving detectors, (c) time dependence of the delay line.

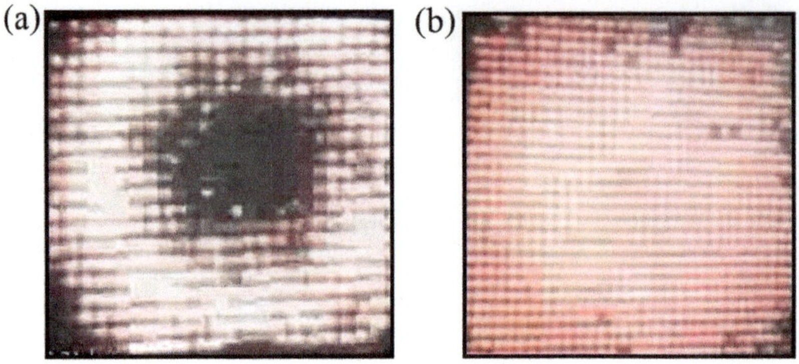

FIG. 9. Typical luminescence on the signal panel of neon tubes under different pulse magnetic field (a)=2.6 Tesla, (b)=1.8 Tesla.

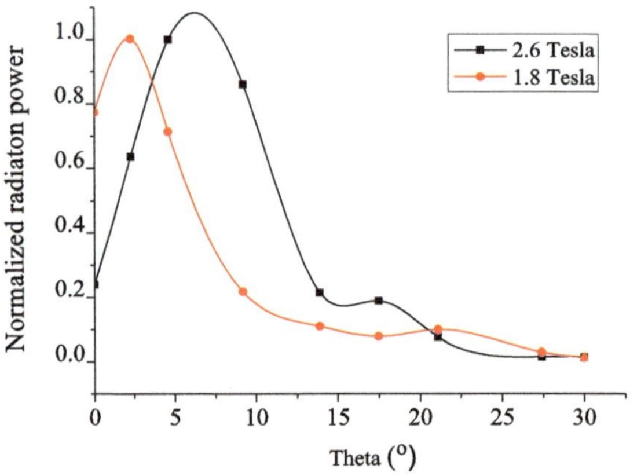

FIG. 10. Power density distribution for different magnetic field intensities.

observed in the experiment. Most of the measured radiation frequencies are in 29~31 GHz range by the dispersive delay line, which is in good accordance with thc simulation results.

It is confirmed by the experiment results that the axial adjustment of the reflector position dramatically improves the performance of the RDG generator. With optimal position of d_2=5mm, over 700 MW peak power is measured at 30.3 GHz. It found that the operation mode changed from TM_{01} to TM_{11} while the constrained magnetic field tuned from 2.6 Tesla to 1.8 Tesla. A probable reason for the phenomenon is that the distance of the beam center and the SWS is an optimum value for TM_{11} mode predominating TM_{01} mode while magnetic field changed. Serving as a mode converter, operation mode TM_{11} can be converted into EH_{11} radiation mode by the reflector. Fig.9 indicates that the light emitting area of neon tube array looks like EH_{11} mode is brighter than the one looks like TM_{01}. Fig.10 illustrates the measured power density distributions of the device. As peaked near axial line center, the power density distribution of 1.8 T indicates a EH_{11} mode. On the contrary, the power density distribution of 2.6 T corresponds to a TM_{01} mode. The measured radiation power for 2.6 T magnetic field is 348 MW, for 1.8T is over 756 MW. We also measured the output powers with a cut-off reflector and without a reflector, which are 300 MW and 200 MW, respectively. Coinciding well with the PIC results, the experiment results verify that the reflector can greatly improve the output power. Remarkably operations at reduced magnetic field are more preferable for system compactness.

V. CONCLUSIONS

In this paper, we presented the theoretical and preliminary experimental results of Ka-band RDG with a tapered coaxial Bragg reflector. PIC reveals that the reflector prevents the backward radiation passing through the diode structure and pre modulates the electron beam. Both the simulation and experiment results show that the reflector can dramatically increase the radiation power. Under 2.6 T magnetic field, the radiation main mode is TM_{01} mode, radiation power is 348 MW, measured frequency is about 29 GHz~30 GHz. Under 1.8 T magnetic field, the radiation main mode is EH_{11} mode, radiation power is 756 MW, measured frequency is about 30 GHz~31 GHz.

ACKNOWLEDGMENTS

This work was supported by the National Natural Science Foundation of China (No. 91438118), State Scholarship Fund of China Scholarship Council (No. 201606075012), and Terahertz Science and Technology Key Laboratory of Sichuan Province.

[1] A. V. Gunin, A. I. Klimov, S. D. Korovin, K. K. Ivan, V. P. Igor, D. P. Sergei, M. R. Albert, V. R. Vladislav, S. S. Aleksei, and M. T. Evgenii, "Relativistic x-band BWO with 3-GW output power," IEEE Trans. Plasma Sci. **26**(3), 326–331 (1998).

[2] S. P. Bugaev, V. I. Kanavets, A. I. Klimov, V. I. Koshelev *et al.*, "Relativistic multiwave Cerenkov generator," IEEE Trans. Plasma Sci. **18**, 525 (1990).

[3] S. P. Bugaev, V. A. Cherepenin, V. I. Kanavets *et al.*, "Investigation of a millimeter wavelength range relativistic diffraction generator," IEEE Trans. Plasma Sci. **18**, 518 (1990).

[4] E. Schamiloglu, *High Power Microwave Sources and Applications* (IEEE MTTS, Fort Worth, Texas, 2004).

[5] L. D. Moreland, E. Schamiloglu, and W. Lemke, "Efficiency enhancement of high-power vacuum BWO's using nonuniform slow wave structures," IEEE Trans. Plasma Sci. **22**, 554–565 (1994).

[6] D. K. Abe, Y. Carmel, S. M. Miller, A. Bromborsky, B. Levush, T. M. Antonsen, and W. W. Destler, "Experimental studies of overmoded relativistic backward-wave oscillators," IEEE Trans. Plasma Sci. **26**(3), 591–604 (1998).

[7] A. I. Slepkov and O. V. Gallyamova, "Self-excitation and establishment of generation in a two-section relativistic diffraction generator," Physics of Wave Phenomena **18**(2), 105–109 (2010).

[8] O. Kazuo, Y. Ryo, Y. Yuichiro, Y. Hosiyuki, K. Kiyofumi, and S. Masakazu, "Study on oscillation starting condition of K-band oversized backward wave oscillator driven by a weakly relativistic electron beam," J. Plasma Fusion Res. SERIES **6**, 703–706 (2004).

[9] A. N. Vlasov, A. G. Shkvarunets, J. C. Rodgers *et al.*, "Overmoded GW-class surface-wave microwave oscillator," IEEE Trans. Plasma Sci. **28**(3), 550–560 (2000).

[10] S. C. Zhang and M. Thumm, "Large-orbit coaxial-structure cyclotron autoresonance maser," Appl Phys Lett. **88**, 033514 (2006).

[11] C. R. Qiu, S. C. Zhang, H. B. Zhang *et al.*, "Nonlinear characteristics of a coaxial-waveguide cyclotron autoresonance maser (CARM) amplifier," J. Phys.D. **39**, 424–428 (2006).

[12] J. J. Barroso, P. J. Castro, J. P. Leite Neto *et al.*, "Experimental characterization of a 6.7 GHz coaxial Bragg reflector," Rev. Sci. Instrum. **78**, 096102 (2007).

[13] Y. X. Lai, S. C. Zhang, and H. B. Zhang, "A coaxial Bragg reflector for cyclotron autoresonance maser oscillators," IEEE Trans. Microw. and Wireless Components Letters **17**(5), 328–330 (2007).

[14] O. Kazuo, M. Yushi, T. Hidekazu, K. Yutaka, A. Satoshi, and S. Akira, "Weakly relativistic K-band oversized backward wave oscillator with Bragg reflector at input end of slow wave structure," *Sixteenth International Toki Conference on Plasma Physics and Controlled Nuclear Fusion*, S1041:1–4, 2007.

[15] A. M. Elfrgani, S. Prasad, M. I. Fuks, and E. Schamiloglu, "Relativistic BWO with linearly polarized Gaussian radiation pattern," IEEE Trans. Plasma Sci. **42**(8), 2135–2140 (2014).

[16] A. M. Elfrgani, S. Prasad, M. I. Fuks, and E. Schamiloglu, "Dual-band operation of relativistic BWO with linearly polarized Gaussian output," IEEE Trans. Plasma Sci. **42**(8), 2141–2145 (2014).

[17] R. Xiao, L. Zhang, T. Liang, Y. Teng, C. Chen, H. Shao, G. Liu, and Y. Lin, "Limitation of cross-excitation instability in a relativistic Cerenkov generator with coaxial slow wave structure," Physics of Plasmas **15**, 053107 (2008).

[18] Z. Q. Yang, Z. Liang, B. Zhang *et.al.*, "Cherenkov oscillator operating at two different wave bands," International Journal of Infrared and Millimeter Waves **20**(1), 83–92 (1999).

[19] H. Y. Wang, Z. Y. Yang, L. X. Zhao *et al.*, "Numerical computation of dispersion curves for symmetrical and asymmetrical modes in an arbitrary cylindrical metal SWS," IEEE Trans. Plasma Sci. **33**, 111 (2005).

[20] Z. J. Shi, Z. Q. Yang, F. Lan, X. Gao, T. Fu, and Z. Liang, "Investigation of a 30 GHz relativistic diffraction generator with a coaxial reflector," Journal of Electromagnetic Waves and Applications **24**, 2453–2462 (2010).

[21] F. Lan, Z. Q. Yang, and Z. J. Shi, "Study on coaxial Bragg reflector with small tapered inner rode corrugated ripples and phase matching section," Acta Phys.Sin. **60**(9), 091101–1 (2011).

[22] J. Di, D. J. Zhu, S. G. Liu *et al.*, *Electromagnetic Field Algorithm of the CHIPIC Code*, 2005.

[23] F. Lan, X. Gao, and Z. J. Shi, "Measurement of radiation power for Ka-band relativistic diffraction generator," Journal of University of Electronic Science and Technology of China **35**(5), 777–779 (2006).

[24] U. Jordan, D. Anderson, M. Backstrom, M. Lisak, and O. Lunden, "Microwave breakdown in slots," IEEE Trans. Plasma Sci. **32**, 2250 (2004).

15

Competition between the inter-valley scattering and the intra-valley scattering on magnetoconductivity induced by screened Coulomb disorder in Weyl semimetals

Xuan-Ting Ji,[1] Hai-Zhou Lu,[2,a] Zhen-Gang Zhu,[1,3,b] and Gang Su[3,4,c]
[1]*School of Electronic, Electrical and Communication Engineering, University of Chinese Academy of Sciences, Beijing 100049, China*
[2]*Institute for Quantum Science and Engineering and Department of Physics, South University of Science and Technology of China, Shenzhen 518055 China*
[3]*Theoretical Condensed Matter Physics and Computational Materials Physics Laboratory, School of Physical Sciences, University of Chinese Academy of Sciences, Beijing 100049, China*
[4]*Kavli Institute of Theoretical Sciences, University of Chinese Academy of Sciences, Beijing 100190, China*

Recent experiments on Weyl semimetals reveal that charged impurities may play an important role. We use a screened Coulomb disorder to model the charged impurities, and study the magneto-transport in a two-node Weyl semimetal. It is found that when the external magnetic field is applied parallel to the electric field, the calculated longitudinal magnetoconductivity shows positive in the magnetic field, which is just the negative longitudinal magnetoresistivity (LMR) observed in experiments. When the two fields are perpendicular to each other, the transverse magnetoconductivities are measured. It is found that the longitudinal (transverse) magnetoconductivity is suppressed (enhanced) sensitively with increasing the screening length. This feature makes it hardly to observe the negative LMR in Weyl semimetals experimentally owing to a small screening length. Our findings gain insight into further understanding on recently actively debated magneto-transport behaviors in Weyl semimetals. Furthermore we studied the relative weight of the inter-valley scattering and the intra-valley scattering. It shows that the former is as important as the latter and even dominates in the case of strong magnetic fields and small screening length. We emphasize that the discussions on inter-valley scattering is out of the realm of one-node model which has been studied.

I. INTRODUCTION

Weyl fermion proposed by Hermann Weyl in 1929 as the massless solution of the Dirac equation is of great significance, and has been searched for several decades from the aspect of particle physics. Neutrino has deemed to be Weyl fermion for a long time before the discovery of the neutrino oscillation,[1] which suggests neutrino is massive. It still has null result in the standard model of the particle physics. Nevertheless, it has been suggested that low energy excitations in some particular condensed matters can be the solutions to the Weyl Hamiltonian, which was understood as Weyl semimetal.[2] Weyl semimetal soon becomes one of central issues of the topological materials.[3–10] Very recently, with the help of angle-resolved photoemission spectroscopy (ARPES), Weyl nodes have been identified in the TaAs family.[11–14] Another candidate material with Weyl nodes is

[a]luhz@sustc.edu.cn
[b]zgzhu@ucas.ac.cn
[c]gsu@ucas.ac.cn

proposed to be $YbMnBi_2$.[15] Moreover the second class of Weyl semimetal has been discovered in $MoTe_2$.[16]

Weyl nodes are the band crossings and behave as monopoles due to the Berry curvature in the three-dimensional (3D) momentum space. The Weyl nodes appear in pairs with opposite chirality. They can only eliminate each other when meet together. Although the Weyl nodes are topologically robust, a chiral charge from one Weyl node can be pumped to the other in the presence of a parallel magnetic and electric field ($\mathbf{E} \cdot \mathbf{B}$ term), thus violating the total chirality. This was named as chiral anomaly.[17] It has been shown that chiral anomaly may induce diverse effects, such as frequency shift on the plasmon mode,[18] anomalous Hall effect[2,3,19–21] and chiral magnetic effect,[22–27] etc. Particularly it may manifests itself by a negative longitudinal magnetoresistance (LMR) which has been theoretically studied[28–30] and has been observed experimentally in the TaAs family.[31–36]

Theoretically, a semiclassical mechanism of the LMR was introduced, wherein the charge transport is dominated by guiding center motion.[37] Pesin *et al.* studied the long-range-disorder-affected density of states in absence of or in presence of weak magnetic field and found that a finite density of states can be induced by the long-range potential fluctuations as the energy approaching the degeneracy point.[38] Klier *et al.*[39] explored the density of states, Landau levels broadening, and the magnetoresistivity in transversal magnetic fields within short-range impurities or charged (Coulomb) impurities. Moreover, δ-potential disorder and random-Gaussian-potential disorder in Weyl semimetal near the Weyl nodes have been studied.[40,41]

Although there are intensive investigations on the theory aspect, it is still insufficient to understand the realistic features observed in experiments because most of previous studies treated one-node model. As known the nodes exist in pairs and the chiral anomaly involves at least one pair. Therefore it is not possible to explore the chiral anomaly in the one-node model. Furthermore, it is noted that the observed LMR in some experiments is not negative but positive; while it is negative in some others. And the Fermi level varies dramatically in different samples. These sensitively sample-dependent features indicate that the charged impurities may be the main type of disorder in the transport measurements. For charged impurities the Coulomb interaction between them and their surrounding electrons is dressed, leading to a screened Coulomb interaction, which induces both intra-valley scattering (when we describe the scattering, we use "valley" rather "node" for convenience) and inter-valley scattering. Therefore, to study the effect of the inter-valley scattering we need start from at least a two-node model.

In this paper, we studied such a two-node model. The inter-valley scattering and the intra-valley scattering are considered simultaneously. We show a negative LMR in magnetic field, consistent with the prediction of the chiral anomaly.[40] It is seen that the transverse magnetoconductivity is enhanced sensitively by increasing the screening length which is just opposite to that of longitudinal magnetoconductivity. Based on this result, we can understand recent observations in experiments and find out a way to facilitate the observation of the negative LMR in the future experiments. We also show that the inter-valley scattering can be as important as the intra-valley scattering, or even more important than the latter. The results on the inter-valley scattering are novel since previously one-node model was mainly explored in which the inter-valley scattering can not be cast.

The paper is organized as follows. We first introduce the two-node model in Sec. II. We then discuss the formulas and analysis of the longitudinal and transverse magnetoconductivities in the presence of the Coulomb screening disorder in Sec. III. The conclusions are given in Sec. IV.

II. MODEL

A. B = 0 case

In a two-node model, the energy dispersion is linear in momentum around the nodes forming cones (or valley), which are labeled by K cone and K' cone. The two-node Hamiltonian[42] in absence of an external magnetic field $\mathbf{B}$ is then given

$$\mathcal{H} = \nu\hbar v_F \left[(\mathbf{k} + \nu\mathbf{k}_c) \cdot \boldsymbol{\sigma} \right], \tag{1}$$

where $\nu = 1$ for K cone and $\nu = -1$ for K' cone, υ_F is the Fermi velocity and $\mathbf{k}_c = (0, 0, \pm k_c)$ represent the positions of the two Weyl nodes in the momentum space. For convenience, we explicitly re-express this Hamiltonian separately,

$$\mathcal{H}_K = \hbar\upsilon_F(\mathbf{k} + \mathbf{k}_c) \cdot \boldsymbol{\sigma},$$
$$\mathcal{H}_{K'} = -\hbar\upsilon_F(\mathbf{k} - \mathbf{k}_c) \cdot \boldsymbol{\sigma}. \tag{2}$$

By a translation, we get two Hamiltonians which only depend on the relative momentum $\mathbf{k}$ with respect to the two nodes as

$$\tilde{H}_K = \mathcal{H}_K - \hbar\upsilon_F \mathbf{k}_c \cdot \boldsymbol{\sigma},$$
$$\tilde{H}_{K'} = \mathcal{H}_{K'} - \hbar\upsilon_F \mathbf{k}_c \cdot \boldsymbol{\sigma}. \tag{3}$$

These two Hamiltonians satisfy the time reversal symmetry (TRS).

For K cone, we have eigenstates of the Hamiltonian in Eq. (3)

$$\Psi_{+K} = \begin{pmatrix} \cos\frac{\theta}{2} \\ \sin\frac{\theta}{2}e^{i\varphi} \end{pmatrix}, \Psi_{-K} = \begin{pmatrix} \sin\frac{\theta}{2} \\ -\cos\frac{\theta}{2}e^{i\varphi} \end{pmatrix}, \tag{4}$$

where $\pm$ represent the conduction band and valence band, respectively. The eigenenergies are thus

$$\tilde{E}_{\pm K} = \pm\hbar\upsilon_F\sqrt{k_x^2 + k_y^2 + k_z^2}, \tag{5}$$

The angle θ and φ are decided by the following relations

$$\cos\theta = \frac{k_z}{|\tilde{E}|}, \tan\varphi = \frac{k_x}{k_y}. \tag{6}$$

The eigenenergies of the original Hamiltonian in Eq. (2) are given by

$$E_K = \tilde{E}_{\pm K} + \hbar\upsilon_F k_c. \tag{7}$$

In K' cone, the relative momentum is expressed by $\mathbf{k}'$ and satisfies the relations, $\mathbf{k}'$=-$\mathbf{k}$, $\theta'=\pi - \theta$, $\varphi'=\pi + \varphi$ due to the TRS. We therefore get the eigenstates for the K' cone

$$\Psi_{+K'} = \begin{pmatrix} \sin\frac{\theta}{2} \\ -\cos\frac{\theta}{2}e^{i\varphi} \end{pmatrix}, \Psi_{-K'} = \begin{pmatrix} \cos\frac{\theta}{2} \\ \sin\frac{\theta}{2}e^{i\varphi} \end{pmatrix}. \tag{8}$$

The corresponding eigenenergies for K' cone are

$$E_{K'} = \tilde{E}_{\pm K'} + \hbar\upsilon_F k_c. \tag{9}$$

where

$$\tilde{E}_{\pm K'} = \mp\hbar\upsilon_F\sqrt{k_x^2 + k_y^2 + k_z^2}. \tag{10}$$

B. B≠0 case

When a magnetic field $\mathbf{B}$ = (0, 0, B) is applied along the z direction, Landau bands form and disperse with k_z. Here we choose the Landau gauge $\mathbf{A}$ = $(-yB, 0, 0)$. Then we have

$$\mathbf{k} = \left(k_x - \frac{eB}{\hbar}y, -i\partial_y, k_z\right). \tag{11}$$

Defining the ladder operators,

$$a = -[(y - l_B^2 k_x)/l_B + l_B\partial_y]/\sqrt{2},$$
$$a^\dagger = -[(y - l_B^2 k_x)/l_B - l_B\partial_y]/\sqrt{2}, \tag{12}$$

where $l_B = \sqrt{\hbar/eB}$ is the magnetic length. For K cone, we have

$$\tilde{H}_K = \hbar v_F \begin{pmatrix} k_z & -\frac{\sqrt{2}}{l_B}a \\ -\frac{\sqrt{2}}{l_B}a^\dagger & -k_z \end{pmatrix}. \tag{13}$$

The corresponding eigenstates of the Landau bands are given by

$$|n \geqslant 1, k_x, k_z, +K\rangle = \begin{bmatrix} \cos\frac{\theta_n^{k_z}}{2}|n-1\rangle \\ \sin\frac{\theta_n^{k_z}}{2}|n\rangle \end{bmatrix} |k_x, k_z\rangle,$$

$$|n \geqslant 1, k_x, k_z, -K\rangle = \begin{bmatrix} \sin\frac{\theta_n^{k_z}}{2}|n-1\rangle \\ -\cos\frac{\theta_n^{k_z}}{2}|n\rangle \end{bmatrix} |k_x, k_z\rangle, \tag{14}$$

where $\cos\theta_n^{k_z} = \frac{k_z}{\sqrt{k_z^2+2n/l_B^2}}$, n is the index of the Landau bands. Please note that $\theta_n^{k_z}$ is the same in K and K' cone, and the TRS has been taken into account already. The eigenenergies for the Hamiltonian (13) are

$$\tilde{E}_{\pm K}^n = \pm\hbar v_F \sqrt{k_z^2 + 2n/l_B^2}. \tag{15}$$

The eigenenergies for the original Hamiltonian are

$$E_{\pm K}^n = \tilde{E}_{\pm K}^n + \hbar v_F k_c. \tag{16}$$

The eigenstate of the $n = 0$ Landau band is obtained

$$|n = 0, k_x, k_z, K\rangle = \begin{bmatrix} 0 \\ |0\rangle \end{bmatrix} |k_x, k_z\rangle, \tag{17}$$

with the eigenenergy

$$E_{k_z}^{0K} = -\hbar v_F k_z + \hbar v_F k_c. \tag{18}$$

The definition of $\theta_{0K}^{k_z}$ is given

$$\cos\left(\theta_{0K}^{k_z}/2\right) = \sqrt{\left(1 + k_z/\sqrt{k_z^2}\right)/2} = \Theta(k_z),$$

$$\sin\left(\theta_{0K}^{k_z}/2\right) = \sqrt{\left(1 - k_z/\sqrt{k_z^2}\right)/2} = \Theta(-k_z), \tag{19}$$

where $\Theta(x)$ is the step-function.

Similarly, the eigenstates for the K' cone are given by

$$|n \geqslant 1, k_x, k_z, +K'\rangle = \begin{bmatrix} \sin\frac{\theta_n^{k_z}}{2}|n-1\rangle \\ -\cos\frac{\theta_n^{k_z}}{2}|n\rangle \end{bmatrix} |k_x, k_z\rangle,$$

$$|n \geqslant 1, k_x, k_z, -K'\rangle = \begin{bmatrix} \cos\frac{\theta_n^{k_z}}{2}|n-1\rangle \\ \sin\frac{\theta_n^{k_z}}{2}|n\rangle \end{bmatrix} |k_x, k_z\rangle. \tag{20}$$

The corresponding eigenenergies are obtained

$$\tilde{E}^n_{\pm K'} = \mp\hbar v_F\sqrt{k_z^2 + 2n/l_B^2},$$
$$E^n_{\pm K'} = \tilde{E}^n_{\pm K'} + \hbar v_F k_c. \tag{21}$$

The eigenstate of the $n = 0$ Landau band is obtained

$$|n = 0, k_x, k_z, K'\rangle = \begin{bmatrix} |0\rangle \\ 0 \end{bmatrix} |k_x, k_z\rangle, \tag{22}$$

with the eigenenergy

$$E^{0K'}_{k_z} = \hbar v_F k_z + \hbar v_F k_c. \tag{23}$$

According to the TRS, $\theta^{k_z}_{0K'} = \pi - \theta^{k_z}_{0K}$, hence we have $\cos(\theta^{k_z}_{0K'}/2) = \sin(\theta^{k_z}_{0K}/2)$ and $\sin(\theta^{k_z}_{0K'}/2) = \cos(\theta^{k_z}_{0K}/2)$.

III. FORMALISM AND RESULTS

A. Longitudinal magnetoconductivity

In the case of low temperature and a Fermi level close to the Weyl nodes, the transport is essentially involved with the 0-th Landau band. When the applied electric and magnetic fields are parallel, the changing rate of density of charge carriers near one node is maximal due to the chiral anomaly. In this case, the semiclassical conductivity of the 0-th Landau band can be derived in terms of the standard Green's function method. Alternatively, it can be simply figured out by using the Einstein relation $\sigma_{zz} = e^2 N_F D$, where the density of state (DOS) can be found as the Landau degeneracy times the DOS of one-dimensional systems, i.e., $N_F = (1/2\pi l_B^2) \times (2/\pi\hbar v_F)$. $D = v_F^2 \tau^{0,\text{tr}}_{k_F}$ is the diffusion coefficient. $\tau^{0,\text{tr}}_{k_F}$ is the transport time, and k_F is the Fermi wave number. Under the first-order Born approximation, for the scattering among the states on the Fermi surface of the 0-th band, the transport time can be obtained

$$\frac{\hbar}{\tau^{0,\text{tr}}_{k_z}} = 2\pi \sum_{k_x',k_z'} \langle |U^{0,0}_{k_x,k_z;k_x',k_z'}|^2\rangle \Lambda\delta(E_F - E^0_{k_z'}) \left(1 - \frac{v^z_{0,k_z'}}{v_F}\right), \tag{24}$$

where $U^{0,0}_{k_x,k_z;k_x',k_z'}$ represents the scattering matrix elements and $\langle \ldots \rangle$ means the impurity average. An extra factor Λ is introduced to correct the unphysical van Hove singularity near the band edge,[43] and the details of Λ can be found in the supplementary material. We note that, for z-axis magnetic field, only the diagonal elements of the velocity matrix with respect to the z-direction contribute to the longitudinal magnetoconductivity. On the contrary, only the off-diagonal elements of the velocity matrix with respect to the x-direction contribute to the transverse magnetoconductivity.

The transport time $\tau^{0,\text{tr}}_{k_z}$ is sensitive to the scattering potential. To reflect the realistic situation, we use the screened Coulomb potential to model the charged impurities,

$$\phi(r) = \frac{e^2}{\alpha\kappa_0 r} e^{-\frac{r}{\lambda}}, \tag{25}$$

where λ measures the screening radius, which can be decided by a finite chemical potential μ, temperature T and the excess electron density $n(\mu, T)$ according to the Eq. (11) in Ref. 44. The $n(\mu, T)$ is defined by

$$n_D - n_A = n(\mu, T), \tag{26}$$

where n_D, and n_A denote the density of the donors and acceptors, respectively. $\alpha\kappa_0$ represents for the dielectric constant, α is a dimensionless constant and e is the charge of the electron.

The longitudinal conductivity σ_{zz} can be found from the formula

$$\sigma_{zz} = \frac{e^2\hbar}{2\pi V} \sum_{k_z,k_x} \text{Tr}\left(\mathbf{v}^z_{0,k_z} \mathbf{G}^R_{0,k_z} \widetilde{\mathbf{v}}^z_{0,k_z} \mathbf{G}^A_{0,k_z}\right), \tag{27}$$

where bold symbols, e.g. **v** and **G**, are all diagonal matrixes in the valley space (i.e. cone space), and the subscript 0 represents the $n = 0$ Landau subspace. We thus have

$$\sigma_{zz} = \frac{e^2\hbar}{2\pi V}\sum_{k_x,k_z,i}\left(v_{0,k_z}^{z,i}\,\tilde{v}_{0,k_z}^{z,i}\,G_{0i}^{R}G_{0i}^{A}\right), \tag{28}$$

where i runs over K and K' cone. Note that, the electrons in different cone move in opposite directions due to the chirality. In our case, we fixed the fermi energy at the zeroth Landau band, and all the Landau bands in the K' cone are fully occupied. Hence, we have

$$\sigma_{zz} = \frac{e^2\hbar}{2\pi V}\sum_{k_x,k_z}\left(v_{0,k_z}^{zK}\,\tilde{v}_{0,k_z}^{zK}\,G_{0K}^{R}G_{0K}^{A}\right), \tag{29}$$

where

$$v_{0,k_z}^{zK} = \partial E_{k_z}^{0K}/\hbar\partial k_z = -v_F, \tag{30}$$

is the $\hat{\mathbf{z}}$-direction velocity of an electron in the 0-th band. $\widetilde{v}_{0,k_z}^{zK}$ is the dressed velocity after taking into account the vertex correction. The retarded/advanced Green's function for the 0-th band are $G_{0K}^{R/A} = 1/(E_F - E_{k_z}^{0K} \pm i\hbar/2\tau)$ where τ is the corresponding momentum relaxation time which is related to the transition probability (induced by the impurities) in K cone at $n = 0$ band. $V = L_xL_yL_z$ is the volume of a box.

In the diffusive regime, one can replace $G_{0K}^{R}\widetilde{v}_{0,k_z}^{zK}G_{0K}^{A}$ by $(2\pi/\hbar)\tau_{k_z}^{0K,\mathrm{tr}}v_{0,k_z}^{zK}\Lambda\delta(E_F - E_{k_z}^{0K})$. We get

$$\sigma_{zz} = \frac{e^2}{V}\sum_{k_z,k_x}(v_{0,k_z}^{zK})^2\tau_{k_F}^{0K,\mathrm{tr}}\Lambda\delta(E_F - E_{k_z}^{0K}). \tag{31}$$

We can expand the Delta function as $\delta(E_F - E_{k_z}^{0K}) = \delta(k_F + k_z - k_c)/\hbar v_F$ since the equation, i.e. $E_F - E_{k_z}^{0K} = 0$, leads us only one root at $k_F = k_c - k_z$. Similarly, expanding the Delta function as $\delta(E_F - E_{k_z}^{0K'}) = \delta(k_F - k_z - k_c)/\hbar v_F$ due to $E_F - E_{k_z}^{0K'} = 0$, leads us only one root at $k_F = k_c + k_z$. This allows us to perform the summation over k_z. And we have

$$\begin{aligned}\sigma_{zz} &= \frac{e^2}{V}\sum_{k_x,k_z}\left(v_{0,k_z}^{zK}\right)^2\Lambda\tau_{k_F}^{0K,\mathrm{tr}}\delta(k_F + k_z - k_c)/\hbar v_F \\ &= \frac{L_ze^2}{V\hbar v_F}\sum_{k_x}\int\frac{dk_z}{2\pi}\left(v_{0,k_z}^{zK}\right)^2\Lambda\tau_{k_F}^{0K,\mathrm{tr}}\delta(k_F + k_z - k_c) \\ &= \frac{e^2}{h}\frac{v_F}{L_xL_y}\sum_{k_x}\tau_{k_F}^{0K,\mathrm{tr}}\Lambda.\end{aligned} \tag{32}$$

We need to calculate the transport time $\tau_{k_F}^{0K,\mathrm{tr}}$ at the Fermi surface, which can be found in the supplementary material. After a straight forward calculation, we obtain

$$\sigma_{zz} = \frac{e^2}{h}\frac{(\hbar v_F)^2}{2\pi n_{\mathrm{imp}}\left(\frac{4\pi e^2}{\alpha\kappa_0}\right)^2 I_1 l_B^4}. \tag{33}$$

In Fig. 1, we plot the longitudinal magnetoconductivity versus B at different screening length λ. For clarity, we treat the λ as a varying parameter and focus on investigation its effect on the relative weight of the inter-valley scattering to the intra-valley scattering. It is noted that σ_{zz} is linear in B. This linear dependence on B was also obtained in Refs. 28, 30, and 40. However the effect of screened Coulomb disorder on σ_{zz} is unknown so far. It shows that the magnitude of σ_{zz} is inversely proportional to λ^4 when λ is small. This λ^4-dependence results from the λ dependence of I_1. In other words, a larger λ induces a larger longitudinal magnetoresistance. This result seems in contradiction to our intuitive expectation that a small λ may induce a stronger inter-valley scattering leading to a larger LMR due to Heisenberg uncertainty principle. However, this effect may be overwhelmed by another effect, i.e. a weak scattering potential for small λ. The overall behavior stems from that the screened Coulomb potential induces not only the intra-valley scattering but also the inter-valley

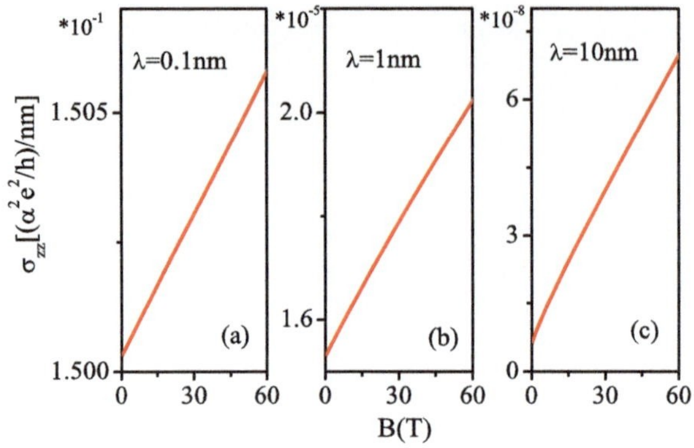

FIG. 1. The longitudinal magnetoconductivity versus the magnetic flux density for different screening lengths. Here $k_c = 0.1\ \mathrm{nm}^{-1}$.

scattering. We should emphasize that the screening length λ in our calculation is taken from 0.1 nm to 10 nm. This interval is consistent with that in Ref. 39 and Ref. 44.

B. Transverse magnetoconductivity

After displaying the results for longitudinal magneto-conductivity, we study in this subsection the transverse magnetoconductivity, i.e. σ_{xx}. For consistency, the Fermi energy is assumed to be near the Weyl nodes, and the conductivity along the $\hat{\mathbf{x}}$ direction can be written as

$$\sigma_{xx} = \frac{e^2\hbar}{2\pi V}\sum_{\mathbf{k}}\mathrm{Tr}\left(\mathbf{G}^R\mathbf{v}^x_{\mathbf{k}}\mathbf{G}^A\mathbf{v}^x_{\mathbf{k}}\right). \tag{34}$$

G and **v** are all matrixes and we have

$$\sigma_{xx} = \frac{e^2\hbar}{\pi v}\sum_{k_x,k_z,i,i'}\Re\left(G^R_{0i}v^x_{0i,1+i'}G^A_{1+i'}v^x_{1+i',0i}\right), \tag{35}$$

where i, i' run over K and K' indexes. A derivation of Eq. (35) is given in supplementary material. As an example, $v^x_{0K,1+K}$ links a scattering from zeroth band in K cone to the first Landau band in K cone.

With the definition of the velocity $v^x = \frac{1}{\hbar}\frac{\partial H}{\partial k_x} = \nu v_F\sigma_x$, the elements are derived

$$\begin{aligned}
v^x_{0K,1+K} &= v_F\sin\frac{\theta^{k_z}_0}{2}\cos\frac{\theta^{k_z}_1}{2} = v_F\Theta(-k_z)\cos\frac{\theta^{k_z}_1}{2},\\
v^x_{0K,1+K'} &= v_F\sin\frac{\theta^{k_z}_0}{2}\sin\frac{\theta^{k_z}_1}{2} = v_F\Theta(-k_z)\sin\frac{\theta^{k_z}_1}{2},\\
v^x_{0K',1+K} &= v_F\cos\frac{\theta^{k_z}_0}{2}\cos\frac{\theta^{k_z}_1}{2} = v_F\Theta(-k_z)\cos\frac{\theta^{k_z}_1}{2},\\
v^x_{0K',1+K'} &= v_F\cos\frac{\theta^{k_z}_0}{2}\sin\frac{\theta^{k_z}_1}{2} = v_F\Theta(-k_z)\sin\frac{\theta^{k_z}_1}{2}.
\end{aligned} \tag{36}$$

The momentum relaxation time for i valley ($i = K, K'$) is defined as

$$\frac{1}{\tau_i} \equiv \frac{1}{\tau_{0i}} + \frac{1}{\tau_{Ii}}, \tag{37}$$

where τ_{0i} (τ_{Ii}) is the momentum relaxation time induced by intra-valley (inter-valley) scattering, and defined by

$$\frac{1}{\tau_{0i}} \equiv \frac{2\pi}{\hbar} \sum_{k'_x,k'_z} |U^{1+i,0i}_{k_x,k_z;k'_x,k'_z}|^2 \delta\left(E_F - E^{0i}_{k_z}\right), \tag{38}$$

$$\frac{1}{\tau_{Ii}} \equiv \frac{2\pi}{\hbar} \sum_{k'_x,k'_z} |U^{1+i,0\bar{i}}_{k_x,k_z;k'_x,k'_z}|^2 \delta\left(E_F - E^{0i}_{k_z}\right), \tag{39}$$

where $\bar{i} = K'$ for $i = K$; while $\bar{i} = K$ for $i = K'$, and one element of U matrix is shown here

$$U^{1+K,0K}_{k_x,k_z;k'_x,k'_z} = \sin \frac{\theta^{k_z}_1}{2} I_{1,0}(\mathbf{R}_i). \tag{40}$$

Other elements of matrix U can be found in supplementary material. The details of the calculation of the momentum relaxation time is given in the supplementary material. Then we obtain the total transverse magnetoconductivity

$$\sigma_{xx} = \sigma^K_{xx,\text{inter}} + \sigma^K_{xx,\text{intra}} + \sigma^{K'}_{xx,\text{inter}} + \sigma^{K'}_{xx,\text{intra}}, \tag{41}$$

where

$$\sigma^K_{xx,\text{inter}} = \frac{e^2}{h}\left(\frac{4\pi e^2}{\alpha\kappa_0}\right)^2 \frac{\Lambda n_{\text{imp}}}{\hbar v_F} F^K_1(k_z) I_1,$$
$$\sigma^K_{xx,\text{intra}} = \frac{e^2}{h}\left(\frac{4\pi e^2}{\alpha\kappa_0}\right)^2 \frac{\Lambda n_{\text{imp}}}{\hbar v_F} F^K_2(k_z) \frac{I_2}{4},$$
$$\sigma^{K'}_{xx,\text{intra}} = \frac{e^2}{h}\left(\frac{4\pi e^2}{\alpha\kappa_0}\right)^2 \frac{\Lambda n_{\text{imp}}}{\hbar v_F} F^{K'}_1(k_z) I_3,$$
$$\sigma^{K'}_{xx,\text{inter}} = \frac{e^2}{h}\left(\frac{4\pi e^2}{\alpha\kappa_0}\right)^2 \frac{\Lambda n_{\text{imp}}}{\hbar v_F} F^{K'}_2(k_z) \frac{I_4}{4}. \tag{42}$$

We call Fs form factors, and derive them in supplementary material, as well as the expressions of I_2, I_3 and I_4.

In Fig. 2, we plot the transverse magnetoconductivities as a function of B at different screening length λ. The most important feature is that the magnitude of transverse magnetoconductivities are all enhanced sensitively in orders by increasing λ. This is just opposite to the behavior of the longitudinal magnetoconductivity which is suppressed by larger λ. We will come back this point later. Another feature is about the B-dependence of the transverse magnetoconductivities. At small λ, σ_{xx} increases with B approximately linearly; while it is inversely proportional to B at large λ. Therefore at some

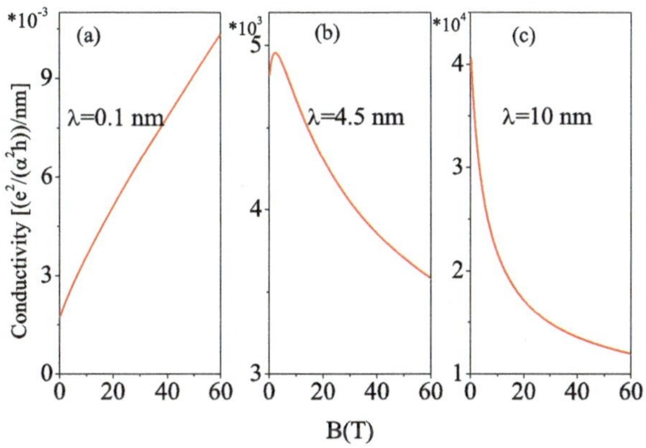

FIG. 2. The transverse magnetoconductivity, σ_{xx}, as a function of B at different screening lengths for (a) λ = 0.1 nm, (b) λ = 4.5 nm, and (c) λ = 10 nm, respectively. Here $k_c = 0.1\ \text{nm}^{-1}$.

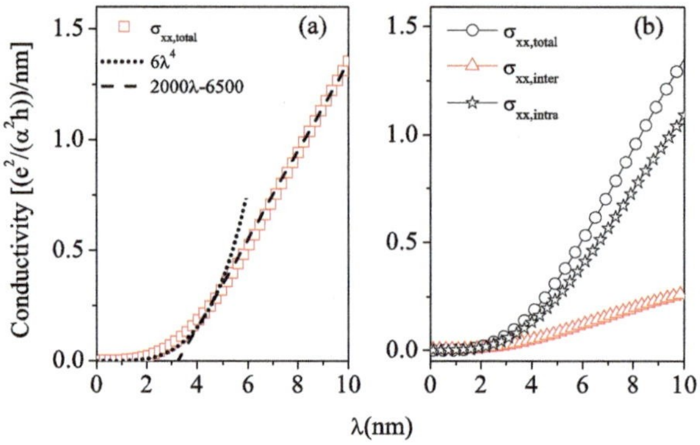

FIG. 3. The transverse magnetoconductivity, σ_{xx}, as a function of screening length λ at $B = 5T$. (a) is shown in logarithmic coordinates, (b) is in linear coordinates, respectively. Here $k_c = 0.1$ nm^{-1}.

specific screening length, the two tendencies compete with each other, giving rise to a non-monotonic behavior of σ_{xx} (see Fig. 2(b)).

In Fig. 3, we show the λ-dependence of the transverse magnetoconductivities. It is seen from Fig. 3(a) that the magnitude of the transverse magnetoconductivities is proportional to λ^4 for small λ, which is also supported by the λ dependence of I_1,I_2,I_3 and I_4. This λ^4-dependence can also be derived analytically at $\lambda \to 0$ limit. For a comparison, we use a linear function to fit the data when $\lambda \in (4$ nm, 10 nm) in Fig. 3(a). It is shown that they match very well. This linearity can not continue for much larger λ but turns to a saturation in this case (not shown in the figure). The inter-valley and intra-valley contributions are shown in Fig. 3(b). We observe a similar λ-dependence to these two contributions as well.

The enhanced transverse magnetoconductivity with λ can be understood by looking at the behavior of screened Coulomb disorder. The screened Coulomb potential can be reformulated $\phi(q) = \frac{4\pi e^2}{\alpha\kappa_0 q^2} S(q,\lambda)$, where $S(q,\lambda) = \frac{(q\lambda)^2}{1+(q\lambda)^2}$ is a defined screening factor. Since we select the parameter $k_c = 0.1$ nm^{-1}, for an inter-valley scattering q is equal to 0.2 nm^{-1}. $S(q,\lambda)$ varies monotonically from 0 to 1 with $q\lambda \in [0,\infty)$. For the fixed q=0.2 nm^{-1}, a larger λ gives rise to a larger screening factor, the magnitude of screened Coulomb potential, the inter-valley scattering (I_1) and intra-valley scattering (I_2). In accordance with this increase, the transverse magnetoconductivity is enhanced because it is proportional to the relaxation times (in I_1 and I_2).

To study the relative weight of the inter-valley scattering and the intra-valley scattering, we define a ratio

$$R = \frac{\sigma_{xx,\text{inter}}}{\sigma_{xx,\text{intra}}}, \tag{43}$$

where $\sigma_{xx,\text{inter(intra)}} = \sigma^{K}_{xx,\text{inter(intra)}} + \sigma^{K'}_{xx,\text{inter(intra)}}$. Therefore, $R = 0$ reflects that there is only intra-valley scattering, and the inter-valley scattering is absent. When $R = \infty$, the situation is just reversed that there only exists inter-valley scattering. For finite R, both scattering channels contribute. When $R > 1$, the inter-valley scattering dominates.

In Fig. 4(a), the dependence of R on B is shown for small screening lengths λ = 0.1 nm, 0.5 nm, and 1 nm, respectively. It is found that for small screening lengths the intra-valley scattering dominates under small B and the ratio tends to a constant for large B. On the contrary, the inter-valley scattering dominates when B is large. In Fig. 4(b), the effect of large screening lengths on R is considered. Although R shows a non-monotonic behavior, R is less than 1/3 in the entire region of B. We may conclude that the intra-valley scattering dominates for a smoother disorder potential.

As λ is reduced, the screened Coulomb potential is more local. Its limiting situation is a δ-potential which is given by

$$U(\mathbf{r}) = \sum_i u_i \delta(\mathbf{r} - \mathbf{R}_i). \tag{44}$$

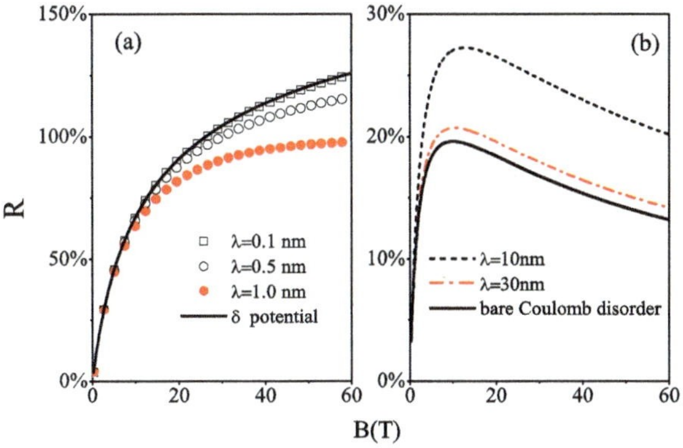

FIG. 4. The ratio R versus B at different screening lengths are studied. In (a), the red spots are for $\lambda = 1.0$ nm; the open spots are for $\lambda = 0.5$ nm, the open squares for $\lambda = 0.1$ nm and the solid line represents the result of δ-potential disorder, respectively. In (b), the short dashed line, dash-dot line and the solid line are for $\lambda = 10$ nm, $\lambda = 30$ nm and the bare Coulomb disorder, respectively. Here $k_c = 0.1\ \text{nm}^{-1}$.

In the same footing and by the virtue of the following two integrals,

$$\tilde{I}_1 \equiv \int \frac{dxdy}{(2\pi)^2} e^{-(x^2+y^2)/2},$$
$$\tilde{I}_2 \equiv \frac{l_B^2}{2} \int \frac{dxdy}{(2\pi)^2} \left(x^2 + y^2\right) e^{-(x^2+y^2)/2}, \tag{45}$$

we calculated the δ-potential case. The results are shown in Fig. 4(a). It is seen that the calculated R for the case of $\lambda = 0.1$ nm and those of δ-potential almost coincide with each other. The δ-potential could serve as the lower boundary of screened Coulomb potential.

On the contrary, a large λ indicates a more extended and smoother potential. Its limit is the bare Coulomb-potential disorder corresponding to $\lambda \to \infty$ in which there is no screening. We also calculated the consequences of this bare Coulomb disorder in Fig. 4(b). It is seen that its results almost coincide with those of $\lambda = 30$ nm. This confirms our expectation. The bare Coulomb disorder can serve as an upper boundary of the screened Coulomb disorder.

In Fig. 5(a), the λ dependence of the ratio R is depicted at different B. It is clear that increasing the screening length has little effect on R for small B. The ratio R remains at low values (see $B = 1$ T case). This means that the inter-valley contribution is tiny in this case, which is consistent with the results in Fig. 4. The ratio R only can be apparently large at a large B. We further note that R tends to a constant after a turning λ^*. With increasing B, λ^* becomes larger. Since in the laboratory range

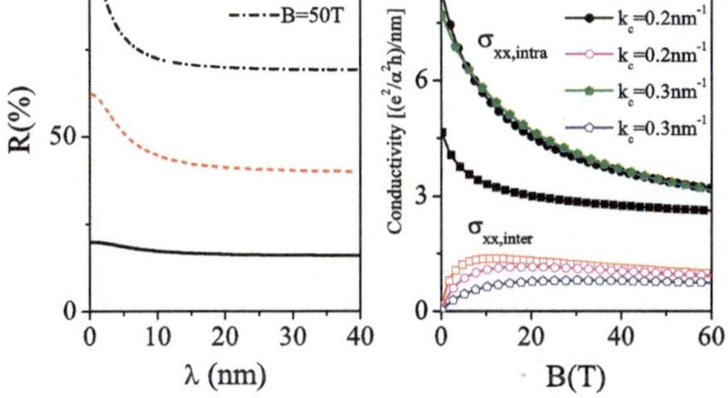

FIG. 5. In (a), the ratio R varies with the screening length at different magnetic flux density, $k_c = 0.1\ \text{nm}^{-1}$. In (b), we show how $\sigma_{xx,\text{inter(intra)}}$ varies with B at different k_c, and $\lambda = 4.5$ nm.

of B, l_B is about several nanometers, i.e. at the order of 10 nm, hence, when $\lambda > \lambda^*$, the impact of the increase of λ becomes invalid.

To know more about the effect of inter-valley and intra-valley scattering, we show how $\sigma_{xx,\text{inter(intra)}}$ varies with B at different k_c in Fig. 5(b). For a fixed k_c, $\sigma_{xx,\text{inter(intra)}}$ increases (decreases) with B. It can be understood that the inter-valley scattering is facilitated at larger B since the magnetic length is shortened and corresponding conductivity channels are enhanced for σ_{xx}. In contrast, $\sigma_{xx,\text{intra}}$ is depressed at larger B. For a fixed B, $\sigma_{xx,\text{inter}}$ is depressed with larger departure of the two nodes in the reciprocal space (larger k_c). This can be simply understood that a larger departure of the two nodes leads to harder transition between the two nodes, making smaller $\sigma_{xx,\text{inter}}$. We also witness a $1/B$-dependence of the intra-valley contribution to the transverse magnetoconductivities, which is consistent with the results in Refs. 38, 39 where only intra-valley scattering is considered. The inter-valley scattering can not be investigated in their one-node model. Therefore, the behavior of the inter-valley contribution to the transverse magnetoconductivities is new. And it is quite different from the intra-valley case. It is seen that the inter-valley scattering is as important as the intra-valley scattering. Without taking into account the inter-valley scattering, it is unlike to get a correct behavior of the total transverse magneto conductivity.

C. Discussions and conclusions

From recent experiments on Weyl semimetals, we observe several key facts. 1) The chemical potential is sample-dependent;[45] 2) The carrier density can be varied in several orders and the mobility can be different in one order;[32,34,45,46] 3) The longitudinal negative MR (related to σ_{zz}) corresponding to the chiral anomaly may not be observed in *every* experiment.[45] From the first and second facts, we may conclude that there exist charged impurities in realistic samples and the density of the impurity is sample-dependent. According to our calculation, it is known that the longitudinal MR is very sensitive to the charged impurities. About the third fact, a reasonable explanation is given that the transverse resistivity (related to σ_{xx}) is usually quite larger than the longitudinal resistivity. In realistic measurements, it is hard to parallelize the magnetic field and electric field. Thus a perpendicular component of magnetic field **B** with respect to the electric field **E** may induce a large transverse MR which may overwhelm the small longitudinal MR. From our study, this may occur for the system with charged impurities of short screening length, for example $\lambda = 0.1$ nm. We thus need to discuss how to observe the negative longitudinal MR in experiments. Due to the opposite effect of the screened Coulomb disorder on the σ_{zz} and σ_{xx}, we may enhance the longitudinal resistivity and decrease the transverse resistivity by increasing the screening length. Eventually, the longitudinal resistivity may be larger than the transverse one. Therefore, the negative longitudinal MR may be easier to be observed in experiments in this case.

IV. SUMMARY

In this work, we study a two-node Weyl semimetal in the presence of magnetic fields. Screened Coulomb potential may be the main source of disorder in Weyl semimetal. We calculated the longitudinal and transverse magnetoconductivity with the screened Coulomb disorder. We find that the longitudinal magnetoconductivity is positive in magnetic field, giving rises to a negative magnetoresistance, which agrees with the results in Ref. 40. In the transverse magnetoconductivity, we disclose a crossover from linear-B dependence in the small screening length limit to $1/B$ dependence in the large screening length limit. We also note that the magnitude of the transverse magnetoconductivity is proportional to the quadruplicate of the screening length λ for small λ. At small λ, the transverse magnetoconductivity σ_{xx} increases with B; while it decreases with B at large λ. At middle λ, σ_{xx} may show a non-monotonic behavior. We define a ratio R quantifying the inter-valley and intra-valley components. It is found that R is quite small for a weak magnetic field, indicating the intra-valley scattering dominates. The inter-valley scattering dominates only in the presence of strong magnetic field and a smaller screening length.

Due to the opposite dependence of longitudinal and transverse magnetoconductivity on the screened Coulomb disorder, we discuss the possible explanation of recent experiments. Here we propose that increasing screening length may facilitate the observation of the negative longitudinal magnetoresistivity induced by the chiral anomaly. Moreover, introducing disorder may provide a helpful way to manipulate the valley degree in the realm of valleytronics of the Weyl semimetal.

ACKNOWLEDGMENTS

This work is supported by Hundred Talents Program of The Chinese Academy of Sciences and the NSFC (Grant No. 11674317). G. S. is supported in part by the MOST (Grant No. 2013CB933401), the NSFC (Grant No. 11474279) and the CAS (Grant No. XDB07010100). H. Z. Lu was supported by the National Key R & D Program (Grant No. 2016YFA0301700), and the National Natural Science Foundation of China (Grant No. 11574127).

[1] B. Pontecorvo, Zh. Eksp. Teor. Fiz. **33**, 2 (1957).
[2] X. Wan *et al.*, Phys. Rev. B **83**, 205101 (2011).
[3] G. Xu *et al.*, Phys. Rev. Lett. **107**, 186806 (2011).
[4] P. Hosur, S. A. Parameswaran, and A. Vishwanath, Phys. Rev. Lett. **108**, 046602 (2012).
[5] P. Hosur, Phys. Rev. B **86**, 195102 (2012).
[6] P. Hosur and X.-L. Qi, C. R. Physique **14**, 857 (2013).
[7] P. Goswami and S. Tewari, Phys. Rev. B **88**, 245107 (2013).
[8] T. Ojanen, Phys. Rev. B **87**, 245112 (2013).
[9] D. Bulmash, C.-X. Liu, and X.-L. Qi, Phys. Rev. B **89**, 081106(R) (2014).
[10] G. Autes *et al.*, Phys. Rev. Lett. **117**, 066402 (2016).
[11] H. M. Weng *et al.*, Phys. Rev. X **5**, 011029 (2015).
[12] B. Q. Lv *et al.*, Phys. Rev. X **5**, 031013 (2015).
[13] S. M. Huang *et al.*, Nat. Commun. **6**, 7373 (2015).
[14] S. Y. Xu *et al.*, Science **349**, 613 (2015).
[15] S. Borisenko *et al.*, arXiv:1507.04847.
[16] K. Deng *et al.*, Nat. Phys. **12**, 1105 (2016).
[17] H. B. Nielsen and M. Ninomiya, Phys. Lett. B **130**, 389 (1983).
[18] J. Zhou, H. R. Chang, and D. Xiao, Phys. Rev. B **91**, 035114 (2015).
[19] K.-Y. Yang, Y.-M. Lu, and Y. Ran, Phys. Rev. B **84**, 075129 (2011).
[20] A. A. Burkov and L. Balents, Phys. Rev. Lett. **107**, 127205 (2011).
[21] S. A. Yang, H. Pan, and F. Zhang, Phys. Rev. Lett. **115**, 156603 (2015).
[22] D. E. Kharzeev and H. U. Yee, Phys. Rev. B **88**, 115119 (2013).
[23] D. T. Son and N. Yamamoto, Phys. Rev. Lett. **109**, 181602 (2012).
[24] M. A. Stephanov and Y. Yin, Phys. Rev. Lett. **109**, 162001 (2012).
[25] K. Landsteiner *et al.*, Phys. Rev. Lett. **107**, 021601 (2011).
[26] M.-C. Chang and M.-F. Yan, Phys. Rev. B **91**, 115203 (2015).
[27] G. Başar, D. E. Kharzeev, and H. U. Yee, Phys. Rev. B **89**, 035142 (2014).
[28] D. T. Son and B. Z. Spivak, Phys. Rev. B **88**, 104412 (2013).
[29] A. A. Burkov, Phys. Rev. Lett. **113**, 247203 (2014).
[30] E. V. Gorbar, V. A. Miransky, and I. A. Shovkovy, Phys. Rev. B **89**, 085126 (2014).
[31] X. Huang *et al.*, Phys. Rev. X **5**, 031023 (2015).
[32] C. Zhang *et al.*, arXiv:1503.02630.
[33] C. Zhang *et al.*, Phys. Rev. B **92**, 041203 (2015).
[34] F. Arnold *et al.*, Nat. Commun. **7**, 11615 (2016).
[35] X. Yang *et al.*, arXiv:1506.02283.
[36] Z. Wang *et al.*, Phys. Rev. B **93**, 121112(R) (2016).
[37] J. C. W. Song, G. Refael, and P. A. Lee, Phys. Rev. B **92**, 180204(R) (2015).
[38] D. A. Pesin, E. G. Mishchenko, and A. Levchenko, Phys. Rev. B **92**, 174202 (2015).
[39] J. Klier, I. V. Gornyi, and A. D. Mirlin, Phys. Rev. B **92**, 205113 (2015).
[40] S.-B. Zhang, H.-Z. Lu, and S.-Q. Shen, New J. Phys. **18**, 053039 (2016).
[41] H.-Z. Lu and S.-Q. Shen, Front. Phys. **12**(3), 127201 (2017).
[42] H.-Z. Lu and S.-Q. Shen, Phys. Rev. B **92**, 035203 (2015).
[43] H.-Z. Lu, S.-B. Zhang, and S.-Q. Shen, Phys. Rev. B **92**, 045203 (2015).
[44] Y. I. Rodionov and S. V. Syzranov, Phys. Rev. B **91**, 195107 (2015).
[45] C.-L. Zhang *et al.*, Nat. Commun. **7**, 10735 (2015).
[46] H. Li *et al.*, Nat. Comm. **7**, 10301 (2016).

16

Optical resonant tunneling in photonic heterostructures containing a tunable dielectric layer

Liyong Cui,[1] Guang Lu,[1] Shan Zhang,[1] Fen Liu,[1] Yanqing Xin,[1] Kunlun Wang,[1] Tianlin Yang,[1] Li Wang,[3] and Guiqiang Du,[1,2,a]
[1]*School of Space Science and Physics, Shandong University at Weihai, Weihai 264209, P. R. China*
[2]*State Key Laboratory of Surface Physics and Department of Physics, Key Laboratory of Micro- and Nano-Photonic Structures (Ministry of Education), Fudan University, Shanghai 200433, P. R. China*
[3]*Beijing Institute of Space Mechanics & Electricity, Beijing 100094, P. R. China*

We demonstrate theoretically and experimentally that complete light transmission can be realized using a photonic heterostructure containing a tunable dielectric layer inserted between two different truncated photonic crystals (PCs). A perfect tunneling state is produced within enlarged photonic band gap (PBG) of the heterostructure by varying the thickness of inserted dielectric layer and the transmittance of the tunneling state depends on the dielectric layer thickness. Additionally, the tunneling state frequency varies with inserted layer thickness but is always located within the small overlapped PBG of two PCs. Therefore, both a perfect tunneling state and an ultrawide PBG can be realized in these heterostructures. The experimental results showed good agreement with theoretical values.

I. INTRODUCTION

Since photonic crystals (PCs) were initially predicted by Yablonovitch in 1987,[1] they have attracted considerable attention for many potential applications, including light-emitting diodes,[2] omnidirectional mirrors,[3] and substrates for efficient antennas. The primary characteristic of a PC is its photonic band gap (PBG), which mainly originates from the multiple Bragg scattering mechanism of the electromagnetic wave. The propagation of electromagnetic waves is strongly prohibited[4,5] within the PBG range, and the PBG can be tuned by selecting different materials and different structural parameters. Based on their spatial dimensions, PCs can be divided into one-dimensional (1D), two-dimensional (2D), and three-dimension (3D) microstructures. Similar to 2D and 3D PCs, 1D PCs also have drawn the attention of researchers because of their well-known properties, including their simple structures and the ease of their fabrication.

In 1958, Anderson found that electronic localization occurred in disordered systems,[6] and the phenomenon was thus named Anderson localization. Later, the concept of Anderson localization was studied in a variety of different fields, such as acoustic waves and optical waves.[7–10] The concept of photonic localization also was proposed in 1987.[11] One important method for realization of photonic localization is that defects are induced in the PCs.[12] The frequency of the defect mode is located in the PBG region and the electromagnetic wave of the defect mode can tunnel through the entire PC. Based on these properties, PCs can be used to fabricate zero-threshold lasers,[13] multichannel filters,[14] monochromatic selectors,[15] and optical sensors.[16] Photonic localization is also found in quasi-periodic PCs[17–20] in a manner similar to the appearance of electron localization in quasi-periodic systems. There is a important related property in that the resonant localized states

[a]E-mail: dgqql@sdu.edu.cn

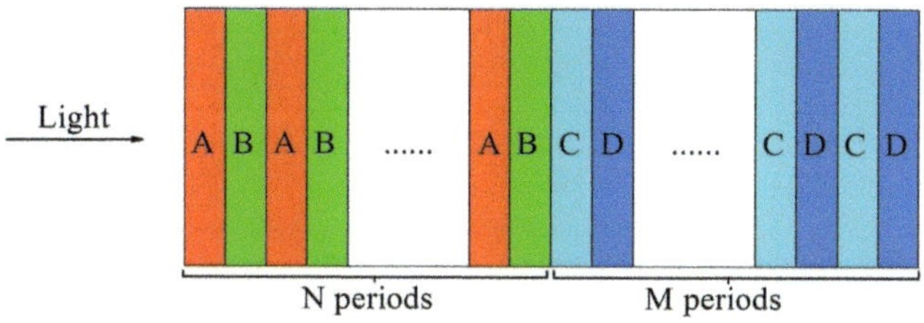

FIG. 1. Schematic of the $(AB)_N(CD)_M$ heterostructure. A and C are TiO_2 layers with $n_A=n_C=n_E=2.123$,[32] while B and D are SiO_2 layers with $n_B=n_D=1.431$,[32] respectively. N and M are periodic numbers.

have perfect transmission if the PC microstructures, including both periodic and quasi-periodic PCs, are symmetrical.[21–26] Structures of this type can be used to fabricate single- and multiple-channel perfect narrowband optical filters. However, high-quality narrowband transmission-type filters usually require the frequencies of the defect modes within a wide PBG. Because they are limited by the availability of dielectrics with appropriate refractive indices, the PBGs of the periodic PCs generally are not large enough. Therefore, many researchers have focused on enlarging the operating regions of PBGs.[27–30] One important way to achieve this aim is to use photonic heterostructures composed of PCs of different materials to realize ultra-wide and omnidirectional PBGs.[27–30] Alternatively, it was found that a tunneling state could emerge in the overlapped and broadened PBG of heterostructures containing two different PCs.[31] Because the maximum electromagnetic fields of the tunneling states are close to the interface of the two PCs, the resulting localized mode is called the interface mode and is also known as the optical Tamm state (OTS).[31] However, it is difficult to realize perfect transmission and a sufficiently broad PBG simultaneously when using heterostructures composed of two different PCs. In this Letter, we propose a type of heterostructure that can realize both complete transmission and an ultra-wide PBG simultaneously that is composed of two different all-dielectric PCs with a tunable inserted dielectric layer.

The paper is organized as follows. In Section II, we discussed the ultra-wide forbidden gap in heterostructures composed of two different all-dielectric PCs. Subsequently, in Section III, by changing the thickness of the inserted dielectric layer, we discussed the conditions to realize perfect transmission and analyzed the physical mechanism of the tunneling state. Then, the results of optical experiments were introduced which are in good agreement with the theoretical values. Finally, the conclusions were draw in Section IV.

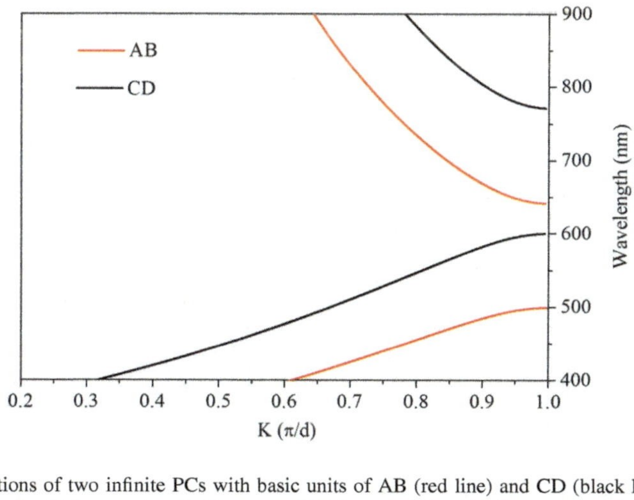

FIG. 2. Dispersion relations of two infinite PCs with basic units of AB (red line) and CD (black line). The thicknesses of layers A, B, C and D are d_A=66.24 nm, d_B=98.27 nm, d_C=79.49 nm and d_D=117.92 nm, respectively. All other parameters are the same as those shown in Fig. 1.

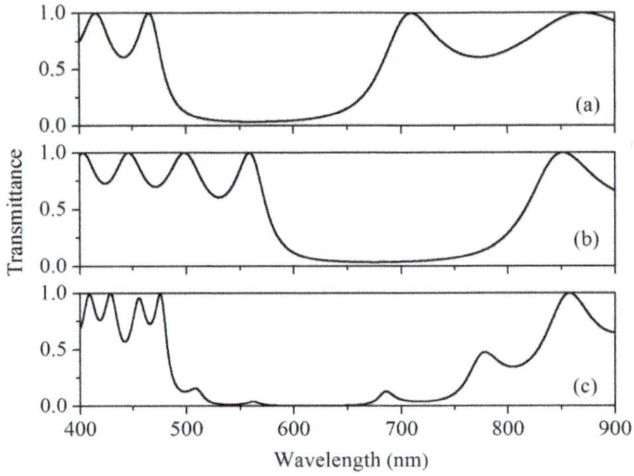

FIG. 3. (a), (b) and (c) show the transmittance characteristics of $(AB)_6$, $(CD)_6$ and $(AB)_6(CD)_6$, respectively. All other layer parameters are the same as those used in Fig. 2.

II. ENLARGED PBG IN HETEROSTRUCTURES WITH TWO DIFFERENT PCS

First, we considered a heterostructure composed of two different PCs, denoted by $(AB)_N$ and $(CD)_M$, without use of the inserted layer to obtain an ultrawide PBG, as shown in Fig. 1, where $(AB)_N$ and $(CD)_M$ represent different truncated all-dielectric PCs with periodic numbers N and M, respectively. A and C represent TiO_2 layers with identical refractive indices of n_A=n_C=n_E=2.123,[32] respectively. B and D are SiO_2 layers with identical refractive indices of n_B=n_D=1.431,[32] respectively.

In order to obtain an ultrawide PBG, where the thicknesses of layers A, B, C and D are set at d_A=66.24 nm, d_B=98.27 nm, d_C=79.49 nm and d_D=117.92 nm, respectively. In this paper, normal incidence conditions are considered. The dispersion relations of the two different PCs are calculated, where AB and CD represent the basic units. As shown in Fig. 2, the PBG with AB (CD) as the basic unit ranges in thickness from 500.15 (600.19) nm to 642.64 (771.17) nm, as indicated by the solid red (black) line. The figure shows that the overlapped PBG of the two different PCs ranges from 600.19 nm to 642.64 nm and that it is very narrow.

Fig. 3 shows the transmittance characteristics of the $(AB)_N$, $(CD)_M$ and $(AB)_N(CD)_M$ structures, respectively, where the periodic numbers are N=M=6, and the other parameters are the same as those used in Fig. 2. The results are calculated using the transfer-matrix method.[33] The characteristics show that a suitably enlarged PBG is realized in the $(AB)_6(CD)_6$ heterostructure.

III. RESONANT TUNNELING STATE IN HETEROSTRUCTURES WITH A TUNABLE DIELECTRIC LAYER

To obtain an optical resonant tunneling state, a tunable dielectric layer denoted by E is inserted between the two different PCs to form the new heterostructure $(AB)_NE(CD)_M$, as shown in Fig. 4. As

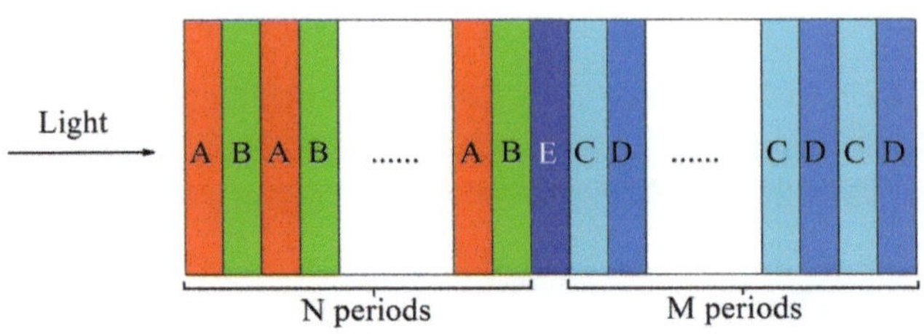

FIG. 4. Schematic of the $(AB)_NE(CD)_M$ heterostructure. E is TiO_2 layer with n_E=2.123.[32] All other layer parameters are the same as those used in Fig. 1.

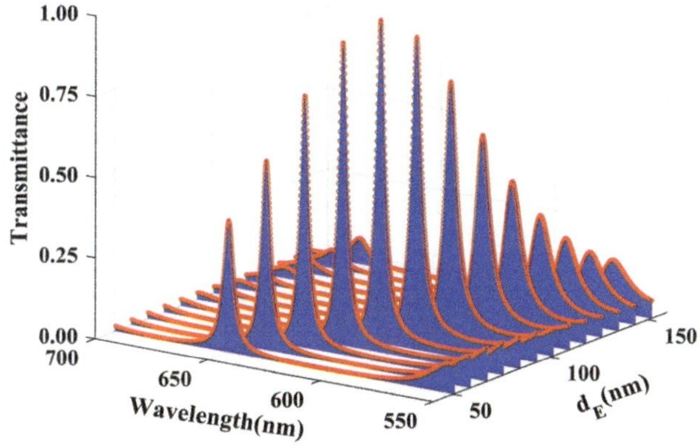

FIG. 5. The transmittance of the $(AB)_6E(CD)_6$ heterostructure varies with the thickness of the inserted dielectric layer E. All other parameters are the same as those used in Fig. 3.

a result, an optical tunneling state is found to appear in the enlarged PBG of the two different PCs, as shown in Fig. 5. Unlike the defect modes that appear in symmetrical structures,[22–26] the transmittance of the tunneling state varies with the thickness of the inserted layer E, and only a specific thickness of E layer corresponds to perfect transmission. The figure also shows that the tunneling wavelength is always located in the small overlapped PBG of the two different PCs. The complete tunneling state is realized when the inserted layer thickness d_E is equal to 84.52 nm, and the corresponding wavelength for this state is 638.13 nm, as shown in Fig. 6. Additionally, the intensity distributions of the electric and magnetic fields, denoted by $|E|^2$ and $|H|^2$, respectively, are shown in Fig. 7, where the incident electric field intensity was given the value 1. The results show that the maximum values of $|E|^2$ and $|H|^2$ are highly localized near the interface between $(AB)_6E$ and $(CD)_6$, and thus the tunneling state with complete transmission can be called an interface-like state or an optical Tamm-like state.

Alternatively, if the positions of the two PCs $(AB)_6$ and $(CD)_6$ are exchanged and a tunable dielectric layer F is inserted between them to form a $(CD)_6F(AB)_6$ heterostructure, a tunneling state also appears in the small overlapped PBG of the two different PCs, and the corresponding transmittance is sensitive to the thickness of inserted layer in a similar manner to the sensitivity

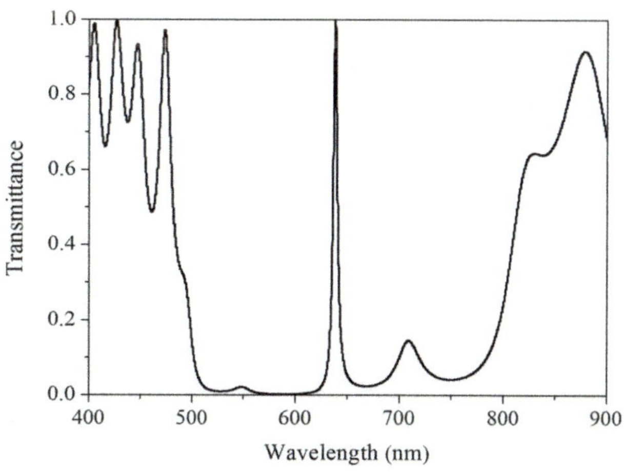

FIG. 6. Transmittance of the $(AB)_6E(CD)_6$ heterostructure with d_E=84.52 nm at the tunneling wavelength of 638.13 nm. All other parameters are the same as those used in Fig. 5.

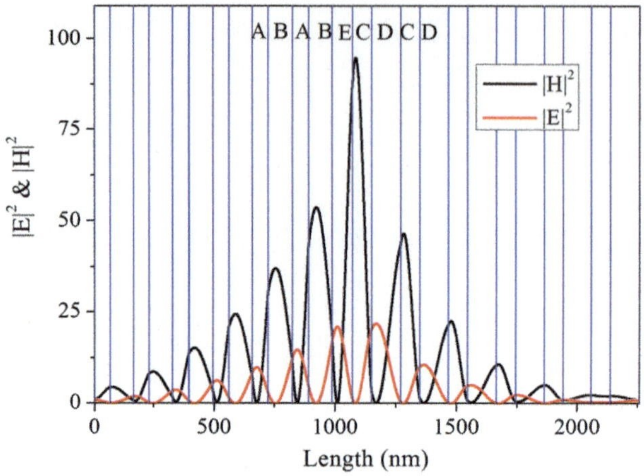

FIG. 7. Simulated intensities of the electric fields (black line) and magnetic fields (red line) in the $(AB)_6E(CD)_6$ heterostructure at the tunneling wavelength of 638.13 nm. All other parameters are the same as those used in Fig. 6.

shown in Fig. 5. The F layer is also selected to be composed of TiO_2 and other parameters are the same as those used in Fig. 3. A perfect tunneling state can be realized at a wavelength of 602.48 nm when the thickness of F is 63.59 nm. The transmission spectrum of the $(CD)_6F(AB)_6$ structure is shown in Fig. 8. Furthermore, the intensity distributions of $|E|^2$ and $|H|^2$ are shown in Fig. 9, where the incident electric field intensity was given a value of 1. The maximum electromagnetic field intensities are also highly localized around the interface between $(CD)_6F$ and $(AB)_6$. Therefore, the tunneling state can also be called an optical Tamm-like state.

Additionally, we experimentally demonstrated these perfect optical tunneling states. The $(AB)_6E(CD)_6$ and $(CD)_6F(AB)_6$ heterostructures were evaporated on the substrate S by the ion-assisted electron-beam evaporation method under high vacuum conditions, giving experimental structures of $(AB)_6E(CD)_6S$ and $(CD)_6F(AB)_6S$, where S is BK7 glass with a refractive index of n_S=1.52. The SiO_2 and TiO_2 layer thicknesses were monitored using quartz crystal sensors. The transmission spectra were measured using a Cary-100 ultraviolet-visible-near-infrared spectrophotometer. Fig. 10(a) and (b) show the measured transmittance characteristics of the $(AB)_6E(CD)_6S$ and $(CD)_6F(AB)_6S$ heterostructures, respectively. In Fig. 10(a), the black line indicates that one near-perfect transmission peak exists at 622.6 nm, which is located within the enlarged PBG, and the

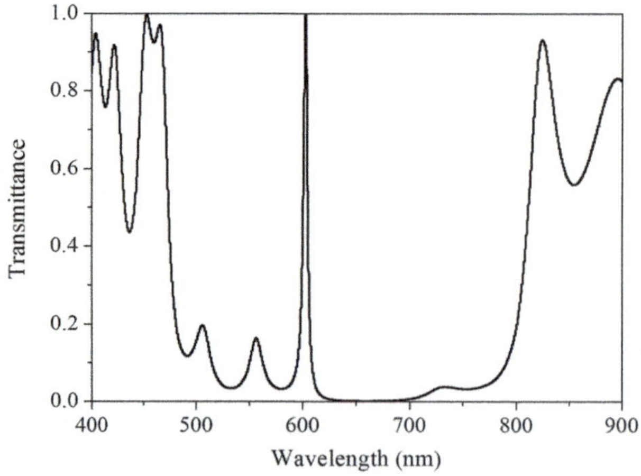

FIG. 8. Transmittance of the $(CD)_6F(AB)_6$ heterostructure with d_F=63.59 nm at the tunneling wavelength of 602.48 nm. All other parameters are the same as those used in Fig. 6.

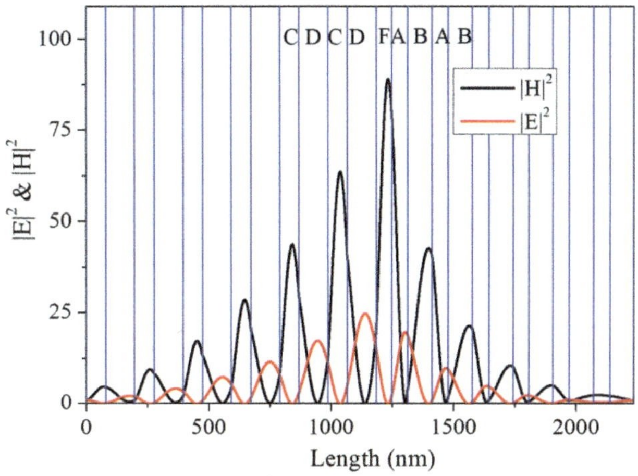

FIG. 9. Simulated intensities of the electric fields (red line) and magnetic fields (blue line) in the $(CD)_6F(AB)_6$ heterostructure at the tunneling wavelength of 602.48 nm. All other parameters are the same as those used in Fig. 8.

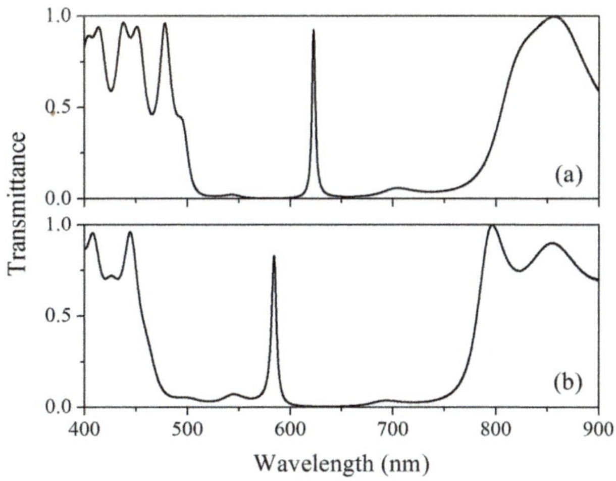

FIG. 10. (a) and (b) show the experimental transmission spectra of the $(AB)_6E(CD)_6$ and $(CD)_6F(AB)_6$ structures, respectively. The maximum transmittances of the two heterostructures are 0.924 at a wavelength of 622.6 nm and 0.823 at a wavelength of 591.0 nm, respectively.

maximum transmittance for $(CD)_6F(AB)_6S$ is 0.924. Fig. 10(b) shows that one near-perfect transmission peak exists at 591.0 nm and the maximum transmittance for the latter heterostructure is 0.823. The small discrepancies between the theoretical and experimental results mainly originate from differences that occur because of errors in layer thickness monitoring during the deposition processes and refractive index discrepancies between the theoretical and experimental materials.

IV. CONCLUSION

In summary, we have presented a type of photonic heterostructure that enables realization of perfect transmission and an ultra-wide PBG simultaneously. The heterostructure is composed of two different PCs separated by a tunable dielectric layer. The perfect transmission is dependent on the inserted dielectric layer thickness. The physical mechanisms that lead to perfect transmission are analyzed in detail and it is found that the perfect tunneling behavior originates from the mechanism of the optical Tamm-like state. Such heterostructures offer a new way to fabricate high-quality optical filters.

ACKNOWLEDGMENTS

This work is supported by the National Natural Science Foundation of China (No. 11264003), by the Natural Science Foundation of Shandong Province (No. ZR2015AM008), by the Basic Research Program of Shandong University at Weihai (No. 2015ZQXM013), open project of State Key Laboratory of Surface Physics and Department of Physics (No. KF2017_01) and by Key Laboratory of Micro- and Nano-Photonic Structures (Ministry of Education).

[1] E. Yablonovitch, Phys. Rev. Lett. **58**, 2059 (1987).
[2] J. J. Wierer, Jr., A. David, and M. M. Megens, Nat. Photonics **3**, 163 (2009).
[3] Y. Fink, J. N. Winn, S. Fan, C. Chen, J. Michel, J. D. Joannopoulos, and E. L. Thomas, Science **282**, 1679 (1998).
[4] M. Plihal and A. A. Maradudin, Phys. Rev. B **44**, 8565 (1991).
[5] P. R. Villeneuve and M. Piche, Phys. Rev. B **46**, 4969 (1992).
[6] P. W. Anderson, Phys. Rev. **109**, 1492 (1958).
[7] A. Figotin, J. Stat. Phys. **76**, 985 (1994).
[8] L. Tsang and A. Ishimaru, J. Opt. Soc. Am. A **2**, 2187 (1985).
[9] M. P. V. Albada, M. B. V. Mark, and A. Lagendijk, Phys. Rev. Lett. **58**, 361 (1987).
[10] P. E. Wolf and G. Maret, Phys. Rev. Lett. **55**, 2696 (1985).
[11] S. John, Phys. Rev. Lett. **58**, 2486 (1987).
[12] K. Inoue and K. Ohtaka, *Photonic Crystals: Physics, Fabrication, and Applications* (Springer-Verlag, Berlin, 2004).
[13] S. Noda, M. Yokoyama, M. Imada, A. Chutinan, and M. Mochizuki, Science **293**, 1123 (2001).
[14] Y. Li, Y. J. Xiang, S. C. Wen, J. H. Yong, and D. Y. Fan, J. Appl. Phys. **110**, 073111 (2011).
[15] I. V. Timofeev, Y. T. Lin, V. A. Gunyakov, S. A. Myslivets, V. G. Arkhipkin, S. Y. Vetrov, W. Lee, and V. Y. Zyryanov, Phys. Rev. E **85**, 011705 (2012).
[16] Mardanih, T. P. Negara, H. Hardhienata, and H. Alatas, AIP Conf. Proc. **1325**, 313 (2010).
[17] P. Hu and C. S. Ting, Phys. Rev. B **34**, 8331 (1986).
[18] Y. Y. Liu and R. Riklund, Phys. Rev. B **35**, 6034 (1987).
[19] M. Kohmoto, B. Sutherland, and K. Iguchi, Phys. Rev. Lett. **58**, 2436 (1987).
[20] W. Gellermann, M. Kohmoto, B. Sutherland, and P. C. Taylor, Phys. Rev. Lett. **72**, 633 (1994).
[21] S. H. Fan, P. R. Villeneuve, and J. D. Joannopoulos, Phys. Rev. Lett. **80**, 960 (1998).
[22] R. W. Peng, X. Q. Huang, F. Qiu, M. Wang, A. Hu, and S. S. Jiang, Appl. Phys. Lett. **80**, 3063 (2002).
[23] W. J. Hsueh, S. J. Wun, Z. J. Lin, and Y. H. Cheng, J. Opt. Soc. Am. B **28**, 2584 (2011).
[24] X. Q. Huang, S. S. Jiang, R. W. Peng, and A. Hu, Phys. Rev. B **63**, 245104 (2001).
[25] R. W. Peng, Y. M. Liu, X. Q. Huang, F. Qiu, M. Wang, A. Hu, S. S. Jiang, D. Feng, L. Z. Ouyang, and J. Zou, Phys. Rev. B **69**, 165109 (2004).
[26] R. Nava, J. T. Martinez, J. A. delRio, and G. G. Naumis, J. Phys.: Condens. **21**, 155901 (2009).
[27] L. Wang, Z. S. Wang, Y. G. Wu, L. Y. Chen, S. W. Wang, X. S. Chen, and L. Y. Chen, J. Appl. Phys. **95**, 424 (2004).
[28] F. Qiao, C. Zhang, J. Wan, and J. Zi, J. Appl. Phys. **77**, 3698 (2000).
[29] X. Wang, X. H. Hu, Y. Z. Li, W. L. Jia, C. Xu, X. H. Liu, and J. Zi, Appl. Phys. Lett. **80**, 4291 (2002).
[30] V. Kumar, M. Anis, K. S. Singh, and G. Singh, Optik-Int. J. Light Electron Opt. **122**, 2186 (2011).
[31] A. V. Kavokin, I. A. Shelykh, and G. Malpuech, Phys. Rev. B **72**, 233102 (2005).
[32] J. C. Manifacier, J. Gasiot, and J. P. Fillard, J. Phys. E: Scientific Instruments **9**, 1002 (1976).
[33] P. Yeh, *Optical Waves in Layered Media* (Wiley, 1988).

17

Magnetic coupling in nonmagnetic metal adsorption on arsenene monolayer: Ab initio study

Luo Min,[1,a] Yu E. Xu,[2] and Xi Yu Song[3]
[1]*Department of Physics, Shanghai Second Polytechnic University, Shanghai 201209, China*
[2]*Department of Electronic Engineering, Shang Hai Jian Qiao University, Shanghai 201306, China*
[3]*Key Laboratory of Polar Materials and Devices, East China Normal University, Shanghai 200241, China*

The structural, electronic and magnetic properties of adsorbed arsenene monolayer with a series of nonmagnetic metal atoms, including Al, Ga, Li, Mg, and Na, are systematically investigated using first-principles calculations. From our calculations, nonmagnetic states appear in the cases of Li, Mg, and Na. The magnetism has been observed for Al and Ga adatoms. Then, we study the magnetic interaction between the moments induced by Al and Ga adatoms. It is found that a *p–d* exchange-like p-p hybridization results in the ferromagnetic (FM) state. However, when the distance of Ga-Ga or Al-Al is increased, it is found out that the FM interaction has been extremely depressed. This phenomenon can be explained by a simple Heisenberg model. The X-adsorbed (X= Al and Ga) arsenene systems are promising candidates for spintronic devices in the future.

I. INTRODUCTION

During these years, the scientists pay much attention to the two-dimensional (2D) materials because of their novel functions. One of the most studied materials is graphene[1] owing to its unique physics, optical, and chemical properties and potential applications in nanoelectronic.[2,3] Furthermore, recently, scientists also try to find out new 2D materials, which have natural band gaps.[4,5] Phosphorene, which has potential to be a 2D atomic crystal, exhibits great properties of electronic, optical, thermoelectric and mechanical aspects.[6–9]

As is known to all, it has been verified that by doping transition metal (TM) atoms into 2D materials at a low concentration could tune the electronic properties obviously and induce magnetism, scattering and superconduction.[10] In addition, another effective and promising method to modify the magnetic and electronic properties is adsorption. According to the latest studies on graphene, blue phosphorene and MoS_2, by using surface adsorption method, we could tune magnetic properties of 2D materials.[11–13] In contrast to the magnetism induced by TM atoms, the magnetic moment based on nonmetal atoms sometimes has much stronger long-range magnetic interactions.[14–16] Recently, a new elemental 2D material has been discovered based on theoretical studies,[17,18] namely grey arsenic, which is the structural counterpart of blue phosphorus. Zhang *et al.*[19] predicted that although the bulk phase of gray arsenic is typically semimetal, it can transform into an indirect band-gap semiconductor as thin as a monolayer (arsenene), which is different from planar graphene, silylene and puckered phosphorene. However, to our knowledge, less interest has been gained to the magnetic properties of nonmagnetic metal elements absorbed arsenene.

In this work, we systematically study the electronic and magnetic properties of nonmagnetic metal atoms X-adsorbed arsenene (X= Al, Ga, Li, Mg, and Na) by using the first principles method.

[a]luomin@sspu.edu.cn

According to our calculations, we found that Al, Ga, Li, Mg, and Na adatoms prefer different adsorption sites. The adsorption of Al and Ga atoms on arsenene can induce the magnetism, which is mainly attributed to 3p/4p states of Al/Ga atom and the 4p states of its neighboring As atoms. Next, we further studied the interaction between two Al atoms. Our results show the *p–p* hybridization mechanism between Al and its neighboring As atoms results in its ferromagnetic (FM) state. Moreover, similar FM behavior appears in the two-Ga-adsorbed system and a long-range ferromagnetic (FM) coupling is observed. Our studies suggest that the Al/Ga adsorbed arsenene might have some potential spintronic applications in the future.

II. METHOD

All calculations are performed with density functional theory (DFT)[20] implemented in the Vienna ab initio simulation package (VASP 5.4). Exchange correlation functions of GGA-PBE[21] and the projector augmented-wave (PAW)[22] pseudopotential are used. An energy cut off of 400 eV, and $6 \times 6 \times 1$ k-points for geometry optimization are used. The lattice parameter and buckling height of arsenene is 3.60 and 1.39 Å, respectively. The vacuum is 15 Å between adjacent supercells. The convergence of force optimization is set at the value of 0.02 eV/Å.

III. RESULTS AND DISCUSSIONS

A. Electronic structures and magnetism of X-adsorbed arsenene

Firstly, we systematically study the magnetic properties of one nonmagnetic metal atom (Al, Ga, Li, Mg, and Na) adsorbed arsenene. For all the cases, three possible adsorption sites are considered. Nonmagnetic metal could bond with the bottom or top As atom, or be in the center of the As hexagon, marking, as shown in Figs. 1a and b (S1, S2 and S3, correspondingly). With the purpose of checking the stability of each case, adsorption energies are calculated, which is estimated as $E_{ads} = E_T - E_P - E_{TM}$, where E_P and E_T is the total energy of perfect and nonmagnetic metal atom adsorbed arsenene, respectively; E_{TM} is the energy of a nonmagnetic metal. Besides, total magnetic moments (μ_{total}) and bond distance have been calculated, as shown in Table I. According to the calculated adsorption energies, as exhibited in Fig. 2, for Al, Ga, and Li cases, the stable adsorption sites is S1 and the Na atom prefers S2. The stable site of Mg is S3, however, adsorption energy of S3 is very low (-0.558 eV), implying the Mg atom could not adsorption on arsenene surface. Moreover, among all the cases, magnetism appears in the Al and Ga cases, and the magnetic moment (μ_{total}) is about 0.89 and 0.74 μB, respectively.

To analyze the origin of magnetism of nonmagnetic metal adatoms on arsenene, we investigated the band structures of each case. As shown in Figs. 3c, d, and e, for Li, Mg, and Na adatoms,

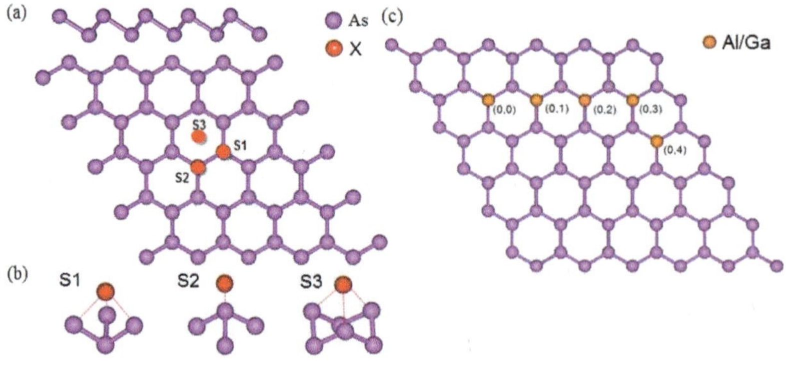

FIG. 1. Schematic structure showing (a) top and side view of three different adsorption sites in the 4×4×1 arsenene ($As_{32}X_1$, X= Al, Ga, Li, Mg, and Na), marking as S1, S2 and S3, respectively; (b) side view of three adsorption sites; (c) the hexagonal structure viewed along the c axis, and the configurations in two Al (Ga) atoms adsorption on the 6×6×1 arsenene.

TABLE I. Adsorption energy (E_{ads}), total magnetic moment (μ_{total}), and X-As bond lengths (L) of the X-adsorbed 4 × 4 × 1 arsenene (X= Al, Ga, Li, Mg, and Na).

Structure		Al	Ga	Li	Mg	Na
E_{ads} (eV)	S1	-1.567♦	-1.516♦	-1.689♦	-0.311	-1.183
	S2	-1.141	-1.076	-1.591	-0.107	-1.334♦
	S3	-1.441	-1.379	-1.510	-0.558♦	-1.169
μ_{total} (μ_B)	S1	0.0♦	0.0♦	0.0♦	0.0	0.0
	S2	0.89	0.74	0.0	0.0	0.0♦
	S3	0.0	0.0	0.0	0.0♦	0.0
L♦ (Å)	X-As	2.728	2.815	2.680	2.891	2.795

there are some impurity levels in the band gap. However, both the spin-up and spin-down parts in the above cases are identical, and there are no magnetic properties in these systems. With the one Al- and Ga-adsorbed arsenene monolayer, as shown in Figs. 3a and b, the impurity states in spin-up band are above the Femi level (E_F) and those are below E_F in spin-down band. Therefore, the spin-up and spin-down parts in the above cases are asymmetric and the systems are magnetic. Next, we further investigate the magnetism distributions of Al (Ga) adsorbed arsenene, as shown in Fig. 4. The magnetic moments are primarily originate from Al (Ga) and its neighboring As27. As shown in Figs. 3a and c, the three second nearest atoms, As11, As12, and As16 still have partial contributions.

We next analyze the origin of the magnetic behavior of Al and Ga, we calculated the total density of states (TDOS) and projected density of states (PDOS) of these two systems, as shown in Figs. 5a and b. As we mentioned above, there are obviously asymmetry spin spitting around the E_F. Corresponding PDOS discussions further reveal that these is a strong hybridization between Al (Ga) and its surrounding As atoms and then induce some spitting states in spin-up part, which are mainly originated from the 3p (4p) electrons of Al (Ga) and the 4p electrons of As atoms.[23]

B. Magnetic coupling in Al- and Ga-adsorbed arsenene

In order to check if the Al- and Ga-adsorbed arsenene show a long-rang magnetism, we investigate magnetic coupling between two Al (Ga) atoms in the 6 × 6 × 1 arsenene. As listed in Fig. 1c, we fix one Al (Ga) at (0,0), and fix another Al atom at (0,1), (0,2), (0,3) and (0,4), respectively. In order to check the ground state, we calculated energy differences between antiferromagnetic (AFM) and FM states ($\Delta E = E_{AFM} - E_{FM}$) of different Al-Al/Ga-Ga distances ($d_{Al\text{-}Al}/d_{Ga\text{-}Ga}$), where E_{FM} and E_{AFM} are the total energies of two Al (Ga) atoms aligned ferromagnetically (FM) and antiferromagnetically (AFM), respectively. The FM and AFM configurations are always fixed and fully relaxed. As listed in Table II, for the two-Al-adsorbed system, with the increasing $d_{Al\text{-}Al}$, ΔE transforms from +0.353 to +0.002 eV, and the Al-Al interaction is always FM. The similar tendency appears in Ga case, as the $d_{Ga\text{-}Ga}$ increases, ΔE transforms from +0.323 to +0.002 eV and the FM is its favor state.

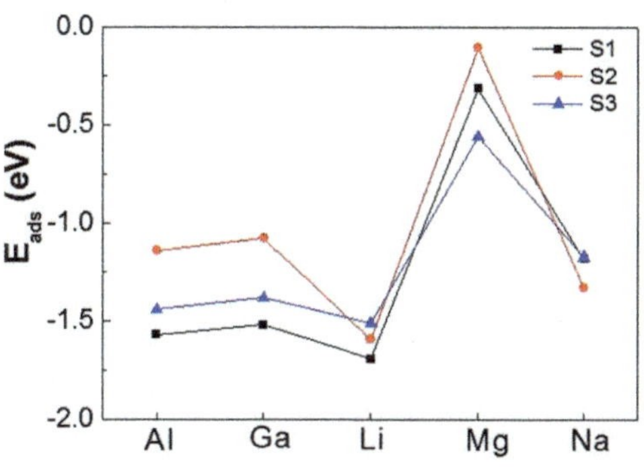

FIG. 2. Adsorption energies of different adsorption sites in the X-adsorbed 4×4×1 arsenene (X= Al, Ga, Li, Mg, and Na).

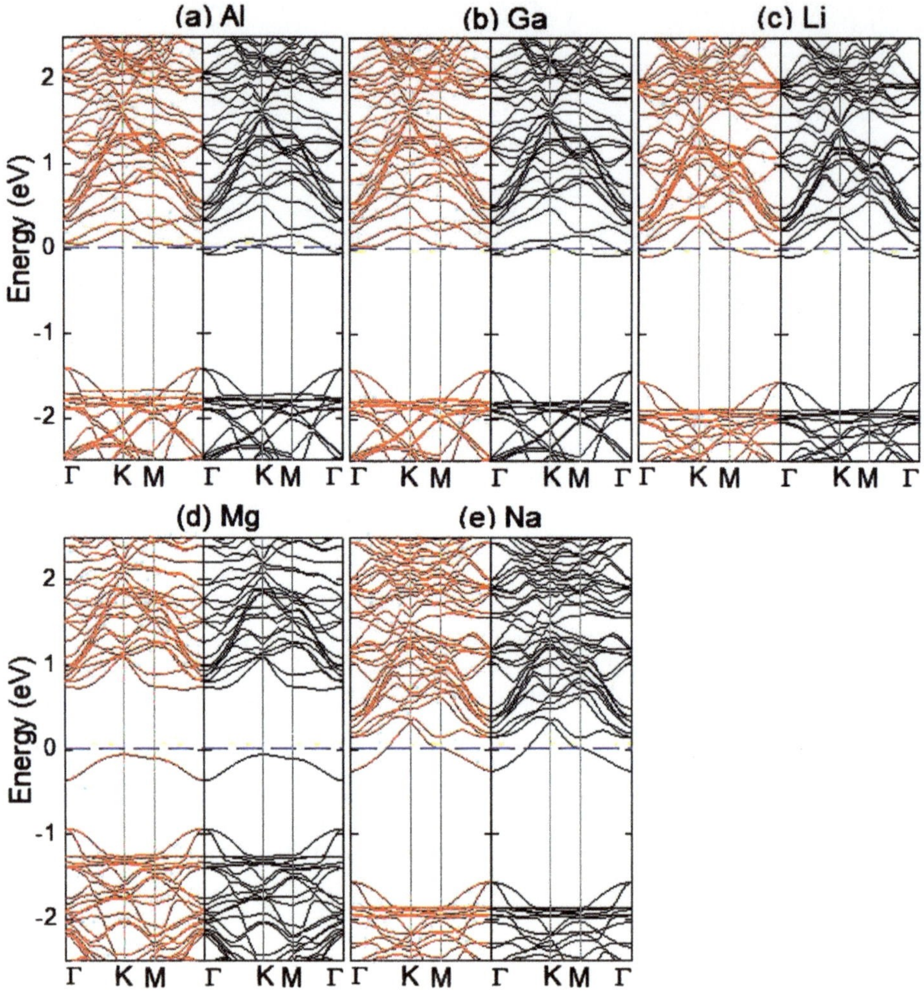

FIG. 3. Band structures of nonmagnetic metal atoms X-adsorbed arsenene (X= Al, Ga, Li, Mg, and Na). The red and black lines represent the spin-up and spin-down components, respectively. The Fermi level is indicated by the dot line.

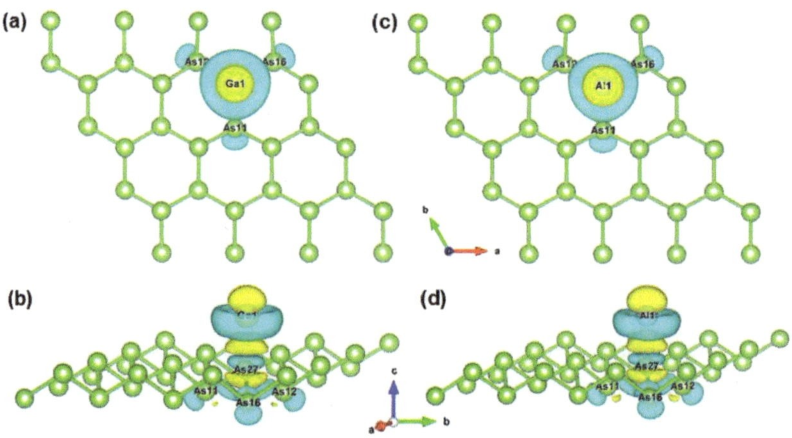

FIG. 4. (a) Top and (b) side view of calculated spin density distribution of one-Ga-adsorbed arsenene; (c) and (d) for one-Al-adsorbed arsenene. Yellow and cyan isosurfaces represent positive and negative spin densities (0.001 e/Å^3), respectively.

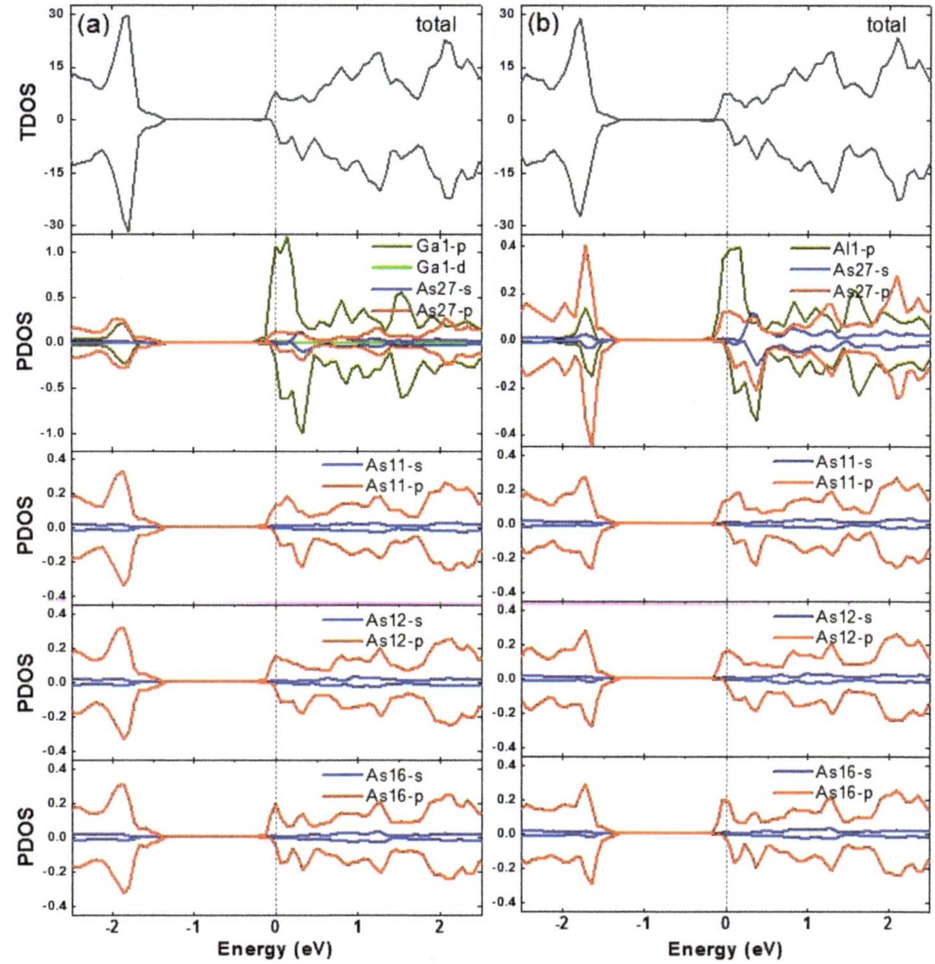

FIG. 5. TDOS and PDOS of (a) Ga and its bonding As atoms in one-Ga-adsorbed system; (b) Al and its bonding As atoms. The vertical dash line indicates the Fermi level.

To understand the FM magnetic coupling in above cases, we study the two-Al-adsorbed system as an example. The spin density distribution and PDOS of case (0,1) are calculated, as shown in Figs, 6 and 7, respectively. Based on the calculated results, the long-range FM coupling between two Al adatoms might be explained as follows. As shown in Fig. 6, the spin polarizations of As atoms neighboring to two Al adatoms are different. The hybridization mainly arises from two Al and their neighboring As18, As47, and As 53. It is well known that FM interaction can be mediated by

TABLE II. Optimized Al-Al (Ga-Ga) distance (*d*), the total magnetic moments (M_{tot}), the total energies (E_{tot}), the energy differences (ΔE) between AFM and FM.

System	Configuration	*d* (Å)	M_{tot} (μ_B)	E_{tot} (eV)	ΔE (eV)	Ground State
$As_{72}Al_2$ (6×6×1)	1	2.773	1.96	-337.582	0.353	FM
	2	7.065	1.48	-335.315	0.015	FM
	3	10.800	1.63	-335.329	0.005	FM
	4	13.670	1.72	-335.344	0.002	FM
$As_{72}Ga_2$ (6×6×1)	1	2.741	1.96	-337.053	0.323	FM
	2	7.170	1.49	-335.226	0.010	FM
	3	10.800	1.60	-335.225	0.004	FM
	4	13.669	1.71	-335.237	0.002	FM

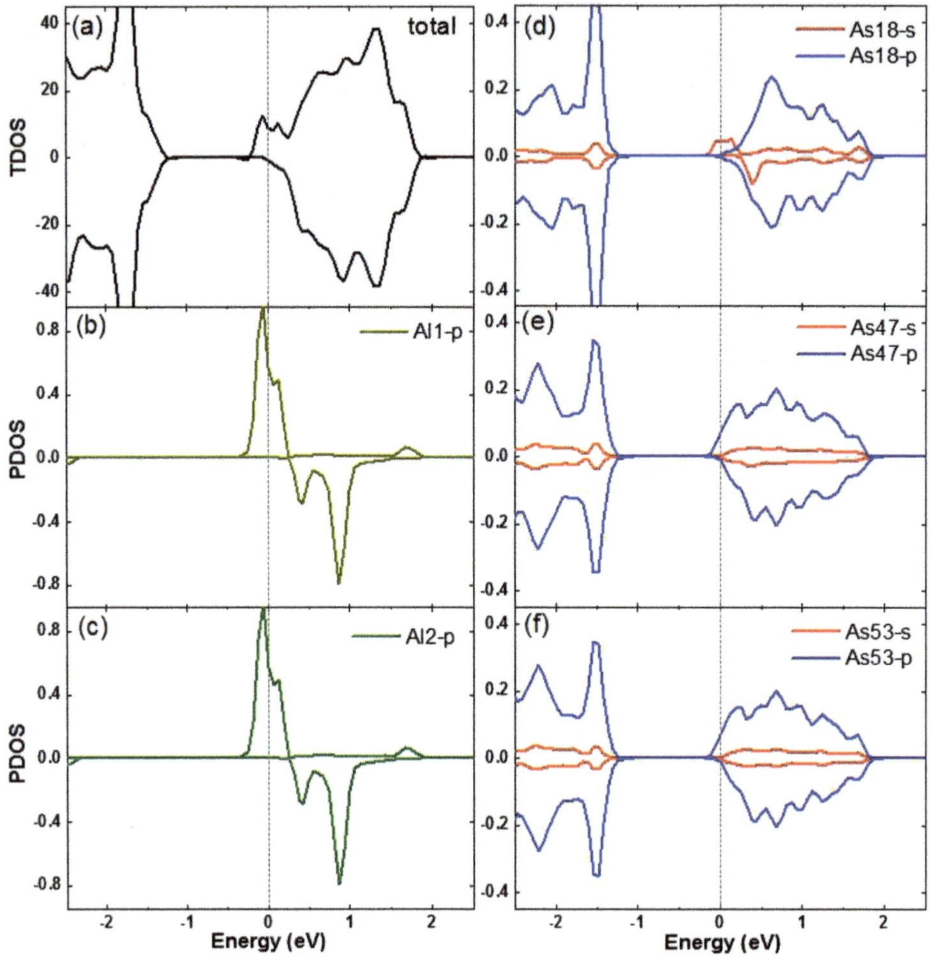

FIG. 6. (a) TDOS of two-Al-adsorbed 6×6×1 arsenene (configuration 1); (b) (c) two Al and (d)-(f) their bonding As atoms. The ground state is FM.

free carriers through the p-d hybridization mechanism.[24–26] These theories have been successfully used. In our cases, we deem that alignment of magnetic moments induced by Al adatoms is achieved via a p-d hybridization-like p-p coupling between Al p states and host As p states. As shown in Fig. 7, a strong p-p interaction between Al (Al1, Al2) and As (As18, As47, As53) atoms shows, which leads to stronger coupling between impurity (Al) and host (As) spin direction. The spin-polarized As atoms can mediate an indirect long-range FM coupling between two Al adatoms effectively. As we discussed above, the nearest As18 atoms between two Al adatoms mediate the magnetic coupling. Therefore, we believe that the polarized As atoms between two Al adatoms are parallel to the spin direction of Al. Accordingly, a long-range FM coupling is expected in current systems.

It is noteworthy that as the Al-Al distance become very large [13,670 Å, case (0,4) in Fig. 1(c)] and the ΔE is extremely small (~2 meV). We could determine that as Al-Al distance becomes more and more large, such FM interaction would be evidently depressed. Similar phenomena have been discussed in TM atoms adsorption on 2D-SiC,[27] therefore, we parameterized the exchange interactions of two Al atoms with a simple Heisenberg model.[28] The exchange interaction coefficient $J(r)$ is parameterized using the functional form

$$J(r) = \begin{cases} \frac{c}{r^3}\exp(-r/r_0) & r \le r_c \\ 0 & otherwise \end{cases} \tag{1}$$

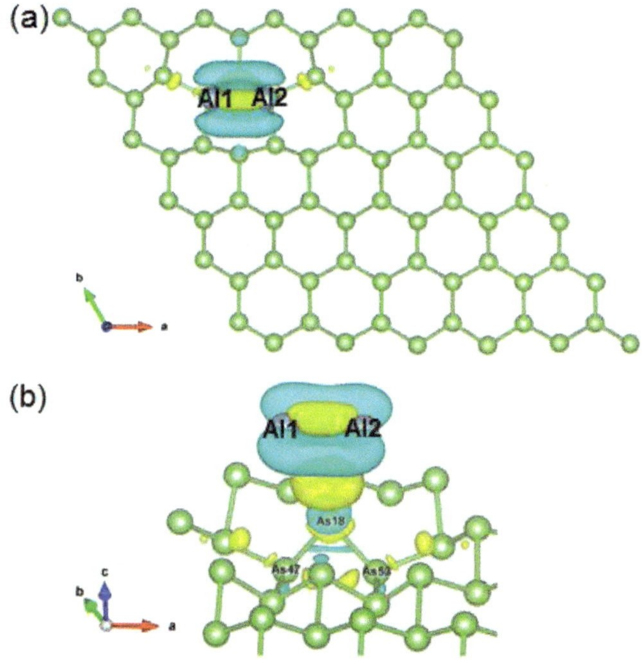

FIG. 7. Spin density distribution of the second nearest Al-Al pair and their bonding three As atoms. Yellow and cyan isosurfaces represent positive and negative spin densities (0.0002 e/Å^3), respectively.

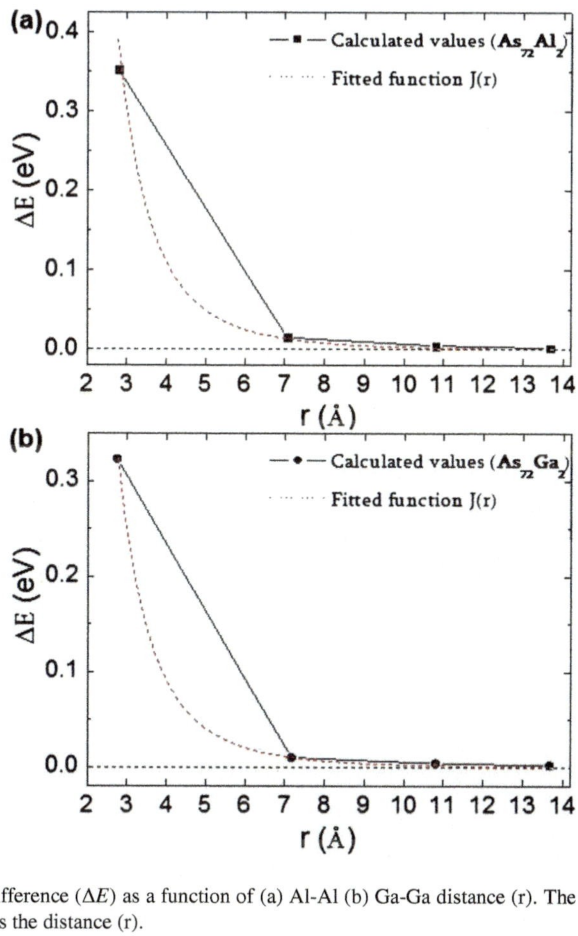

FIG. 8. The total energy difference (ΔE) as a function of (a) Al-Al (b) Ga-Ga distance (r). The dashed line is the Magnetic correlation function [J(r)] vs the distance (r).

where r is equal to $d_{Al\text{-}Al}$, r_0 is the screening length, r_c is the cutoff and lies on the interaction range, and c is a constant of proportionality.[29] For the Al case, $r_c = 14.40$ Å, as the fourth nearest-neighbor shell, $c = 12.05$ eV/Å^3 and $r_0 = 7.55$Å. We simulate $J(r)$ to $d_{Al\text{-}Al}$ (r), as listed in Fig. 8a. Numerical fitting presents the relations between calculated ΔE and $d_{Al\text{-}Al}$ (r), reasonably. Moreover, similar magnetic coupling in the two-Ga-adsorbed system has also been studied and a long-rang FM coupling appears, as shown in Table II. The magnetic correlation function $J(r)$ is estimates as $J(r) = \frac{10.05}{r^3} \exp(-r/7.35)$, as shown in Fig. 8b. From our results, we suggest that while the hybridization is weak, the long range FM interaction between nonmagnetic metal atoms could be explained by the Zener-RKKY theory.[30]

IV. CONCLUSIONS

In summary, the adsorption of the nonmagnetic metal elements X (X=Al, Ga, Li, Mg, and Na) adsorption on arsenene sheets has been investigated by using DFT calculations. Different magnetic properties of arsenene show based on nonmagnetic metal adatoms. Al and Ga adatoms can induce magnetism in arsenene, which mainly results from the 3p/4p electrons of Al/Ga atoms, and 4p states of surrounding As atoms also have significant contributions. Then, the FM coupling of two Al adatoms is studied. Our results verify that such FM states are originated by the *p–d* exchange-like p-p hybridization between Al and its adjacent As. However, such FM coupling was extremely depressed with increasing Al-Al distance, which can be explained by a simple Heisenberg model. Moreover, a similar phenomenon is observed in the two-Ga-adsorbed arsenene.

ACKNOWLEDGMENTS

We thank the Key Laboratory of Polar Materials and Devices, Ministry of Education, East China Normal University (ECNU). Our work is supported by the Supercomputer Center of ECNU.

[1] K. S. Novoselov, A. K. Geim, S. V. Morozov, D. Jiang, Y. Zhang, S. V. Dubonos, I. V. Grigorieva, and A. A. Firsov, Science **306**, 666 (2004).
[2] K. Geim, Science **324**, 1530 (2009).
[3] A. H. C. Neto, F. Guinea, N. M. R. Peres, K. S. Novoselov, and A. K. Geim, Rev. Mod. Phys. **81**, 109 (2009).
[4] K. F. Mak, C. Lee, J. Hone, J. Shan, and T. F. Heinz, Phys. Rev. Lett. **105**, 136805 (2010).
[5] B. Radisavljevic, A. Radenovic, J. Brivio, V. Giacometti, and A. Kis, Nature Nanotech. **6**, 147 (2011).
[6] S. Rodin, A. Carvalho, and A. H. C. Neto, Phys. Rev. Lett. **112**, 176801 (2014).
[7] T. Low, A. S. Rodin, A. Carvalho, Y. Jiang, H. Wang, F. Xia, and A. H. C. Neto, Phys. Rev. B **90**, 075434 (2014).
[8] R. Fei, A. Faghaninia, R. Soklaski, J. A. Yan, C. Lo, and L. Yang, Nano Lett. **14**, 6393 (2014).
[9] A. Ramasubramaniam and A. R. Muniz, Phys. Rev. B **90**, 085424 (2014).
[10] A. Hashmi and J. Hong, J. Phys. Chem. C **119**, 9198 (2015).
[11] T. P. Kaloni, G. Schreckenbach, and M. S. Freund, J. Phys. Chem. C **119**, 3979 (2015).
[12] J. W. Chang, S. Larentis, E. Tutuc, L. F. Register, and S. K. Banerjee, Appl. Phys. Lett. **104**, 141603 (2014).
[13] Y. Ding and Y. Wang, J. Phys. Chem. C **119**, 10610 (2015).
[14] O. V. Yazyev and L. Helm, Phys. Rev. B **75**, 125408 (2007).
[15] L. Shen, R. Q. Wu, H. Pan, G. W. Peng, M. Yang, Z. D. Sha, and Y. P. Feng, Phys. Rev. B **78**, 073306 (2008).
[16] J. J. Attema, G. A. de Wijs, G. R. Blake, and R. A. de Groot, J. Am. Chem. Soc. **127**, 16325 (2005).
[17] C. Kamal and M. Ezawa, Phys. Rev. B **91**, 085423 (2015).
[18] L. Kou, Y. Ma, X. Tan, T. Frauenheim, A. Du, and S. Smith, J. Phys. Chem. C **119**, 6918 (2015).
[19] S. L. Zhang, Z. Yan, Y. F. Li, Z. F. Chen, and H. B. Zeng, Angew. Chem. Int. Ed. **54**, 3112 (2015).
[20] G. Kresse and J. Furthmüller, Phys. Rev. B **54**, 11169 (1996).
[21] J. P. Perdew, K. Burke, and M. Ernzerhof, Phys. Rev. Lett. **77**, 3865 (1996).
[22] G. Kresse and D. Joubert, Phys. Rev. B **59**, 1758 (1999).
[23] P. Wu, M. Huang, W. J. Cheng, and F. L. Tang, Physica E **81**, 7 (2016).
[24] T. Dietl, H. Ohno, F. Matsukura, J. Cibert, and D. Ferrand, Science **287**, 1019 (2000).
[25] L. Liu, P. Y. Yu, Z. Ma, and S. S. Mao, Phys. Rev. Lett. **100**, 127203 (2008).
[26] D. Kitchen, A. Richardella, J. M. Tang, M. E. Flatte, and A. Yazdani, Nature **442**, 436 (2006).
[27] M. Luo, Y. H. Shen, and T. L. Yin, Solid State Commun. **252**, 1 (2017).
[28] A. Ramasubramaniam and D. Naveh, Phys. Rev. B **87**, 195201 (2013).
[29] K. Sato, L. Bergqvist, J. Kudrnovský, P. H. Dederichs, O. Eriksson, I. Turek, B. Sanyal, G. Bouzerar, H. Katayama-Yoshida, V. A. Dinh, T. Fukushima, H. Kizaki, and R. Zeller, Rev. Mod. Phys. **82**, 1633 (2010).
[30] J. Kanamori and K. Terakura, J. Phys. Soc. Jpn. **70**, 1433 (2001).

18

Impurity-induced modulation of terahertz waves in optically excited GaAs

A. S. Kurdyubov,[1,a] A. V. Trifonov,[1] I. Ya. Gerlovin,[1] I. V. Ignatiev,[1] and A. V. Kavokin[1,2,3]
[1]*Spin Optics Laboratory, St. Petersburg State University, 1 Ul'anovskaya, Peterhof, St. Petersburg 198504, Russia*
[2]*CNR-SPIN, Viale del Politecnico 1, I-00133 Rome, Italy*
[3]*School of Physics and Astronomy, University of Southampton, SO17 1NJ Southampton, United Kingdom*

The effect of the photoinduced absorption of terahertz (THz) radiation in a semi-insulating GaAs crystal is studied by pulsed THz transmission spectroscopy. We found that a broad-band modulation of THz radiation may be induced by a low-power optical excitation in the spectral range of the impurity absorption band in GaAs. The measured modulation factor achieves 80%. The amplitude and frequency characteristics of the resulting THz modulator are critically dependent on the carrier density and relaxation dynamics in the conduction band of GaAs. In semi-insulating GaAs crystals, the carrier density created by the impurity excitation is controlled by the rate of their relaxation to the impurity centers. The relaxation rate and, consequently, the frequency characteristics of the modulator can be optimized by an appropriate choice of the impurities and their concentrations. The modulation parameters can be also controlled by the crystal temperature and by the power and photon energy of the optical excitation. These experiments pave the way to the low-power fast optically-controlled THz modulation, imaging, and beam steering.

I. INTRODUCTION

The electromagnetic radiation belonging to the terahertz spectral range is important field of active research promising for a wide range of applications.[1–5] Ability of THz radiation to penetrate most part of non-conductive materials makes it a promising tool for a non-destructive control of organic materials,[6] real-time THz imaging,[7,8] and security applications.[9,10] Very broad band of THz radiation can be used for the ultrafast THz communications and data processing.[5,11,12] Realization of these applications requires, in particular, compact and reliable THz modulators.

Modulators of electromagnetic waves in the optical spectral range are well developed by now. These devices are able to control the amplitude, phase, polarization, and other characteristics of optical waves. Realization of similar modulators for THz radiation is a challenging problem because of the much longer THz wavelength, which would require either the use of bulky macroscopic modulators or the use of new physical principles of modulation in compact solid state devices. One of the promising ways to realize a THz modulator is considered to be a photo-induced modification of the THz characteristics of semiconductor media by an optical excitation.

Several modulation methods based on the optically-induced physical processes in semiconductors are proposed so far.[7,13–18] In particular, pure Si and GaAs crystals have been used for this purpose.[15] The interband optical excitation of the crystals creates free electrons and holes, which efficiently absorb the THz radiation. Due to the fast electron-hole recombination in the

[a]Electronic mail: kurdyubov@yandex.ru

direct-gap semiconductors, e.g., GaAs, such modulators are characterized by the high operation rate (the switching time is of the order of units of ns).[16] At the same time, the fast recombination hampers the accumulation of the carriers and, consequently, the deep modulation of THz absorption. In the indirect-gap semiconductors, e.g., silicon, or in the type II quantum wells,[13] the carrier lifetime is much larger and the required carrier density is easier to accumulate compared to the direct band gap systems.[8,15,19] However, the indirect-gap nature of this material requires relatively strong excitation power (Watts per cm^2) to achieve a noticeable THz absorption.[13,16] Besides, the operation rate is of the order of tens of kHz limited by the carrier lifetime.

In Ref. 20 it was found that, in a semi-insulating (SI) GaAs crystal, a significant concentration of the free carriers in the conduction band can be created by an optical excitation of the electrons bound to acceptor centers. The relaxation rate of the photo-excited electrons strongly depends on the binding energy of electrons at these centers, E_a. At low temperatures, the corresponding relaxation time varies from fractions of microsecond for shallow centers with E_a of order of several tens of meV to fractions of millisecond for much deeper centers.[21] The recombination is accelerated at the temperature rise. So, a variation of the type of centers to be optically excited and the crystal temperature allows one, in principle, to strongly vary the operation rate of a THz modulator.

II. EXPERIMENTAL

In the present work we study the spectral, amplitude, and frequency responses of a THz modulator based on a SI GaAs crystal at different temperatures. We demonstrate the evidence of the giant photomodulation effect in the spectral range corresponding to the impurity bands in SI GaAs crystals. Samples under study are the thin plates (thickness $t = 0.4$ mm) of SI GaAs produced as substrates for epitaxial technologies.

The THz experiments were performed by the use of a standard THz time-domain spectroscopy technique.[2] The block diagram of the setup is shown in Fig. 1. The THz pulses were focused on the end face of the GaAs plate, which length along the THz propagation is of about few mm. This geometry of experiments allowed us to optimize the length of the THz absorbing material. The sample was mounted at the cold metal finger of a closed-cycle helium cryostat. To block the THz radiation

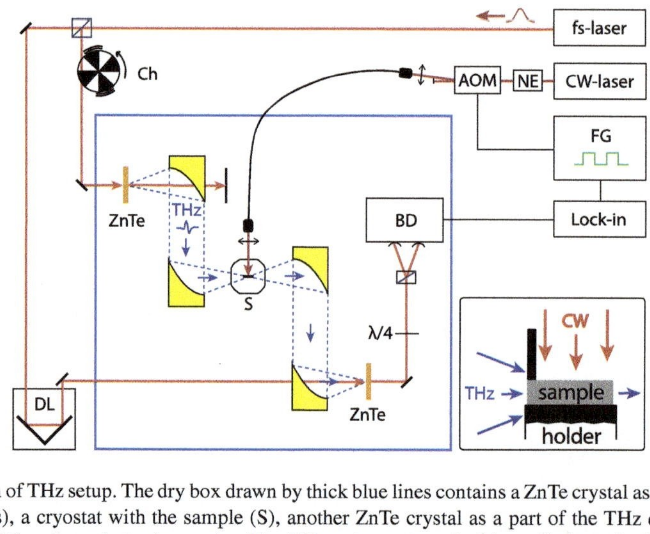

FIG. 1. Block diagram of THz setup. The dry box drawn by thick blue lines contains a ZnTe crystal as a THz source, parabolic mirrors (yellow figures), a cryostat with the sample (S), another ZnTe crystal as a part of the THz detector containing also a balanced detector (BD) and a polarization optics. The THz pulses are excited by radiation of a femtosecond Ti:sapphire laser (fs-laser) modulated by a chopper (Ch). A part of the radiation is directed to the ZnTe crystal included into the THz detection system. Pulses of this laser beam are delayed by a translated stage (DL) to measure the temporal profile of the THz pulses. Radiation of a tunable CW Ti:sapphire laser stabilized in intensity by a noise eater (NE) and modulated by an acousto-optical modulated (AOM) is directed to the sample via a fiber. A functional generator (FG) controls the modulation frequency. A lock-in amplifier (Lock-in) is used to increase sensitivity of the signal detection. Inset shows the geometry of the experiments.

propagating outside the sample, a thin metal plate was attached to the sample perpendicularly to its surface (see inset in Fig. 1). The transmitted THz radiation was detected by a system measuring the temporal profile of electromagnetic field of the THz pulses. A Fourier transformation of the profile gives the spectrum of the transmitted pulses. The THz part of the setup, from the THz source up to the THz detector, was placed into a dry box to reduce the THz absorption by the water vapor.

To induce the THz absorption, the front surface of the plate was illuminated by radiation of a continuous-wave (CW) tunable laser. The geometry of the experiment is shown in Fig. 1. Variation of the laser spot size on the sample allowed us to vary the length of the THz absorbing volume. We studied the photo-induced THz absorption as a function of the photon energy and power of the CW excitation and of the sample temperature. To study the frequency dependence of the absorption, we modulated the laser radiation by an acousto-optical modulator in the frequency range from 5 kHz to 1 MHz.

III. EXPERIMENTAL RESULTS

Fig. 2 shows the spectra of THz pulses transmitted through the sample illuminated by the CW radiation with different excitation densities. Each spectrum is a broad band of about 1.5 THz width centered at 1.7 THz. The spectral dip observed at the central frequency is caused by absorption of THz radiation in the box by the residual water vapor. The CW excitation in the spectral range of the sample transparency (photon energy $E < 1510$ meV) is found to result in a decrease of the transmission of the THz radiation, which is uniform over the whole THz spectrum. The transmission gradually decreases with the increase of the excitation power as it is shown in the insets of Fig. 2. However, at some excitation density, the photo-induced THz absorption is saturated, see low inset in Fig. 2.

The effect of the photo-induced THz absorption can be characterized by a modulation factor (MF) defined as[22,23]

$$MF = 1 - \frac{\int E^2_{\text{CW-on}} d\omega}{\int E^2_{\text{CW-off}} d\omega}, \tag{1}$$

where E is the electric field of the THz wave measured in the presence ($E_{\text{CW-on}}$) and absence ($E_{\text{CW - off}}$) of the CW excitation. As seen in the low inset of Fig. 2, the MF reaches 80% at relatively low excitation density of about 1 W/cm.[2]

At low sample temperature, the dependence of the THz absorption coefficient on the photon energy of the CW radiation (the excitation spectrum of the THz absorption), is characterized by a relatively narrow spectral band centered at 20 meV below the band gap of GaAs [see Fig. 3(a)]. The excitation above this band, in particular into the spectral range of the fundamental absorption of GaAs,

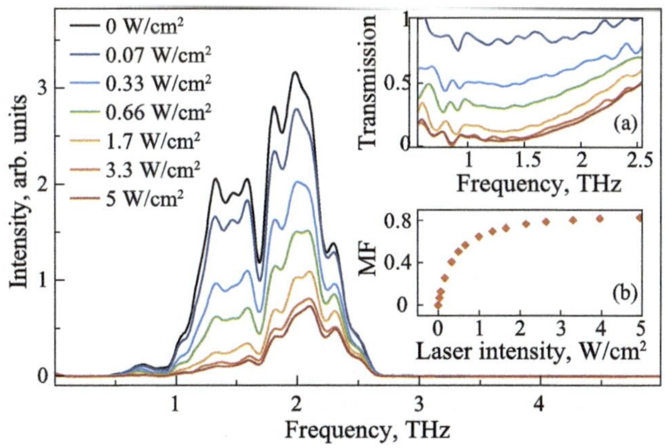

FIG. 2. Fourier spectra of the THz radiation transmitted through the SI GaAs at different excitation densities with photon energy, $E_{\text{CW}} = 1498$ meV. The laser spot diameter along the THz passway, $D = 2$ mm. The upper inset shows the THz transmission spectra normalized to the reference spectrum measured with no optical excitation. The low inset shows the power dependence of the modulation factor calculated using Eq. (1). The sample temperature $T = 5$ K.

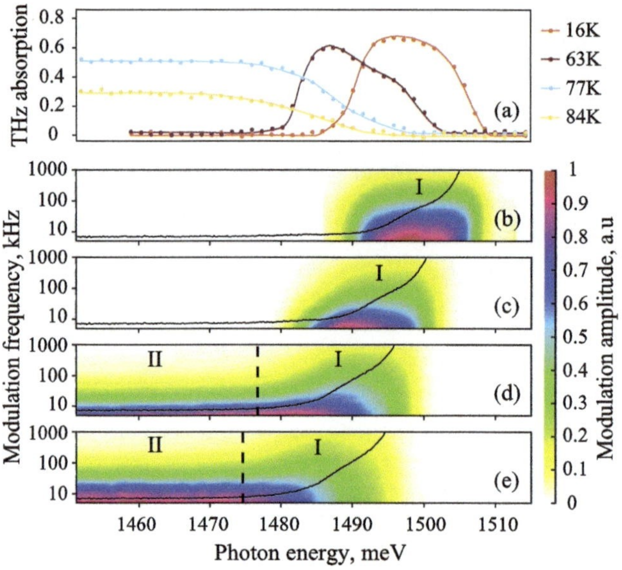

FIG. 3. (a) The photoexcitation spectra of THz absorption (action spectra) in SI GaAs at different sample temperatures. The excitation density is ≈ 2 W/cm^2. Points are the experimental data, solid lines are guides to the eye. (b) - (e) Relative modulation amplitude of the THz radiation as a function of the photon energy and modulation frequency of optical excitation at temperatures T = 16 K (b), T = 63 K (c), T = 77 K (d), T = 84 K (e). The vertical dashed lines separate action bands I and II. The color bar shows the scale of the THz modulation amplitude normalized at its maximal value for each temperature. The optical excitation density ≈ 3 W/cm^2. The black solid curves in panels (b) - (d) show the spectral dependences of optical density of the sample under study at respective temperatures.

does not affect the transmitted THz radiation. This unusual, at the first glance, result indicates that such optical excitation does not create any noticeable carrier density in spite of the large absorption of the optical radiation.

An increase of the sample temperature is followed by a shift of the excitation spectrum of the THz absorption (referred hereafter to as the "action band I") to the lower photon energies. This shift is similar to that of the GaAs band gap. The magnitude of the photo-induced THz absorption in the maximum of the excitation band decreases with the temperature rise. At $T > 80$ K, this band virtually disappears. The energy position of the action band I and its temperature variation points out that it is related to the photo-ionization of the negatively charged acceptor centers.[24,25] The 20-meV energy shift of this band relative to the fundamental absorption edge corresponds to the acceptors created by the carbon impurities.[26] These are the most common impurities related to peculiarities of the GaAs growth technology.

At the larger sample temperatures, $T > 77$ K, another mechanism of the photo-induced absorption appears. Its efficiency is virtually constant in the spectral range of 1450 – 1480 meV. The further rise of the sample temperature is followed by a gradual decrease of the THz absorption. We suggest that the action band II observed at these temperatures is related to the deeper acceptors, whose energy is spread out over a wide range.[24] For some samples of GaAs studied by us, the action band II of the photo-induced THz absorption has been observed even at the low sample temperatures of about 4 K. We assume that this effect is caused by the much larger density of the deep acceptors in these samples. A discussion of these differences in the spectral behavior of different samples of GaAs grown by different technologies is out of scope of the present paper.

For the study of temporal characteristics of the photo-induced THz absorption, we have measured the dependence of the modulation amplitude of the THz radiation on the amplitude modulation frequency of the CW excitation. The modulation amplitude is the difference between the THz signals when the CW excitation is switched on and off normalized to the same difference measured at low modulation frequency. A general behavior of the modulation effect is illustrated in Fig. 3(b–e). One can see that the increase of the modulation frequency gives rise to a decrease of the THz modulation amplitude. At the relatively low temperatures, $T < 20$ K, when only the action band I is observed, the THz modulation amplitude becomes twice lower at the frequency of about 100 kHz. At $T \geq 77$ K where the action band II is observed, similar suppression of the THz modulation occurs at the considerably smaller frequencies [see Fig. 3(d,e)].

To quantitatively describe the frequency dependences of the THz modulation, we have simulated the processes of the excitation and relaxation of the free carriers induced by the amplitude modulation of the CW laser beam forming a meander with period T. It is assumed that the relaxation of free carriers responsible for the THz absorption is a single-exponential decay with characteristic time τ. When the optical excitation is switched on, the carried density increases as $F(t) = F_0[1 - \exp(-t/\tau)]$ where F_0 is the carrier density at the steady state excitation. At the periodical excitation with period T, the density increases from $(F_0 - a)/2$ to $(F_0 + a)/2$ and back during the first and second half periods, respectively. Here a is the amplitude of the alternative component of the density, $a = F_0[1 - \exp(-T/(2\tau))]/[1 + \exp(-T/(2\tau))]$. The temporal dependence of the carrier density is described by:

$$F(t) = F_0 \times \begin{cases} 1 - \dfrac{e^{-t/\tau}}{1 + e^{-T/(2\tau)}}, & \text{for } 0 < t < T/2, \\ \dfrac{e^{-(t-T/2)/\tau}}{1 + e^{-T/(2\tau)}}, & \text{for } T/2 < t < T. \end{cases} \tag{2}$$

We assume that the THz absorption amplitude is proportional to the carrier density in the conduction band.

The THz signal is measured by a system including a lock-in amplifier so that the detected signal is:

$$X(\nu) = \frac{1}{T} \int \tilde{F}(t) \cos(2\pi \nu t) dt, \tag{3}$$

$$Y(\nu) = \frac{1}{T} \int \tilde{F}(t) \sin(2\pi \nu t) dt,$$

$$R(\nu) = \sqrt{X(\nu)^2 + Y(\nu)^2},$$

where $\nu = 1/T$ is the modulation frequency, $X(\nu)$ and $Y(\nu)$ are the signals detected in the two channels with the $\pi/2$ phase shift; $R(\nu)$ is the amplitude of the signal. Quantity $\tilde{F}(t)$ is the alternative component of the carrier density. The integration is fulfilled over the modulation period T.

The frequency dependencies of the THz modulation obtained at temperatures $T > 80$ K for the action band II are well reproduced by Eqs. (2,3) as it is shown in Fig. 4. The obtained agreement with the experiment allows us to estimate the corresponding relaxation rate, $1/\tau$. We found

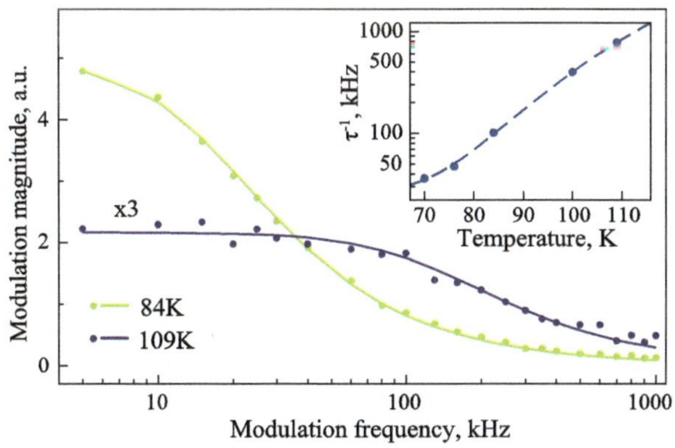

FIG. 4. Amplitude-frequency characteristics, $R(\nu)$, of the THz modulator measured at two temperatures under the optical excitation into the action band II ($E_{exc} = 1450$ meV). Points are the experimental data, solid lines are the results of modelling by Eqs. (2, 3). Inset shows the temperature dependence of the relaxation rate (points) obtained from the fits of the amplitude-frequency characteristics. Dashed line is the fit by Eq. (4). Note the logarithmic scale for τ^{-1}.

that this rate rapidly rises with the sample temperature, see inset in Fig. 4. The physical origin of this strong acceleration of the relaxation can be in the thermally-activated population of the deep centers with electrons coming from the valence band. As a result, the electrons optically excited to the conduction band may recombine with the holes appearing in the valence band. This process is much faster than the electron recombination with the holes at the deep centers. The thermally-induced increase of the relaxation rate is well described by an activation function:

$$\tau^{-1} = \tau_0^{-1} + \frac{\tau_1^{-1}}{\exp(\Delta E/kT) - 1}, \quad (4)$$

where τ_0^{-1}, τ_1^{-1}, and ΔE are the fitting parameters. As seen in the inset of Fig. 4, this function describes well the experimental data with activation energy $\Delta E = 74$ meV corresponding to the energy of two LO phonons in GaAs.[27]

The similar study of the action band I observed at the lower temperatures ($T < 80$ K) shows that its amplitude-frequency characteristics cannot be simulated by Eqs. (2, 3). This means that the electron relaxation to the shallow centers cannot be described by a single exponential law. A cutoff frequency corresponding to the twofold decrease of the THz absorption is of about 100 kHz and weakly depends on the sample temperature.

IV. DISCUSSION

The obtained results demonstrate that the THz absorption in the SI GaAs crystals can be initiated by a relatively weak optical excitation in the spectral range of the impurity absorption bands. This excitation is more efficient from point of view of the THz absorption than the excitation within the spectral range of the fundamental GaAs absorption in spite of the much lower optical density of the impurity bands. This unusual, at the first glance, effect is due to large lifetime of the electrons excited from the impurity centers that is beneficial for electron accumulation in the conduction band. In the case of the interband excitation, fast electron-hole recombination (of the order of ns) prevents the electron accumulation and, consequently, a noticeable THz absorption.

The weak THz absorption in the case of the interband excitation would be, in principle, related to the very thin near-surface layer of the order of 1 μm where electrons in the conduction band are created. The THz radiation propagating along the crystal plate may not "feel" this layer. To verify the origin of the weak THz absorption, we have changed the geometry of the experiment and directed the THz radiation along the optical excitation beam that is perpendicular to the GaAs plate. In this case, the THz waves pass through the optically excited layer and the photocreated electrons may absorb the THz radiation. This experiment has shown that the THz absorption is weak, as before, in the case of the interband excitation and it is much stronger in the case of the optical excitation into the impurity band. So, this experiment confirms that the main reason for the low THz absorption in the case of the interband optical excitation is the fast electron-hole recombination.

The lifetime of the electrons created by the photo-ionization of impurity centers is determined by the reverse relaxation of electrons back to the impurities. The rate of this relaxation depends on the density of the impurity centers and on their binding energy. The relatively slow relaxation of electrons to the impurity centers at low temperatures limits the THz modulation rate to about 100 kHz in the samples studied.

At the elevated crystal temperatures, the deep impurity centers can be populated with electrons due to the thermally-activated ionization of the shallow centers. This process explains the appearance of the action band II and the disappearance of the action band I. In turn, the shallow centers can be repopulated with electrons that leads to the generation of holes in the valence band. The fast electron-hole recombination strongly accelerates the electron relaxation that allows one to control the cutoff frequency of the THz modulation in a wide range, as it is shown in the inset of Fig. 4. In the same time, the fast recombination reduces the electron density and, consequently, the THz absorption at the fixed optical excitation power.

V. CONCLUSION

The experimental study of the main characteristics of the impurity-related photoinduced THz absorption in SI GaAs shows that there are real prospects for creating the broad-band THz modulators on the basis of this effect. The cutoff modulation frequency achievable for such modulators is of the order of several tens of kHz at the liquid nitrogen temperature and it rapidly increases up to 100 kHz and more with the crystal temperature rise. The required optical excitation power is a fraction of Watt and the wavelength (of about 850 nm) falls within the operating range of cheap, commercially available, semiconductor lasers. It is essential that the excitation in the spectral range of the impurity band allows one to reach nearly the 100% MF of the THz radiation by the use of GaAs crystal plates with thickness of several mm. The growth of the GaAs crystals required for the modulators is the well developed area of industry because SI GaAs crystal plates are widely used as substrates for the epitaxial growth of heterostructures.

The effect of the impurity-induced THz absorption at the optical excitation allows one to vary the modulation parameters in the wide range. In particular, the frequency characteristics of the modulators, which is limited by the relaxation rate of electrons to the impurity centers, can be optimized by the appropriate choice of the type and density of the impurities. The reduction of the relaxation time broadens the modulation frequency band but it reduces, other parameters being equal, the density of photocreated carriers controlling the THz MF. To improve the MF without degrading the frequency response, one can use the stronger optical excitation and/or the larger crystal thickness along the THz propagation path. Additional methods of control of the modulation parameters are the variation of the crystal temperature and of the spectrum of the excitation light.

An important feature of such modulators is the opportunity of employing them as the dynamic two-dimensional amplitude filters, which are required for the parallel data transmission systems operating in the THz spectral range[18] as well as for the THz beam steering[8] by means of the dynamic holograms, Fresnel optics etc.

ACKNOWLEDGMENTS

The authors thank the Resource Center "Nanophotonics" and "Interdisciplinary Resource Centre for Nanotechnology" St. Petersburg State University for the technical support of the experiments. This work was supported by the Russian Foundation for Basic Research, project No. 15-59-30406 RT. AVK acknowledges the support by the EPSRC in the framework of the Programme grant on Hybrid Polaritonics.

[1] M. Tonouchi, Nature Photon. **1**, 97 (2007).
[2] X.-C. Zhang and J. Xu, *Introduction to THz Wave Photonics* (Springer Science and Business Media, LLC, 2010).
[3] G. Chattopadhyay, IEEE Trans. THz Sci. Techn. **1**, 33 (2011).
[4] M. A. Belkin and F. Capasso, Phys. Scr. **90**, 118002 (2015).
[5] T. Nagatsuma, G. Ducournau, and C. C. Renaud, Nat. Phot. **10**, 371 (2016).
[6] F. Rutz, M. Koch, S. Khare, M. Moneke, H. Richter, and U. Ewert, Int. Journal Infrared and Millimeter Waves **27**, 547 (2006).
[7] S. Busch, B. Scherger, M. Scheller, and M. Koch, Opt. Lett. **37**, 1391 (2012).
[8] M. I. B. Shams, Z. Jiang, S. M. Rahman, L.-J. Cheng, J. L. Hesler, P. Fay, and L. Liu, IEEE Transact. THz Sci. Techn. **7**, 310 (2017).
[9] P. U. Jepsen, D. G. Cooke, and M. Koch, Laser Photonics Rev. **5**, 124 (2011).
[10] K. B. Cooper and G. Chattopadhyay, IEEE Microwave Mag. **15**, 51 (2014).
[11] C. Jastrow, K. Munter, R. Piesiewicz, T. Kurner, M. Koch, and T. Kleine-Ostmann, Electron. Lett. **44**, 213 (2008).
[12] H.-J. Song and T. Nagatsuma, IEEE Trans. THz Sci. Techn. **1**, 256 (2011).
[13] I. H. Libon, S. Baumgärtner, M. Hempel, N. E. Hecker, J. Feldmann, M. Koch, and P. Dawson, Appl. Phys. Lett. **76**, 2821 (2000).
[14] T. Okada and K. Tanaka, Scientific Reports **1**, 121 (2011).
[15] L.-J. Cheng and L. Liu, Optics Express **21**, 028657 (2013).
[16] M. Rahm, J.-S. Li, and W. J. Padilla, J. Infrared Milli Terahertz Waves **34**, 1 (2013).
[17] T. P. Steinbusch, H. K. Tyagi, M. C. Schaafsma, G. Georgiou, and J. Gomez Rivas, Opt. Expr. **22**, 26559 (2014).
[18] M. I. B. Shams *et al.*, Electron. Lett. **50**, 801 (2014).
[19] T. Nozokido, H. Minamide, and K. Mizuno, RIKEN Review **11**, 11 (1995).
[20] A. S. Kurdyubov, A. V. Trifonov, I. Ya. Gerlovin, I. V. Ignatiev, and A. V. Kavokin, Phys. Solid St. **59**, 1298 (2017) [Fiz. Tv. Tela **59**, 1274 (2017)].

[21] W. P. Dumke, Phys. Rev. **132**, 1998 (1963).
[22] G. Wang, B. Zhang, H. Ji, X. Liu, T. He, L. Lv, Y. Hou, and J. Shen, Appl. Phys. Lett. **110**, 023301 (2017).
[23] T. Matsui, R. Takagi, K. Takano, and M. Hangyo, Opt. Lett. **38**, 4632 (2013).
[24] S. M. Sze and J. C. Irvin, Solid State Electron. **11**, 599 (1968).
[25] P. A. Schultz and O. A. von Lilienfeld, Modelling Simul. Mater. Sci. Eng. **17**, 084007 (2009).
[26] R. Heilman and G. Oelgart, Semicond. Sci. Technol. **5**, 1040–1045 (1990).
[27] D. Strauch and B. Dorner, J. Phys.: Condens. Matter. **2**, 1457 (1990).

19

Effect of in-plane strain anisotropy on (011) epitaxial $BaTiO_3$ and $PbTiO_3$ thin films

Wenhui Ma[a] and Fei Wang
Department of Physics, Shantou University, Shantou, Guangdong 515063, People's Republic of China

Misfit strain phase diagrams of single-domain (011) epitaxial $BaTiO_3$ and $PbTiO_3$ thin films are computed using a thermodynamic phenomenological approach. In the case of (001) epitaxial films, the two orthogonal in-plane misfit strains are equivalent. For (011) epitaxial films, however, the two substrate constraints exert quantitatively different influences due to crystal anisotropy. Effects of preferential strain relaxation and arbitrary in-plane strains are investigated, and similarity and contrast of $BaTiO_3$ and $PbTiO_3$ phase diagrams are revealed. Impact of anisotropy of in-plane constraint on polarization, dielectric and piezoelectric properties is explored, and the calculated structural transitions and out-of-plane lattice strain are compared with relevant theoretical and experimental data reported in the literature.

I. INTRODUCTION

Both Experimental and theoretical investigations have clearly demonstrated that epitaxial strain is capable of strongly manipulating phase transition and properties of ferroelectric thin films.[1–10] It is noted that most of the published research are about <001>-oriented epitaxial films. In the recent years, however, epitaxial films of other crystallographic orientations such as <011> and <111> have started to draw increasing attention.[11–22] Both thermodynamic phenomenological theory and first-principles calculations have been used to make relevant calculations.[23–31] Due to crystal anisotropy, parent phases for <001>- and <011>-oriented epitaxial perovskites have different symmetry even under the same biaxial in-plane constraint arisen from equilateral lattice misfit between the films and substrates. When cooling down from elevated growth temperatures, <001>- and <011>-oriented epitaxial thin films may undergo vastly different phase transition sequences.

As for <011>-oriented films, the existing calculations are essentially centered on equal biaxial constraints. In the real situations, however, epitaxial films can be subject to uniaxial or other anisotropic biaxial in-plane constraints.[11–14,17–19,22] <011>-oriented epitaxial films subject to various specially-manipulated substrate constraints may permit new sequence and order of phase transitions, and occurrence of new phases forbidden in bulk crystals and <001>-oriented epitaxial films, thus providing new opportunities for strain engineering structure and physical properties of technologically important perovskite ferroelectrics. In this work, misfit strain phase diagrams are computed and the effect of in-plane strain anisotropy on phase stability and properties of <011>-oriented epitaxial $BaTiO_3$ and $PbTiO_3$ thin films is investigated using a thermodynamic phenomenological approach.

In the literature, orientation of the epitaxial films is usually simply stated as (110), and the coordinate systems used are sometimes not explicitly described, making proper interpretation of

[a]Corresponding author. E-mail: whma@stu.edu.cn

the results elusive. To facilitate comparison with relevant theory and experiment, in this work the coordinate system and phase notations are elaborated first before presenting our calculations on polarization states, lattice structure, and properties.

This paper is organized as follows. Sec. II introduces coordinate systems and phase notations, and thermodynamic phenomenological theory of single-domain <011> epitaxial films. Sec. III presents and compares phase diagrams of epitaxial $BaTiO_3$ and $PbTiO_3$, and reveal the effect of strain anisotropy. Sec. IV presents properties of the single-domain epitaxial films. Evolution of polarization and out-of-plane lattice strain for the epitaxial films on (110) $SrTiO_3$ is discussed in Sec. V. Finally, a brief summary will be given in Sec. VI.

II. METHOD

A. Coordinate systems and phase notations

In the case of <001>-oriented epitaxial ABO_3 perovskite films, a conventional pseudo-cubic coordinate system is commonly used, in which $^0x_1$, $^0x_2$, $^0x_3$ correspond to $(100)_{pc}$, $(010)_{pc}$ and $(001)_{pc}$, respectively. For <011>-oriented epitaxial films, two kinds of coordinate systems may be used: one is the conventional pseudo-cubic coordinate system, the other is a new coordinate system formed by rotation about any of three pseudo-cubic <001> axes. Even in the simplest case of equal biaxial in-plane substrate constraint, parent phase for <011>-oriented films is of orthorhombic *mmm* symmetry, in contrast to that for $(001)_{pc}$-oriented ones which is of tetragonal 4/*mmm*. The orthorhombic lattice cell has a volume approximately two times that of the pseudocubic or tetragonal one.

In the present study for <011>-oriented epitaxial films, we adopt a new system created by rotating the conventional coordinate system by 45° about $[100]_{pc}$. In the new orthorhombic reference frame x_1, x_2, x_3 correspond to coordinates along $(100)_{pc}$, $(01\text{-}1)_{pc}$ and $(011)_{pc}$, which can also be referred to as $(100)_{ortho}$, $(010)_{ortho}$ and $(001)_{ortho}$, respectively, in the orthorhombic system. In the new coordinate system, the 3 or *c* direction (i.e., surface normal of the epitaxial films) is parallel to $(011)_{pc}$. As a comparison, in the first-principles calculations Gui et al.[30] chose to rotate the reference frame about $(001)_{pc}$, so the coordinate system used by them is an alternative pseudocubic one, where x_1', x_2', x_3' correspond to $(001)_{pc'}$, $(1\text{-}10)_{pc'}$ and $(110)_{pc'}$, respectively.

Choice of coordinate system is arbitrary and should not affect the general conclusions about phase stability, despite difference in the notations for various phases. In fact, what really matters are symmetry and structural evolution of the polarization states for the respective phases. In accordance with conventionally-accepted usage, we use the term $(011)_{pc}$-oriented epitaxial film throughout the paper. As for phase notations, however, for the sake of clarity in symmetry analysis we adopt the above-discussed new system based on the orthorhombic lattice cell, similar to that used by Mtebwa et al.[28] Conversion of phase notations in various coordinate systems is given in Table I.

TABLE I. Conversion of various phase notations for $(011)_{pc}$ epitaxial films.

Orthorhombic coordinate system (x_1, x_2, x_3)	Pseudocubic coordinate system ($^0x_1, {}^0x_2, {}^0x_3$)	Alternative pseudocubic coordinate system ($^0x_1', {}^0x_2', {}^0x_3'$)	Crystal system and point group
This work and Ref. 28	Refs. 9, 26, and 27	Ref. 30	
a_1: $P_1 \neq 0, P_2=P_3=0$	$O(a_1)$: $^0P_1 \neq 0, {}^0P_2={}^0P_3=0$	c: $^0P_{1'}={}^0P_{2'}=0, {}^0P_{3'} \neq 0$	Orthorhombic mm2
a_2: $P_2 \neq 0, P_1=P_3=0$	N/A	N/A	Orthorhombic mm2
c: $P_1=P_2=0, P_3 \neq 0$	$O(a_2c)$: $^0P_1=0, {}^0P_2={}^0P_3 \neq 0$	aa: $^0P_{1'}={}^0P_{2'} \neq 0, {}^0P_{3'}=0$	Orthorhombic mm2
a_1a_2: $P_1 \neq P_2 \neq 0, P_3=0$	$M_A(r)$: $^0P_1>{}^0P_2={}^0P_3 \neq 0$ $M_B(r)$: $^0P_2={}^0P_3>{}^0P_1 \neq 0$	r_c: $^0P_{3'}>{}^0P_{1'}={}^0P_{2'} \neq 0$	Monoclinic m
a_1c: $P_2=0, P_1 \neq P_3 \neq 0$	$M_C(a_1c)$: $^0P_2=0, {}^0P_1 \neq {}^0P_3 \neq 0$	r_{aa}: $^0P_{1'}={}^0P_{2'}>{}^0P_{3'} \neq 0$	Monoclinic m
a_2c: $P_1=0, P_2 \neq P_3 \neq 0$	$M_C(a_2c)$: $^0P_1=0, {}^0P_2 \neq {}^0P_3 \neq 0$	ab: $^0P_{1'} \neq {}^0P_{2'} \neq 0, {}^0P_{3'}=0$	Monoclinic m
r^*: $P_1 \neq P_2 \neq P_3 \neq 0$	$Tr(a_1a_2c)$: $^0P_1 \neq {}^0P_2 \neq {}^0P_3 \neq 0$	abc: $^0P_{1'} \neq {}^0P_{2'} \neq {}^0P_{3'} \neq 0$	Triclinic 1

B. Thermodynamic free energy function for strained single-domain $(011)_{pc}$ epitaxial films

In the pseudo-cubic coordinate system (${}^0x_1$, ${}^0x_2$, ${}^0x_3$), to determine phase stability of a stressed state the elastic Gibbs function of the form $G({}^0P, {}^0\sigma_{ij}) = G_2 + G_4 + G_6 + G_8 + G_{el} + G_{em}$ is usually used, where ${}^0P$ and ${}^0\sigma_{ij}$ refer to polarization and stress components in the pseudo-cubic system, G_2, G_4, G_6, G_8 the second-, fourth-, sixth- and eighth-order polarization terms, G_{el} and G_{em} the elastic and electromechanical coupling terms, respectively. To derive free energy function suitable for calculating phase stability of $(011)_{pc}$ epitaxial films, we use an approach similar to that used by Tagantsev et al.,[23] i.e., polarization and stress (${}^0P$ and ${}^0\sigma_{ij}$) in the old system are changed to those (P and σ_{ij}) in the new system (x_1, x_2, x_3) via coordinate transformation $x_i = A_{ij} \cdot {}^0x_j$, in which A_{ij} is the corresponding transformation matrix.[23,26–28] In the new orthorhombic coordinate system, the free energy can thus be written as $G(P, \sigma_{ij}) = G_2^{\#} + G_4^{\#} + G_6^{\#} + G_8^{\#} + G_{el}^{\#} + G_{em}^{\#}$, where $G_2^{\#}$, $G_4^{\#}$, $G_6^{\#}$ and $G_8^{\#}$ are explicit functions of P_j, $G_{el}^{\#}$ a pure elastic term and an explicit function of σ_{kl}, and $G_{em}^{\#}$ an electromechanical term and an explicit function of σ_{kl} and P_j. To facilitate calculations in strained $(011)_{pc}$ epitaxial films, a Legendre transformation[5] is performed: $F(P_i, u_k) = G(P_i, \sigma_k) + \sigma_1 u_1 + \sigma_2 u_2$, where σ_k and u_k are reduced notations for the in-plane stress and strain, respectively. u_1 and u_2 are the in-plane misfit between lattices of the substrate and paraelectric free-standing film along $(100)_{pc}$ and $(01\text{-}1)_{pc}$, respectively. Besides equal biaxial constraints ($u_1 = u_2 = u_m$), there can be uniaxial constraints ($u_1 \neq 0$ or $u_2 \neq 0$), and anisotropic or unequal biaxial constraints (e.g., $u_1 = -u_2$, or more arbitrarily, $|u_1| \neq |u_2|$).

After the Legendre transformation, the free energy function now has the following form,

$$\begin{aligned} F = {} & \alpha_1^* P_1^2 + \alpha_2^* P_2^2 + \alpha_3^* P_3^2 + \alpha_{11}^* P_1^4 + \alpha_{22}^* P_2^4 + \alpha_{33}^* P_3^4 \\ & + \alpha_{12}^* P_1^2 P_2^2 + \alpha_{13}^* P_1^2 P_3^2 + \alpha_{23}^* P_2^2 P_3^2 + G_6^{\#} + G_8^{\#} + F_{el} \end{aligned}, \tag{1}$$

where α_i^* and α_{ij}^* are the renormalized dielectric stiffness coefficients. $\alpha_i^* = \alpha_i + f_i(u_1, u_2, Q_{ij}, s_{ij})$, in which α_i is the dielectric stiffness, and Q_{ij} and s_{ij} are electrostrictive coefficients and elastic compliances at constant polarization, respectively. F_{el} is a polarization-independent pure elastic term whose value does not affect the exact phase determination. $G_6^{\#}$ and $G_8^{\#}$ remain unchanged after the Legendre transformation, their expressions are shown below,

$$\begin{aligned} G_6^{\#} = {} & \alpha_{111} P_1^6 + \alpha_{222}^{\#} P_2^6 + \alpha_{222}^{\#} P_3^6 + \alpha_{112} P_1^4 (P_2^2 + P_3^2) + P_2^4 (\alpha_{122}^{\#} P_1^2 + \alpha_{223}^{\#} P_3^2) \\ & + P_3^4 (\alpha_{122}^{\#} P_1^2 + \alpha_{223}^{\#} P_2^2) + \alpha_{123}^{\#} P_1^2 P_2^2 P_3^2 \end{aligned}, \tag{2}$$

$$\begin{aligned} G_8^{\#} = {} & \alpha_{1111} P_1^8 + \alpha_{2222}^{\#} P_2^8 + \alpha_{2222}^{\#} P_3^8 + \alpha_{1112} P_1^6 (P_2^2 + P_3^2) + P_2^6 (\alpha_{1222}^{\#} P_1^2 + \alpha_{2223}^{\#} P_3^2) \\ & + P_3^6 (\alpha_{1222}^{\#} P_1^2 + \alpha_{2223}^{\#} P_2^2) + \alpha_{1122}^{\#} P_1^4 P_2^4 + \alpha_{1122}^{\#} P_1^4 P_3^4 + \alpha_{2233}^{\#} P_2^4 P_3^4 \\ & + \alpha_{1123}^{\#} P_1^4 P_2^2 P_3^2 + \alpha_{1223}^{\#} P_2^4 P_3^2 P_1^2 + \alpha_{1223}^{\#} P_3^4 P_1^2 P_2^2 \end{aligned}, \tag{3}$$

where α_{ijk}, $\alpha_{ijk}^{\#}$, α_{ijkl}, and $\alpha_{ijkl}^{\#}$ are high-order dielectric stiffness coefficients at constant stress: $\alpha_{222}^{\#} = (\alpha_{111} + \alpha_{112})/4$, $\alpha_{122}^{\#} = (2\alpha_{112} + \alpha_{123})/4$, $\alpha_{223}^{\#} = (15\alpha_{111} - \alpha_{112})/4$, $\alpha_{123}^{\#} = (6\alpha_{112} - \alpha_{123})/2$, $\alpha_{2222}^{\#} = (2\alpha_{1111} + 2\alpha_{1112} + \alpha_{1122})/16$, $\alpha_{1222}^{\#} = (\alpha_{1112} + \alpha_{1123})/4$, $\alpha_{2223}^{\#} = (14\alpha_{1111} + 2\alpha_{1112} - \alpha_{1122})/4$, $\alpha_{1122}^{\#} = (2\alpha_{1122} + \alpha_{1123})/4$, $\alpha_{2233}^{\#} = (70\alpha_{1111} - 10\alpha_{1112} + 3\alpha_{1122})/8$, $\alpha_{1123}^{\#} = (6\alpha_{1122} - \alpha_{1123})/2$, $\alpha_{2213}^{\#} = (15\alpha_{1112} - \alpha_{1123})/4$.

C. Calculations of structure and properties for single-domain $(011)_{pc}$ epitaxial films

Using the free energy function shown in Eq. (1), we can deal with $(011)_{pc}$ epitaxial films subject to arbitrary in-plane constraints. Table I lists possible polarization states and various notations for possible ferroelectric phases that may occur on cooling from the paraelectric (p) phase of orthorhombic mmm symmetry. Since the previous calculations on $(011)_{pc}$ epitaxial films mainly concern equal biaxial in-plane constraint, a_2 phase ($P_2 \neq 0$, $P_1 = P_3 = 0$) has seldom been reported so far.[24] At present we show that this symmetry-allowed ferroelectric orthorhombic state with only one in-plane polarization component along $(01\text{-}1)_{pc}$ direction may occur under various anisotropic strain conditions.

TABLE II. Free energy coefficients for $BaTiO_3$ and $PbTiO_3$ (T in °C).

Constant	$BaTiO_3$[32]	$PbTiO_3$[35,36]
α_1 ($10^5C^{-2}m^2N$)	4.124 × (T-115)	3.765 × (T-478.8)
α_{11} ($10^8C^{-4}m^6N$)	-2.097	-7.252
α_{12} ($10^8C^{-4}m^6N$)	7.974	7.5
α_{111} ($10^9C^{-6}m^{10}N$)	1.294	2.606
α_{112} ($10^9C^{-6}m^{10}N$)	-1.950	6.1
α_{123} ($10^9C^{-6}m^{10}N$)	-2.5	-3.66
α_{1111} ($10^{10}C^{-8}m^{14}N$)	3.863	
α_{1112} ($10^{10}C^{-8}m^{14}N$)	2.529	
α_{1122} ($10^{10}C^{-8}m^{14}N$)	1.637	
α_{1123} ($10^{10}C^{-8}m^{14}N$)	1.367	

The free energy coefficients used for the present studies are listed in Table II, and the materials constants (Q_{ij} and s_{ij}) used are listed in Table III. Some details on the present choice of free energy coefficients and materials constants will be introduced in Section III. In Table IV we show numerical values of $\Delta\alpha_i^*$ [i.e., $f_i(u_1,u_2,Q_{ij},s_{ij})$] for $BaTiO_3$ and $PbTiO_3$. Clearly, even for equal biaxial constraints, $\alpha_1^* \neq \alpha_2^* \neq \alpha_3^*$, indicating that substrate clamping lowers the symmetry of the $(011)_{pc}$ epitaxial films to orthorhombic. In Table IV effect of strain anisotropy on α_i^* coefficients can be seen as well. The 4th-order coefficients α_{ij}^* do not vary with in-plane strain anisotropy. For simplicity, we do not give their explicit expressions, but instead present the numerical values, as shown in Table V.

In this work, all the physical quantities are defined based on the new coordinate system. Dielectric tensor components η_{ij} can be calculated from dielectric stiffness (χ_{ij}) matrices: $\eta_{ij} = 1/\chi_{ij}$, in which $\chi_{ij} = \partial^2F/\partial P_i\partial P_j$. Components of lattice strain S_i in the epitaxial films can be deduced from $S_i = -\partial F/\partial\sigma_i$. Longitudinal ($d_{33}$) and shear piezoelectric constants (d_{35} and d_{35}) can be obtained from E-field dependence of longitudinal (or out-of-plane) strain component (S_3) and shear strain components (S_4 and S_5) of the epitaxial films, respectively. In the present studies, we separate the out-of-plane lattice strain into two parts: $S_3 = S_3^0 + \Delta S_3$, where $S_3^0 = f(u_1,u_2)$ depends only on the misfit strain, and $\Delta S_3 = f(P_1,P_2,P_3)$ is the polarization-associated part.

TABLE III. Materials constants (Q_{ij} and s_{ij}) used in the calculations for $(011)_{pc}$ epitaxial $BaTiO_3$ and $PbTiO_3$ (SI units, T in °C).

Constant	$BaTiO_3$[33,34]	$PbTiO_3$[5,35,36]
Q_{11} (m^2/C)	0.10	0.089
Q_{12} (m^2/C)	-0.034	-0.026
Q_{44} (m^2/C)	0.029	0.0675
s_{11} ($10^{-12}m^2/N$)	8.33	8.0
s_{12} ($10^{-12}m^2/N$)	-2.68	-2.5
s_{44} ($10^{-12}m^2/N$)	9.24	9.0

TABLE IV. List of coefficients $\Delta\alpha_i^*$ used in the calculations for $(011)_{pc}$ epitaxial $BaTiO_3$ and $PbTiO_3$ (Both equal biaxial and uniaxial constraint situations are included). $\Delta\alpha_i^* = \alpha_i^* - \alpha_i$. For uniaxial u_2, $\Delta\alpha_i^*$ equals the number in the table multiplied by 10^9u_2; for the other two situations, $\Delta\alpha_i^*$ equals the number in the table multiplied by 10^9u_1.

Material	Biaxial $u_1=u_2$ ($\times10^9u_m$)			Uniaxial $u_1\neq0$ ($\times10^9u_1$)			Uniaxial $u_2\neq0$ ($\times10^9u_2$)		
	$\Delta\alpha_1^*$	$\Delta\alpha_2^*$	$\Delta\alpha_3^*$	$\Delta\alpha_1^*$	$\Delta\alpha_2^*$	$\Delta\alpha_3^*$	$\Delta\alpha_1^*$	$\Delta\alpha_2^*$	$\Delta\alpha_3^*$
$BaTiO_3$	-11.440	-4.985	-0.500	-11.867	1.875	2.966	0.428	-6.860	-3.466
$PbTiO_3$	-11.689	-9.272	1.228	-11.259	0.269	2.769	-0.430	-9.541	-1.541

TABLE V. List of coefficients $\Delta\alpha_{ij}^*$ ($\times10^8$) used in the calculations for $(011)_{pc}$ epitaxial $BaTiO_3$ and $PbTiO_3$. $\Delta\alpha_{11}^*=\alpha_{11}^*-\alpha_{11}$, $\Delta\alpha_{12}^*=\alpha_{12}^*-\alpha_{12}$, $\Delta\alpha_{13}^*=\alpha_{13}^*-\alpha_{13}$, $\Delta\alpha_{22}^*=\alpha_{22}^*-(2\alpha_{11}+\alpha_{12})/4$, $\Delta\alpha_{33}^*=\alpha_{33}^*-(2\alpha_{11}+\alpha_{12})/4$, $\Delta\alpha_{23}^*=\alpha_{23}^*-(6\alpha_{11}-\alpha_{12})/2$, where α_{ij} and α_{ij}^* are the 4th-order dielectric stiffness coefficients at constant stress and constant strain, respectively.

Material	$\Delta\alpha_{11}^*$	$\Delta\alpha_{12}^*$	$\Delta\alpha_{13}^*$	$\Delta\alpha_{22}^*$	$\Delta\alpha_{23}^*$	$\Delta\alpha_{33}^*$
$BaTiO_3$	6.006	-3.752	-4.145	1.699	2.404	0.951
$PbTiO_3$	4.955	-0.188	-2.865	2.343	1.465	0.473

III. PHASE DIAGRAMS OF $(011)_{PC}$ SINGLE-DOMAIN EPITAXIAL FILMS

In this section we present and compare misfit vs. temperature phase diagrams for $(011)_{pc}$ $BaTiO_3$ and $PbTiO_3$ under four typical kinds of epitaxial strain conditions, namely, equal biaxial $u_1=u_2=u_m$, uniaxial $u_1\neq0$, and uniaxial $u_2\neq0$, and equal but opposite biaxial in-plane strain ($u_1=-u_2\neq0$). We also present and discuss the room-temperature misfit-misfit phase diagrams.

For the present studies of $BaTiO_3$ we choose the 8th-order free energy function[32] which, combined with the carefully chosen electrostrictive coefficients[33] and elastic constants,[34] has been shown to be able to reproduce well the electric-field induced phase transitions, and dielectric and piezoelectric properties reported in literature. This 8th-order free energy function has also been shown to work well for $(001)_{pc}$ epitaxial films at high strain level.[3] Variation of free energy coefficients and materials constants certainly can modify several features of the misfit strain phase diagrams, and those results will be published in a separate paper. The calculated phase diagrams for $(011)_{pc}$ epitaxial $BaTiO_3$ under four different substrate constraints, and the room-temperature polarization components as a function of in-plane strain are presented in Fig. 1 and Fig. 2, respectively.

Regarding the T-u_m phase diagram for $BaTiO_3$ *c* phase is much stabilized on the compressive side. Under biaxial constraints, there exists a nearly flat *p*-*c* phase boundary as shown in Fig. 1(a), with the Curie transition point T_c decreasing slightly with increasing compressive strain. Such result is rather rare for epitaxial films, in particular for biaxial compression. Wu et al.[27] reported a T-u_m biaxial misfit phase diagram for $(011)_{pc}$ epitaxial $BaTiO_3$ using a similar thermodynamic approach, general features of the respective phase regions are basically similar to our present studies,

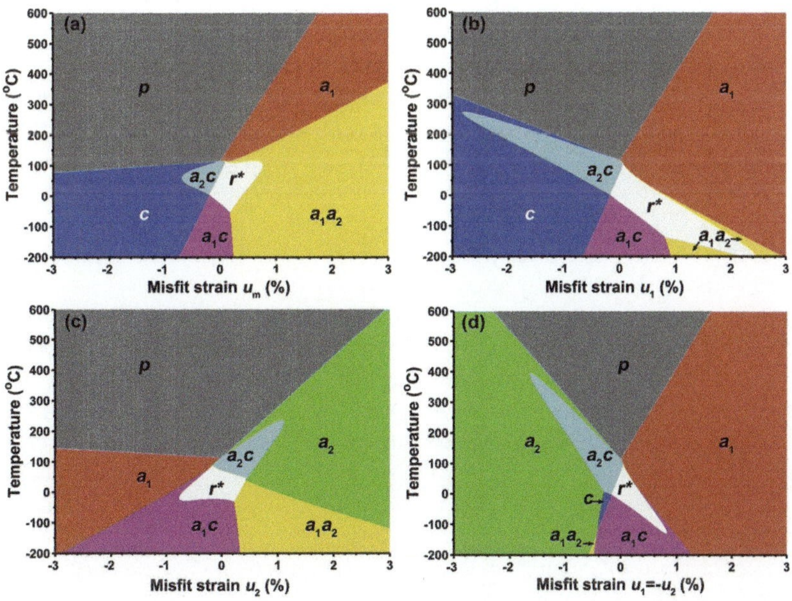

FIG. 1. Temperature vs. misfit strain phase diagrams for $(011)_{pc}$ epitaxial $BaTiO_3$ under four different in-plane constraints: (a) biaxial, $u_1=u_1=u_m$, (b) uniaxial, u_1, (c) uniaxial, u_2, (d) equal but opposite, $u_1=-u_2$.

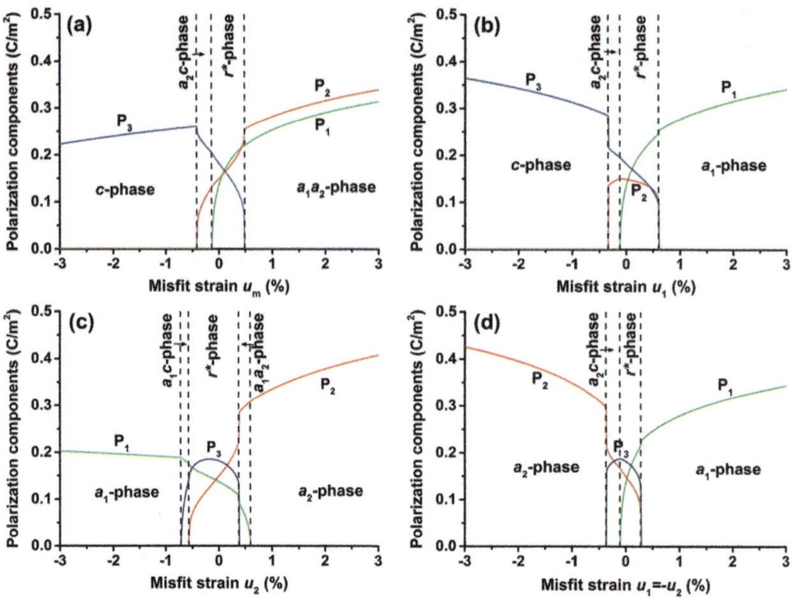

FIG. 2. Polarization components vs. in-plane strain for $(011)_{pc}$ epitaxial $BaTiO_3$ at 25 °C: (a) biaxial u_m, (b) uniaxial u_1, (c) uniaxial u_2, (d) $u_1=-u_2$.

albeit with some differences in the detailed features, for which the most striking one lies in their *p-c* phase boundary line that points to higher temperature with increasing compressive strain. Wu et al.[27] did not reveal the exact free energy coefficients and materials constants used, but according to our unpublished studies, in their calculations the 8th-order potential proposed by Li et al.[32] and the materials constants provided by Pertsev et al.[5] were most probably used. In $(011)_{pc}$ epitaxial $BaTiO_3$ subject to equal biaxial compression, Gui et al.[30] obtained using first-principles calculations a temperature-independent Curie transition line (*p-c* phase boundary) on the compressive side of the T-u_m phase diagram, which is in qualitative agreement with our present calculations.

For $BaTiO_3$, equal biaxial compression lowers *p-c* transition temperature [see Fig. 1(a)], as $\Delta\alpha_3^*$ is a relatively small negative value (see Table IV). As a comparison, uniaxial compression along $(01\text{-}1)_{pc}$ (i.e., u_2) can just slightly enhance *p*-a_1 transition temperature [see Fig. 1(c)], as $\Delta\alpha_1^*$ is a relatively small positive value (see Table IV). Provided that $\Delta\alpha_i^*$ is negligibly small, the Curie transition point would then become essentially independent of epitaxial strain. In strong contrast to the above two cases, uniaxial compression along $(100)_{pc}$ (i.e., u_1) considerably broadens a_2c phase region in $BaTiO_3$, as shown in Fig. 1(b). Even for a magnitude of u_1 as large as ~ -2.8%, our calculations indicate that, below T_c the film transforms first to *c* phase and then to a_2c phase at a lower temperature., T_c is ~274 °C at a uniaxial strain of u_1= -2.2%, in well agreement with the experimental data for the $(011)_{pc}$ epitaxial $BaTiO_3$ on (110) $SrTiO_3$,[17] which is preferentially relaxed along x_2 and therefore only uniaxially strained.

In the present studies of $(011)_{pc}$ epitaxial $PbTiO_3$ we adopt the widely-accepted 6th-order free energy coefficients and electrostrictive coefficients proposed by Haun et al.[35] The phase diagrams calculated for four typical kinds of in-plane constraints are shown Fig. 3. Fig. 4 plots the room-temperature polarization components vs. in-plane strain. In contrast to the case of $BaTiO_3$, for $PbTiO_3$, a_1c phase is absent under biaxial constraints. Furthermore, a_2c phase region is much broader in the T-u_m diagram, which is probably due to very high stability of tetragonal state in the stress-free state of $PbTiO_3$. The biaxial T-u_m phase diagram of Fig. 3(a) is very similar to that obtained by Mtebwa et al.[28] using the same thermodynamic potential and the same material constants except Q_{44}, for which Mtebwa et al. used the value of 0.096, rather than that of 0.0675 reported by Haun et al.[36] (see Table II and Fig. 3(d) of Ref. 28).

As for uniaxial constraints, on the compressive side of T-u_1 phase diagrams, in $BaTiO_3$ *c* phase is much stabilized, but in $PbTiO_3$ it is a_2c phase that is much stabilized. However, on the tensile side of

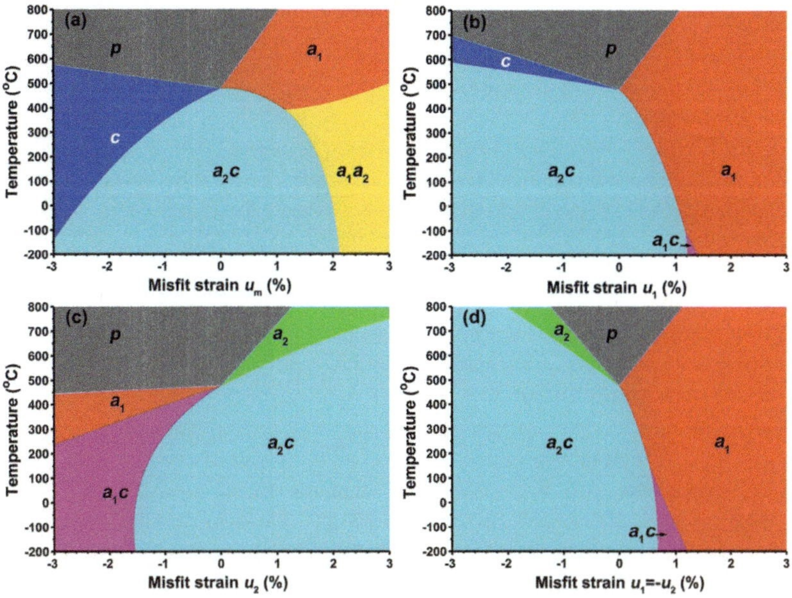

FIG. 3. Temperature vs. misfit strain phase diagrams for $(011)_{pc}$ epitaxial $PbTiO_3$ under four different in-plane constraints: (a) biaxial u_m, (b) uniaxial u_1, (c) uniaxial u_2, (d) u_1=-u_2.

T-u_1 phase diagrams, a_1 phase is stabilized for both $BaTiO_3$ and $PbTiO_3$. On the compressive side of T-u_2 phase diagrams, for $BaTiO_3$ the p-a_1 phase transition shifts to slightly higher temperature with increasing strain, whereas for $PbTiO_3$ it shifts to slightly lower temperature; in both cases overall the p-a_1 phase boundary line is nearly flat, showing that the Curie transition is nearly independent of the compressive strain along $(010)_{ortho}$ or $(01\text{-}1)_{pc}$ direction. Similar to the case of $BaTiO_3$ subject to biaxial u_m, slight downshift of p-a_1 transition temperature in $PbTiO_3$ [see Fig. 3(c)] by uniaxial compression u_2 is related to a relatively small negative $\Delta\alpha_1^*$ value (see Table IV).

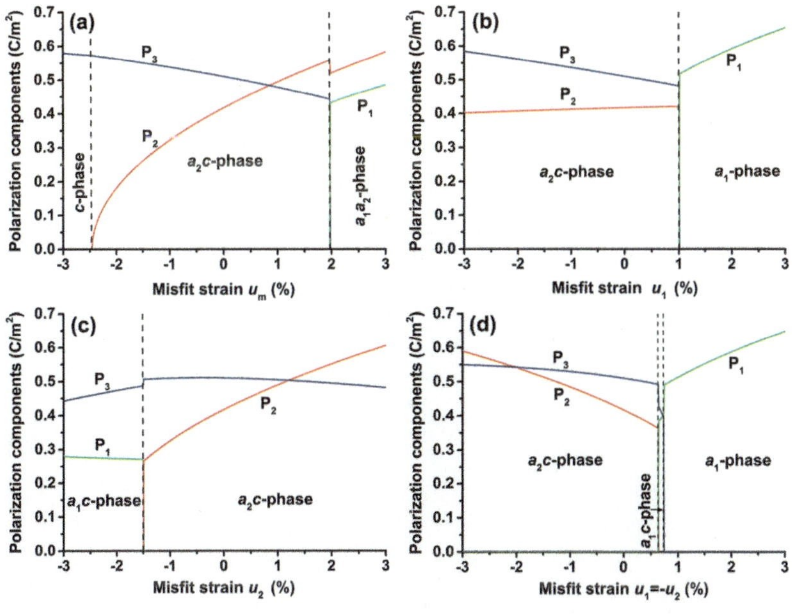

FIG. 4. Polarization components vs. in-plane strain for $(011)_{pc}$ epitaxial $PbTiO_3$ at 25 °C: (a) biaxial u_m, (b) uniaxial u_1, (c) uniaxial u_2, (d) u_1=-u_2.

The above results presented for $BaTiO_3$ and $PbTiO_3$ reveal that for $(011)_{pc}$ epitaxial films the two orthogonal in-plane strains u_1 and u_2 are not equivalent. Regarding these two conditions, the former substrate constraint is along pseudocubic $(100)_{pc}$, the later along $(01\text{-}1)_{pc}$. Definitely, the T-u_2 phase diagrams are much different from the T-u_1 diagrams. Such situation is unlike that for $(001)_{pc}$ epitaxial films, in which two orthogonal in-plane constraints along $(100)_{pc}$ and $(010)_{pc}$ are equivalent and the corresponding uniaxial misfit strain phase diagrams (T-u_1 and T-u_2) are exactly the same. Origin for such situation can be attributed to the unique strain dependence of α_i^* for $(011)_{pc}$ epitaxial films. As can be seen in Table IV, the two orthogonal in-plane strains u_1 and u_2 lead to differing values for $\Delta\alpha_i^*$ coefficients.

It is seen that a_2 phase cannot be stabilized by equal biaxial constraints for both $BaTiO_3$ and $PbTiO_3$. For $PbTiO_3$ subject to uniaxial compressive strain [see Fig. 3(b)], a_2c phase region replaces most of the c-phase region shown to be stable in the biaxial T-u_m diagram [Fig. 3(a)]. On the other hand, for uniaxial tensile strain u_2 [see Fig. 3(c)], a_2 is stabilized at high temperature, while a_2c is stabilized at low temperatures. The a_1c phase occurs on the compressive side of T-u_2 diagram, which transforms to a_1 phase at high temperatures.

In $(001)_{pc}$ epitaxial films, the two orthogonal in-plane strains u_1 and u_2 are equivalent, but in $(011)_{pc}$ epitaxial films, they are remarkably different. This point is further shown by the u_1=-u_2 phase diagrams [Fig. 1(d) and Fig. 3(d)]. On the tensile side of T vs. u_1=-u_2 phase diagrams, a_1 phase is much stabilized for both $BaTiO_3$ and $PbTiO_3$. On the compressive side, however, for $BaTiO_3$ a_2 phase is much stabilized, while for $PbTiO_3$ it is a_2c phase that is much stabilized. In contrast, for $(001)_{pc}$ epitaxial films the T vs. u_1=-u_2 phase diagrams must be symmetrical.[9]

Uniaxial misfit strain and equal but opposite misfit strain u_1=-u_2 are two special anisotropic in-plane strains. In general, for arbitrary biaxial constraints, both u_1 and u_2 can be non-zero and different, i.e., $u_1 \neq u_2 \neq 0$. Fig. 5 and Fig. 6 display room-temperature misfit strain vs. misfit strain (u_1-u_2) phase diagrams for $(011)_{pc}$ epitaxial $BaTiO_3$ and $PbTiO_3$, respectively. For $BaTiO_3$ a considerably sized region of paraelectric phase occurs at 25 °C but a_2c phase region is very small. In contrast, for $PbTiO_3$ no paraelectric phase can be stabilized at 25 °C, and c and a_1a_2 phase regions are severely shrunk, whereas a_2c phase region is greatly broadened. Notably for $BaTiO_3$, at room temperature triclinic r^* phase and a_2 phase can be readily induced by anisotropic epitaxial strain, but it is not the case for $PbTiO_3$.

In comparison to the case of $(011)_{pc}$ epitaxial $BaTiO_3$ for which the triclinic phase region can be very large, for $PbTiO_3$ it is found that triclinic phase is absent in all the four typical strain situations examined in the present calculations. Mtebwa et al.[28] found a narrow triclinic phase region at the a_2c-a_1a_2 phase boundary of the biaxial T-u_m diagram for $(011)_{pc}$ $PbTiO_3$ using some different sets of material constants. In fact, according to our calculations triclinic phase does not occur even in the

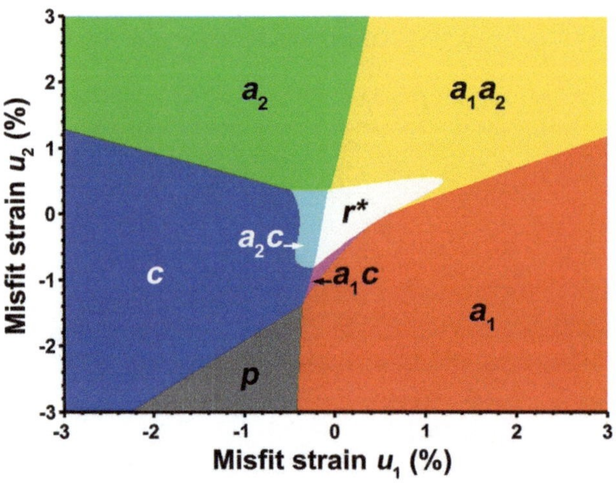

FIG. 5. Misfit strain vs. misfit strain (u_1-u_2) phase diagram for $(011)_{pc}$ epitaxial $BaTiO_3$ at 25 °C.

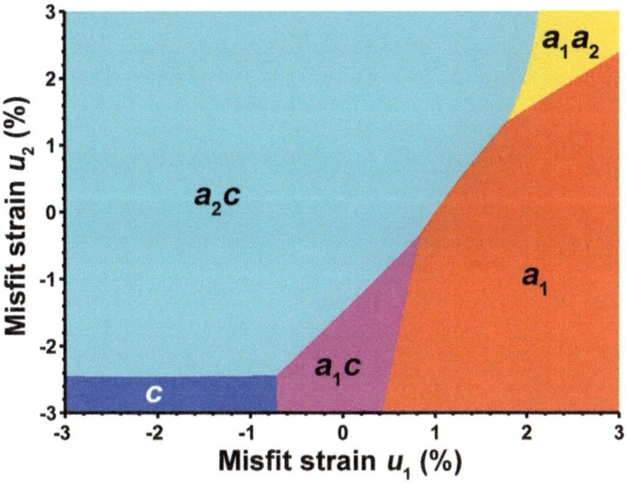

FIG. 6. Misfit strain vs. misfit strain (u_1-u_2) phase diagram for $(011)_{pc}$ epitaxial $PbTiO_3$ at 25 °C.

room-temperature u_1-u_2 phase diagram (see Fig. 6). It is worthwhile mentioning that, in the case of $(001)_{pc}$ epitaxial $PbTiO_3$, Ma et al.[9] found triclinic phase under various anisotropic in-plane strain conditions.

IV. PROPERTIES OF $(011)_{PC}$ SINGLE-DOMAIN EPITAXIAL FILMS

In this section we present dielectric and piezoelectric properties for $(011)_{pc}$ $BaTiO_3$ and $PbTiO_3$ under four typical kinds of epitaxial strain conditions, namely, equal biaxial u_1=u_2=u_m, uniaxial $u_1 \neq 0$, uniaxial $u_2 \neq 0$, and equal but opposite in-plane strain u_1=-u_2. The strain dependence of dielectric and piezoelectric properties corresponds well with the strain dependence of phase transitional behaviors presented in the preceding section.

Fig. 7 shows dielectric constant η_{33} vs. in-plane strain for $(011)_{pc}$ epitaxial $BaTiO_3$ at 25 °C. Some observations of the dielectric behaviors under different epitaxial constraints are listed as follows: 1) c-phase always has lower η_{33}; 2) under zero misfit strain, the order of magnitude of η_{33} lies between 1000 and 2000 at room temperature; 3) near the phase boundaries of c-a_2c, r^*-a_1a_2, r^*-a_1, and a_1-a_1c, η_{33} tends to reach very high values. Overall, high dielectric η_{33} is usually expected when epitaxial strain induces polarization rotation towards in-plane directions.

Room-temperature longitudinal (d_{33}) and shear piezoelectric constant (d_{34} and d_{35}) as a function of in-plane strain for $BaTiO_3$ are shown in Fig. 8 and Fig. 9, respectively. As shown by the evolution of polarization in Fig. 2(a), for equal biaxial constraints, both a_2c-r^* and r^*-a_1a_2 transitions are clearly of second-order, while the c-a_2c transition exhibits drastic polarization change, similar to a weak first-order transition near a critical point. Accordingly, in the dielectric spectra of Fig. 7(a), η_{33} has a sharp peak around u_m=0.005, and in the piezoelectric spectra of Figs. 8(a) and 9(a) there are a d_{33} peak around u_m=-0.004, and high positive d_{35} and ultrahigh d_{34} at the r^*-a_1a_2 transition, but high negative d_{34} at the c-a_2c transition.

In the case of uniaxial u_1, the c-a_2c transition is of first-order, the a_2c-r^* second-order, and the r^*-a_1 very weak first-order. Correspondingly, in Fig. 7(b), η_{33} has only a jump at u_1=-0.33%, but exhibits a sharp peak around u_1=0.6%. Similar to the case of biaxial u_m, for a_2c-r^* transition there is no dielectric anomaly. At the c-a_2c transition there is a small negative jump of d_{34}, while at the r^*-a_1 transition a positive peak d_{34} and a fairly big jump of d_{35} occur.

In the case of uniaxial u_2, notably, at the a_1-a_1c transition a very broad η_{33} peak occurs, along with two lower sharp peaks at a_1c-r^* and r^*-a_1a_2 transitions, respectively. d_{34} is very high at the r^*-a_1a_2 transition, while d_{33} and d_{35} are ultrahigh at the a_1-a_1c transition.

For equal but opposite in-plane strain (u_1=-u_2), ultrahigh and relatively broad η_{33} peak occurs at r^*-a_1 transition, and the very high d_{35} peak is remarkable, but at this transition the d_{33} peak is

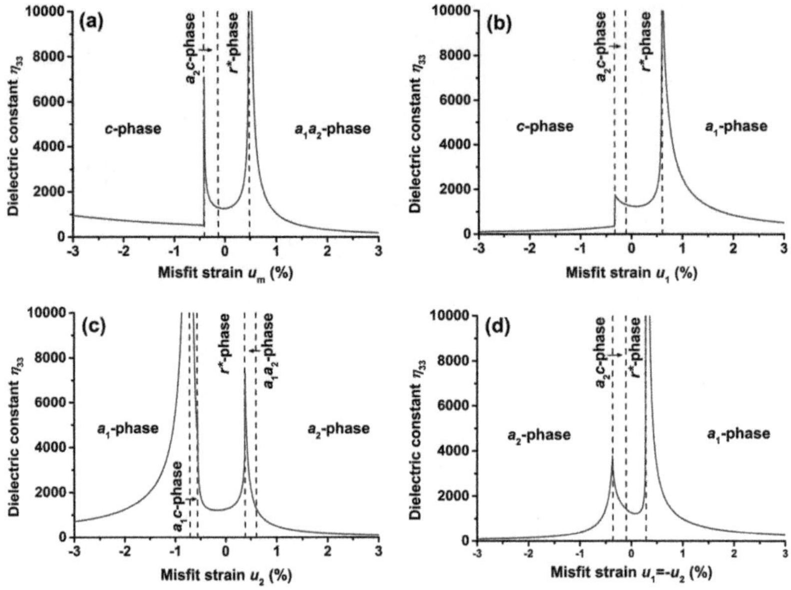

FIG. 7. Dielectric constant η_{33} vs. in-plane strain for $(011)_{pc}$ epitaxial $BaTiO_3$ at 25 °C: (a) biaxial u_m, (b) uniaxial u_1, (c) uniaxial u_2, (d) u_1=-u_2.

not so high as compared to that at the a_1-a_1c transition in the case of uniaxial u_2. It is noted that for both uniaxial u_2 and anisotropic u_1=-u_2, positive d_{34} peaks are found at the compressive side of the diagrams, in contrast to the cases of biaxial u_m and uniaxial u_1.

Now we turn to $PbTiO_3$. Fig. 10 shows dielectric constant η_{33} vs. in-plane strain for $(011)_{pc}$ epitaxial $PbTiO_3$ at 25 °C. As shown by the polarization spectra in Fig. 4(a), for biaxial u_m, the c-a_2c transition is clearly of second-order, but the a_2c-a_1a_2 one is of first-order. In the dielectric spectra of Fig. 10(a), however, η_{33} exhibits no anomaly at the c-a_2c transition, but has a low peak around

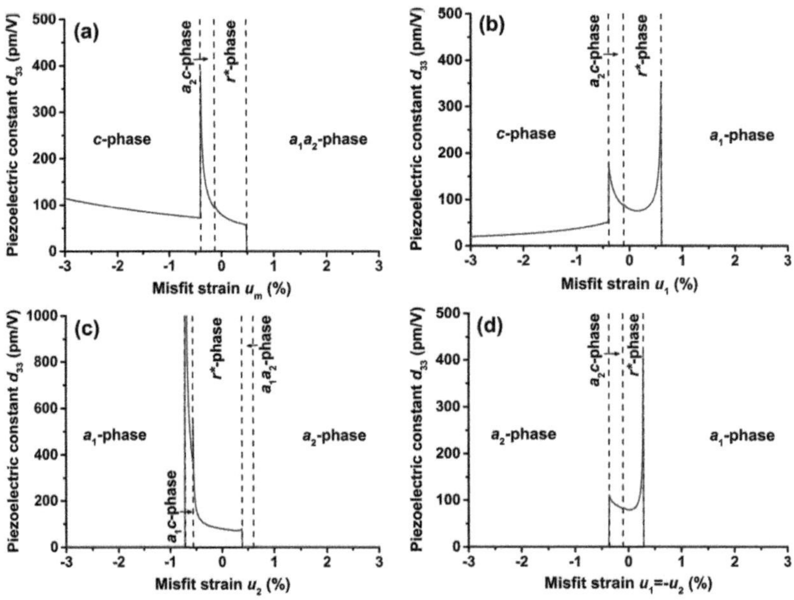

FIG. 8. Piezoelectric constant d_{33} vs. in-plane strain for $(011)_{pc}$ epitaxial $BaTiO_3$ at 25 °C: (a) biaxial u_m, (b) uniaxial u_1, (c) uniaxial u_2, (d) u_1=-u_2.

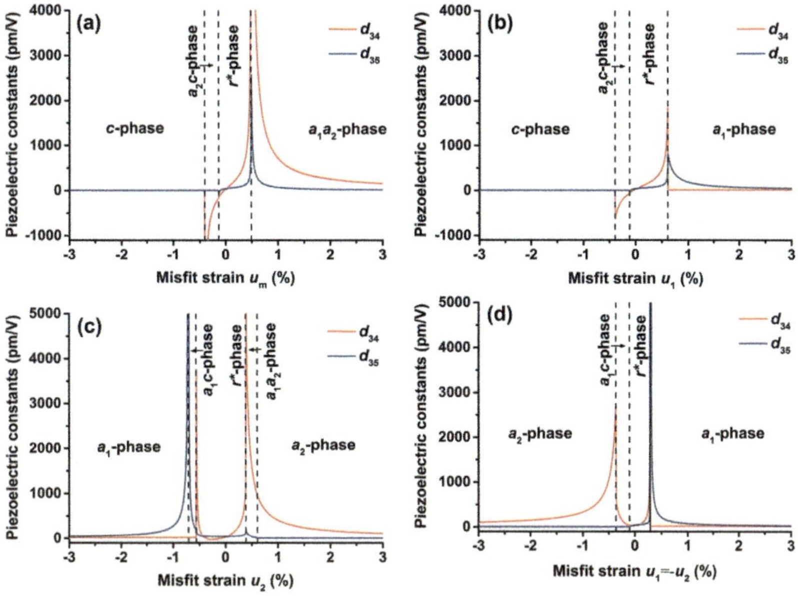

FIG. 9. Piezoelectric constants d_{34} and d_{35} vs. in-plane strain for $(011)_{pc}$ epitaxial $BaTiO_3$ at 25 °C: (a) biaxial u_m, (b) uniaxial u_1, (c) uniaxial u_2, (d) u_1=-u_2.

u_m=0.02. Fig. 11 and Fig. 12 show strain dependence of room-temperature longitudinal (d_{33}) and shear (d_{34} and d_{35}) piezoelectric constants, respectively. In the piezoelectric spectra of Figs. 11(a) and 12(a), the longitudinal d_{33} basically remain flat until disappearance at the a_2c-a_1a_2 transition where the shear d_{35} shows only a low jump while the shear d_{34} shows a positive peak and a negative peak at the tension and compression sides, respectively.

In the case of uniaxial u_1, at room temperature there is a first-order a_2c-a_1 transition as shown in Fig. 4(b). Correspondingly, in Fig. 10(b) η_{33} displays a fairly big jump around u_1=1%. However, for

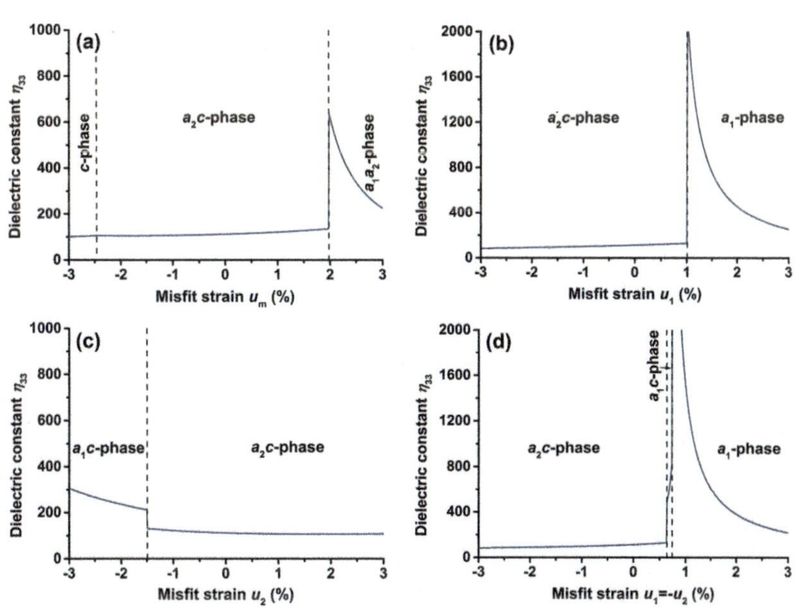

FIG. 10. Dielectric constant η_{33} vs. in-plane strain for $(011)_{pc}$ epitaxial $PbTiO_3$ at 25 °C: (a) biaxial u_m, (b) uniaxial u_1, (c) uniaxial u_2, (d) u_1=-u_2.

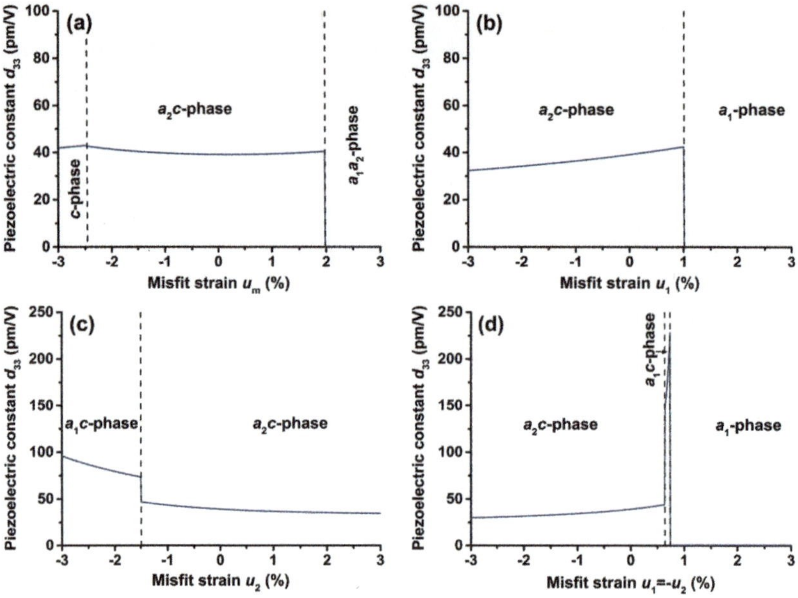

FIG. 11. Piezoelectric constant d_{33} vs. in-plane strain for $(011)_{pc}$ epitaxial $PbTiO_3$ at 25 °C: (a) biaxial u_m, (b) uniaxial u_1, (c) uniaxial u_2, (d) u_1=-u_2.

$PbTiO_3$ there is no anomaly in d_{33}, as seen in Fig. 11(b). Similarly, Fig. 12(b) shows that d_{34} exhibits merely a trivial jump at the transition, in contrast to a fairly big jump for d_{35}.

In the case of uniaxial u_2, corresponding to the first-order a_1c-a_2c transition [see Fig. 4(b)] at room temperature, both dielectric η_{33} and piezoelectric d_{33} increase with compressive u_2, but are almost independent of tensile u_2, as can be seen in Figs. 10(c) and 11(c), while non-zero d_{34} and d_{35} occur at the tension and compression side of the diagrams, respectively, as shown in Fig. 12(c).

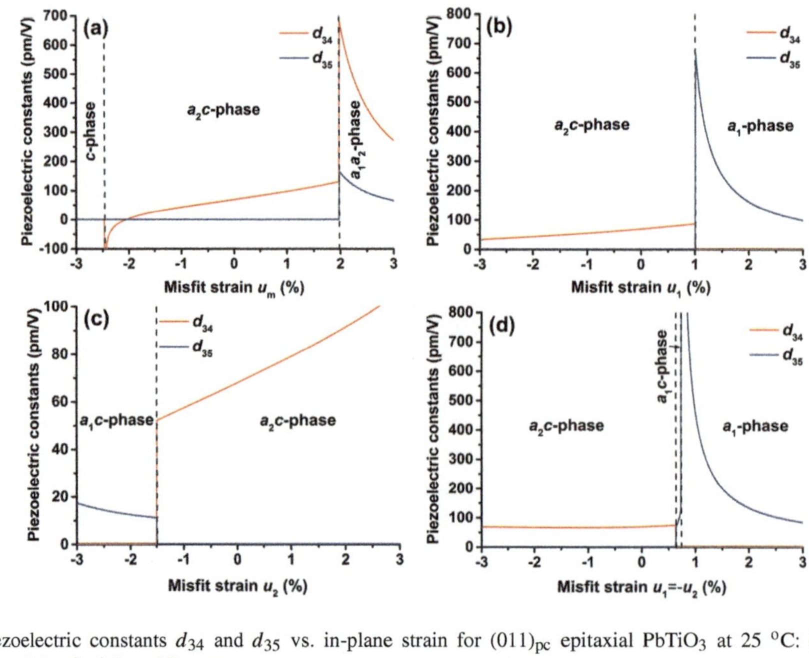

FIG. 12. Piezoelectric constants d_{34} and d_{35} vs. in-plane strain for $(011)_{pc}$ epitaxial $PbTiO_3$ at 25 °C: (a) biaxial u_m, (b) uniaxial u_1, (c) uniaxial u_2, (d) u_1=-u_2.

In the case of equal but opposite in-plane strain (u_1=-u_2), at room temperature for $PbTiO_3$ there is a very narrow a_1c phase region in between a_2c and a_1 phases, for which polarization vector rotates quickly to in-plane positions. The η_{33} peak is sharp and ultrahigh at the a_1c-a_1 transition, where d_{35} peak is very high, but it is not the case for d_{33} and d_{34}.

Theoretically great enhancement is predicted near the phase boundaries. Experimentally, however, order of magnitudes of the expected enhancement is not always achievable. One reason for this is that the available substrates are usually limited. As a result, the required critical misfits at the phase boundaries might be hardly accessible. Another reason is associated with domain formation. When analyzing epitaxial ferroelectrics using phenomenological thermodynamic approaches, phase stability is determined by calculating polarization states with the lowest free energy. The stable states can be either single-domain or poly-domain, depending on the coupling and delicate balance of elastic and electrostatic energies in the films with various thicknesses, interfacial states and processing conditions. The balance in the whole free energy for strained epitaxial films of over a few tens of nanometers thickness processed under conventional conditions often gives rise to poly-domain states. Although in some cases, domain engineering does bring about enhancement in piezoelectric properties (see e.g., Ref. 13), in other cases domain clamping may lead to considerable reduction in the electromechanical response.

V. OUT-OF-PLANE LATTICE STRAIN FOR $(011)_{PC}$ EPITAXIAL FILMS ON (110) $SrTiO_3$

In this section we present temperature dependence of polarization and out-of-plane lattice strain calculated under various epitaxial strain conditions, (i.e., equal biaxial u_m, uniaxial $u_1 \neq 0$, and uniaxial $u_2 \neq 0$). For $(011)_{pc}$ epitaxial $BaTiO_3$ and $PbTiO_3$ grown on (110) $SrTiO_3$, values of the misfit strain are -2.2% and -1.6%, respectively. Initially the epitaxial strain should be of equal biaxial in nature. However, after preferential relaxation the strain state may turn into uniaxial $u_1 \neq 0$, or uniaxial $u_2 \neq 0$, depending on the exact conditions. We perform phenomenological calculations for the three typical epitaxial strain states, and discuss the theoretical results in comparison with relevant experimental data available in the literature.

Fig. 13 presents polarization components as a function of temperature and evolution of polarization-associated part (ΔS_3) of the out-of-plane lattice strain for $(011)_{pc}$ epitaxial $BaTiO_3$. The epitaxial strain $u_1=u_2=u_m$ is assumed to be -2.2%, corresponding to the films grown on (110) $SrTiO_3$. Apart from equal biaxial substrate constraint, we also take into account two extreme conditions of in-plane anisotropic strain relaxation in the epitaxial films, i.e., uniaxial u_1=-2.2% or u_2=-2.2%. Although for biaxial u_m and uniaxial u_2 the calculated room-temperature P_3 are close to the measure one of about 0.18 C/m^2,[18] the estimated T_c for these two strain conditions are far different than the one derived from the measured temperature-dependent out-of-plane lattice strain.[17] As compared to equal biaxial constraint, clearly, preferential relaxation of u_2 enhances T_c in $BaTiO_3$ considerably (see Fig. 13(b)), and the calculated T_c is ~274 °C agrees with the experimental data for anisotropically relaxed epitaxial $(011)_{pc}$ $BaTiO_3$ on $SrTiO_3$.[17] But unfortunately, in this case the a_2c phase occurs only in a narrow temperature window between 200 and 250 °C, and the calculated room-temperature P_3 far exceeds the measure one. On the other hand, when u_1 is fully relaxed, the ferroelectric-paraelectric transition in the $BaTiO_3$ film would occur between a_1 phase (i.e., only non-zero P_1) and paraelectric phase, resulting in a negligible negative ΔS_3, as shown in Fig. 13(d). It appears that the experimental data of ΔS_3 for the 100 nm thick $(011)_{pc}$ $BaTiO_3$ subject to a uniaxial strain (u_1=-2.2%)[17] do not agree with our theoretical calculations for uniaxially strained single-domain film, rather they seem to lie in between the present theoretical data for biaxial u_m and uniaxial u_1 strain conditions. To resolve this issue, we relax the single-domain supposition and assume that the uniaxially strained $(011)_{pc}$ $BaTiO_3$ film in fact takes a special monoclinic structure (i.e., $P_2=P_3$). It turns out that now the calculated temperature-dependent ΔS_3 are in better agreement with the measured ones,[17] and the calculated room-temperature P_3 agrees with the measure one[18] as well. But interestingly the assumption leads to an unexpected first-order Curie phase transition, distinct from phase transitions of the other strained epitaxial states, which are typically of second-order. If the assumption of $P_2=P_3$ is indeed the case, smearing of the Curie phase transition may originate from inhomogeneity effect or relaxation-caused strain gradient, which is indeed present in the 100 nm-thick film.[17] Full modeling

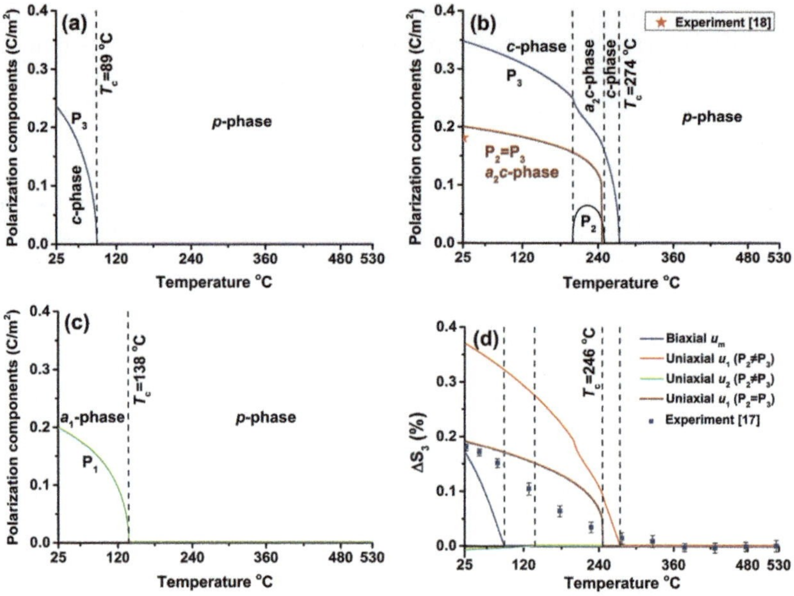

FIG. 13. Polarization components (Panels a-c) and out-of-plane lattice strain ΔS_3 (Panel d) vs. temperature for strained $(011)_{pc}$ epitaxial $BaTiO_3$. (a) biaxial, u_m=-2.2%; (b) uniaxial, u_1=-2.2%; (c) uniaxial, u_2=-2.2%. The P_3 data point (~0.18 C/m^2) at 25 °C in Panel (b) refers to the experimental result for $(011)_{pc}$ epitaxial $BaTiO_3$ on (110) $SrTiO_3$ taken from Ref. 18. The experimental data of ΔS_3 in Panel (d) come from Ref. 17. For the uniaxially strained state (see Panels b and d) we also present the calculations on a special monoclinic a_2c state assuming P_2=P_3.

of the phase transitional behavior should consider the balance of polarization-related Landau free energy, elastic energy due to mechanical constraint, surface or interface effects, and electrostatic energy terms associated with specific charge conditions.

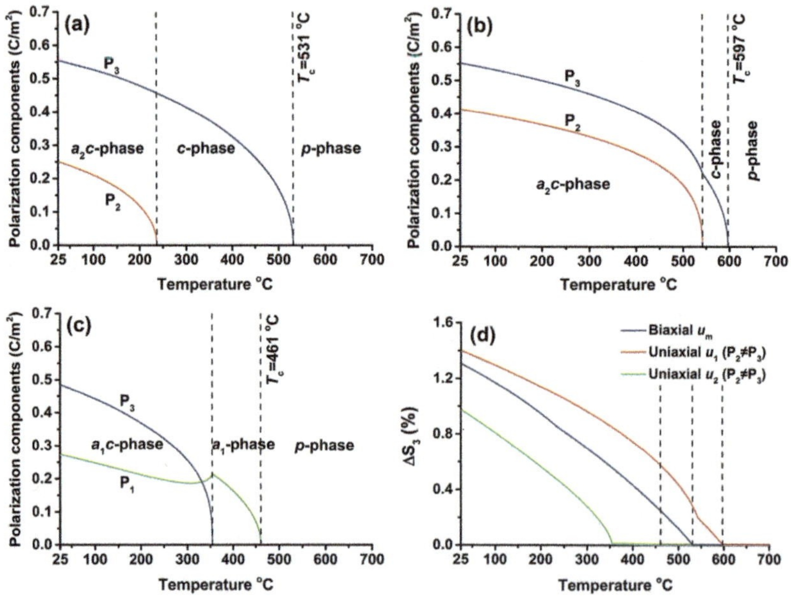

FIG. 14. Polarization components (Panels a-c) and out-of-plane lattice strain ΔS_3 (Panel d) vs. temperature for strained $(011)_{pc}$ epitaxial $PbTiO_3$. (a) biaxial u_m=-1.6%, (b) uniaxial u_1=-1.6%, (c) uniaxial u_2=-1.6%. For $(011)_{pc}$ $PbTiO_3$ epitaxially grown on (110) $SrTiO_3$, the initial misfit strain is about -1.6%. Similar to the situation for $(011)_{pc}$ $BaTiO_3$,[17] after preferential strain relaxation the equal biaxial in-plane constraint in $(011)_{pc}$ $PbTiO_3$ may turn into uniaxial one as well.

Fig. 14 presents polarization components as a function of temperature and evolution of polarization-associated part (ΔS_3) of the out-of-plane lattice strain for $(011)_{pc}$ epitaxial $PbTiO_3$. The epitaxial strain $u_1=u_2=u_m$ is assumed to be -1.6%, corresponding to the films grown on (110) $SrTiO_3$. Apart from equal biaxial substrate constraint, we also take into account the other two extreme conditions of anisotropic in-plane strain relaxation, i.e., uniaxial u_1=-1.6% (u_2=0) or uniaxial u_2=-1.6% (u_1=0). For $PbTiO_3$, a_2c-c phase transition temperature is remarkably enhanced if compressive strain is preferentially relaxed along $(01\text{-}1)_{pc}$ direction, as shown by comparing Fig. 14(a) and Fig. 14(b), whereas T_c is only slightly increased in this case. For $PbTiO_3$, occurrence of a_1 phase contributes a tiny positive ΔS_3, compared to the equally trivial negative ΔS_3 in the case of $BaTiO_3$.

As for $(011)_{pc}$ epitaxial films of pure $PbTiO_3$ so far there have been no suitable experimental data for comparison. Mtebwa et al.[28] reported out-of-plane lattice parameters for $(011)_{pc}$-oriented $Pb(Zr_{0.05}Ti_{0.95})O_3$ epitaxial films. According to their reported data, the films seem to be subject to biaxial constraints. Based upon the results garnered in our present theoretical studies, it is reasonable to expect that, similar to the case of $BaTiO_3$, a special monoclinic a_2c structure with equal a_2- and c-oriented polarization components would be favored, if the in-plane substrate constraint along $(01\text{-}1)_{pc}$ is preferentially relieved. We await experimental verification of our present theoretical calculations.

VI. SUMMARY

Using a thermodynamic phenomenological approach, effect of the anisotropy of in-plane substrate constraint on phase stability and properties of $(011)_{pc}$ epitaxial $BaTiO_3$ and $PbTiO_3$ thin films has been investigated. Below Curie point, usually the films transform first into orthorhombic, and subsequently into a variety of monoclinic structures. For $BaTiO_3$, triclinic phase can occur even under equal biaxial strains. For $PbTiO_3$, monoclinic a_2c phase is rather stable, while monoclinic a_1a_2 phase seems difficult to be stabilized except for equal biaxial strain conditions. Polarization states and dielectric and piezoelectric constants accompanying the phase transitions have been computed. Unlike $(001)_{pc}$ epitaxy, for $(011)_{pc}$ epitaxial films effects of the two orthogonal substrate constraints differ significantly, and preferentially relaxed films may have distinguished structure and properties as a result of the in-plane strain anisotropy. These findings shed light on the particular phase transition features of strained $(011)_{pc}$ epitaxial films and may stimulate further research for possible future applications.

ACKNOWLEDGMENTS

The financial support from the National Natural Science Foundation of China (NSFC Grant No. 10874109) and the Program for New Century Excellent Talents in University (Grant No. NCET-08-0648) is gratefully acknowledged.

[1] J. F. Scott, Science **315**, 954 (2007).
[2] D. G. Schlom, L. Q. Chen, C. B. Eom, K. M. Rabe, S. K. Streiffer, and J. M. Triscone, Annu. Rev. Mater. Res. **37**, 589 (2007).
[3] K. J. Choi, M. Biegalski, Y. L. Li, A. Sharan, J. Schubert, R. Uecker, P. Reiche, Y. B. Chen, X. Q. Pan, V. Gopalan, L. Q. Chen, D. G. Schlom, and C. B. Eom, Science **306**, 1005 (2004).
[4] W. Ma and D. Hesse, Appl. Phys. Lett. **84**, 2871 (2004).
[5] N. A. Pertsev, A. G. Zembilgotov, and A. K. Tagantsev, Phys. Rev. Lett. **80**, 1988 (1998).
[6] O. Diéguez, S. Tinte, A. Antons, C. Bungaro, J. B. Neaton, K. M. Rabe, and D. Vanderbilt, Phys. Rev. B **69**, 212101 (2004).
[7] O. Diéguez, K. M. Rabe, and D. Vanderbilt, Phys. Rev. B **72**, 144101 (2005).
[8] Y. L. Li and L. Q. Chen, Appl. Phys. Lett. **88**, 072905 (2006).
[9] W. Ma and A. Hao, J. Appl. Phys. **116**, 214110 (2014).
[10] W. Ma, AIP Advances **6**, 045310 (2016).
[11] J. Slutsker, A. Artemev, and A. L. Roytburd, Acta Mater. **52**, 1731 (2004).
[12] J. Ouyang and A. L. Roytburd, Acta Mater. **54**, 5565 (2006).
[13] J. Ouyang, J. Slusker, I. Levin, D. M. Kim, C. B. Eom, R. Ramesh, and A. L. Roytburd, Adv. Funct. Mater. **17**, 2094 (2007).
[14] W. K. Simon, E. K. Akdogan, A. Safari, and J. A. Bellotti, Appl. Phys. Lett. **87**, 082906 (2005).
[15] S. Utsugi, T. Fujisawa, Y. Ehara, T. Yamada, M. Matsushima, H. Morioka, and H. Funakubo, Appl. Phys. Lett. **96**, 102905 (2010).
[16] T. Yamada, T. Kamo, H. Funakubo, D. Su, and T. Iijima, J. Appl. Phys. **109**, 091605 (2011).
[17] T. Shimizu, D. Suwama, H. Taniguchi, T. Taniyama, and M. Itoh, J. Phys.: Condens. Matter **25**, 132001 (2013).

[18] T. Shimizu, D. Suwama, H. Taniguchi, T. Taniyama, and M. Itoh, Appl. Phys. Express **6**, 015803 (2013).
[19] H. Liu, P. Yang, Z. Fan, A. Kumar, K. Yao, K. P. Ong, K. Zeng, and J. Wang, Phys. Rev. B **87**, 220101(R) (2013).
[20] X. Wan, E. P. Houwman, R. Steenwelle, R. van Schaijk, M. D. Nguyen, M. Dekkers, and G. Rijnders, Appl. Phys. Lett. **104**, 092902 (2014).
[21] W. Zhang, H. Cheng, Q. Yang, F. Hu, and J. Ouyang, Ceram. Int. **42**, 4400 (2016).
[22] J. H. Lee, K. Chu, K. E. Kim, J. Seidel, and C. H. Yang, Phys. Rev. B **93**, 115142 (2016).
[23] A. K. Tagantsev, N. A. Pertsev, P. Muralt, and N. Setter, Phys. Rev. B **65**, 012104 (2001).
[24] G. Akcay, I. B. Misirlioglu, and S. P. Alpay, J. Appl. Phys. **101**, 104110 (2007).
[25] J. X. Zhang, Y. L. Li, Y. Wang, Z. K. Liu, L. Q. Chen, Y. H. Chu, F. Zavaliche, and R. Ramesh, J. Appl. Phys. **101**, 114105 (2007).
[26] R. Xu, J. Zhang, Z. Chen, and L. W. Martin, Phys. Rev. B **91**, 144106 (2015).
[27] H. Wu, X. F. Ma, Z. Zhang, J. Zeng, J. Wang, and G. Z. Chai, AIP Advances **6**, 015309 (2016).
[28] M. Mtebwa, A. K. Tagantsev, T. Yamada, P. Gemeiner, B. Dkhil, and N. Setter, Phys. Rev. B **93**, 144113 (2016).
[29] J. H. Qiu, Z. H. Chen, X. Q. Wang, N. Y. Yuan, and J. N. Ding, Solid State Commun. **236**, 1 (2016).
[30] Z. Gui, S. Prosandeev, and L. Bellaiche, Phys. Rev. B **84**, 214112 (2011).
[31] Z. Gui and L. Bellaiche, Phys. Rev. B **91**, 020102(R) (2015).
[32] Y. L. Li, L. E. Cross, and L. Q. Chen, J. Appl. Phys. **98**, 064101 (2005).
[33] T. Yamada, J. Appl. Phys. **43**, 328 (1972).
[34] D. Berlincourt and H. Jaffe, Phys. Rev. **111**, 143 (1958).
[35] M. J. Haun, E. Furman, S. J. Jang, H. A. McKinstry, and L. E. Cross, J. Appl. Phys. **62**, 3331 (1987).
[36] M. J. Haun, Z. Q. Zhuang, E. Furman, S. J. Jang, and L. E. Cross, Ferroelectrics **99**, 45 (1989).

20

Characteristics of energy deposition from 1-1000 MeV proton and neutron induced nuclear reactions in silicon

Jinhua Han[1,2,a] and Gang Guo[1,2]
[1]*Department of Nuclear Physics, China Institute of Atomic Energy, Beijing 102413, China*
[2]*National Innovation Center of Radiation Application, Beijing 102413, China*

The characteristics of the energy deposition of the secondary particles from the proton and neutron induced elastic and nonelastic reactions in silicon over a wide energy range of 1-1000 MeV, which are of great value for the proton and neutron single event effect (SEE) study, are investigated systematically in detail with Geant4 simulations. According to our simulation results, some important issues about the proton and neutron SEEs, including the energy dependence of the SEE cross sections, the equivalence of protons and neutrons in SEEs, the effects of the critical charge on SEEs, the geometric effect of the sensitive volumes, and the role of the elastic scattering mechanism in SEEs, are discussed. And some significant conclusions are tried to be drawn. With our systematic study, a comprehensive and profound understanding about the proton and neutron SEEs is reached from the physical perspective.

I. INTRODUCTION

Single event effects (SEEs) induced by energetic particles have been identified as a major reliability concern for microelectronic systems in radiation environments.[1,2] Protons and neutrons, which feature high penetrating ability in matter, are the dominant radiation components with broad energy distribution in space and in the atmosphere, respectively. And thus they are the major causes of SEEs in microelectronic devices used in spaceflight and aviation, respectively.[3] Neutrons induce SEEs through the ionization of the secondary ion products from their nuclear reactions with the device material, and protons induce SEEs through the same mechanism in most cases.[4] The species, yields, energies, ranges and linear energy transfer (LET) values of the secondary ion products from the proton and neutron induced nuclear reactions in silicon (the most important device material) have been investigated in many papers such as Refs. 5–9. Several models such as the burst generation rate (BGR)[4,10,11] and the analytical microdosimetry model[12–14] for predicting the proton and neutron SEE cross sections from the heavy ion ones have also been developed based on the proton/neutron SEE mechanism.

At present, the feature size of microelectronic devices has shrunken to the nano-scale following the technological roadmap, and the lateral dimensions of a heavy ion track become comparable to or even much larger than it. Therefore, multiple bit upsets (MBUs) pose severe challenges to advanced nano-devices.[15–17] The realistic ion track structure, i.e. the spatial structure of the ion energy deposition, thus becomes of great significance for the SEE study.[18–20] As reported in Refs. 21–25, protons and neutrons are also capable of triggering MBUs. Clearly, the above typical studies on the characteristics of the species, yields, energies, ranges and LETs of the reaction products[5–9] are also valuable for the proton/neutron MBUs. In this paper, we try to give some implications from the other perspective, i.e. the spatial characteristics of the energy deposition from nuclear

[a]Electronic mail: jinhua.h@outlook.com.

reactions, following the heavy ion case. Kobayashi et al. in Ref. 19 gave some radial characteristics of the proton energy deposition in silicon including the contribution of both ionization and nuclear reactions based on Geant4 simulations,[26] and the most energetic events in their simulations were thought to be dominated by the contribution of nuclear reactions. Here we focus on the characteristics of the energy deposition from only the proton/neutron nuclear reactions using a special simulation technique.

Several previous studies about the proton and neutron SEEs conducted by experiments or/and simulations showed a discrepancy about the incipient equivalent energy, above which the proton and neutron SEE cross sections can be seen as equal; it was considered to be 20, 30 and 50 MeV in Refs. 27, 3 and 28, respectively. We believe that the related investigations on the energy deposition from proton/neutron reactions will help to clarify this discrepancy. The contribution of elastic and nonelastic reactions to the proton SEE cross sections was discussed in Ref. 29, and a further study should be helpful. Based on the above considerations, the characteristics of the energy deposition from the 1-1000 MeV proton and neutron induced elastic and nonelastic reactions in silicon are investigated systematically in detail with Geant4 simulations, and some conclusions of great value for the proton and neutron SEE study are tried to be drawn. The proton/neutron SEE cross sections for most devices turn up around a few MeV and saturate far below 1000 MeV, so the considered energy range is rather wide and meaningful for the SEE study.

II. SIMULATION DETAILS

For heavy ions, the track structure in matter is taken as radially symmetric in general.[15–19] For nuclear reactions, every reaction event can induce a unique energy deposition structure that is not radially symmetric with respect to the primary particle path. As shown in Fig.1, the products of different species and energies from different kinds of reaction channels induced by neutrons move in different directions and have their own tracks. It is clear that the corresponding spatial energy deposition structure is much more complex than heavy ions. Referring to the research method of the ion track structure in Refs. 18–19 with Geant4 simulations, we try to get the general characteristics of the radial energy deposition from proton/neutron reactions in a similar way.

Geant4 version 10.2 was employed to perform our simulations in this work. As shown in Fig.2, a cylindrical target detector of natural silicon material was defined. Its height was set at 1 μm that is on the sensitive volume thickness scale.[3] Its radius was set at 10 μm that is much larger than the feature size of advanced nano-devices. In order to get the radial energy deposition distribution, the detector was divided into an innermost cylinder of radius 1 nm (much smaller than the feature size of advanced devices) and 43 concentric cylindrical shells of increasing thickness. Considering protons and neutrons of 17 different energies (1, 2, 5, 8, 10, 14, 20, 30, 50, 80, 100, 150, 200, 300, 500, 800 and 1000 MeV) over the 1-1000 MeV range, 34 simulations were carried out with 10^6 particles irradiating normally at the center of one of the detector circular surfaces in each simulation.

The pre-packaged physics list QGSP_BIC_HP including the hadronic and electromagnetic physics[30] was employed in the simulations. In this physics list, for neutrons with energy

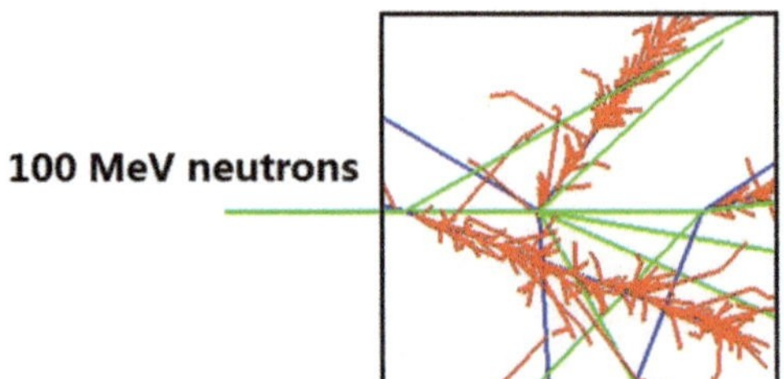

FIG. 1. Geant4-simulated tracks of the secondary particles from three nuclear reactions induced by three neutrons at 100 MeV in a 25-nm silicon cube. Here, a biasing technique is used to increase the cross sections of neutron reactions. The green, blue, and red lines are for neutrons, ions, and δ electrons of energy above 100 eV, respectively.

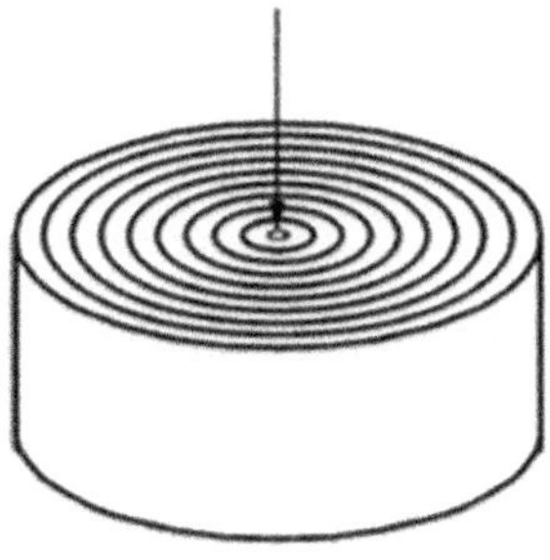

FIG. 2. Schematics of the silicon target detector.

below 20 MeV, the G4NeutronHP package using evaluated nuclear data libraries is employed; for neutrons over the 20 MeV - 10 GeV energy range and protons with energy below 10 GeV, the binary intranuclear cascade model is employed. To save computing time and decrease variance, the cross sections were intentionally increased by a biasing factor of 100000. For our special research aim, the biasing was applied to the hadronic processes but not to the electromagnetic processes, to the primary particles but not to the secondary particles, to their first reactions but not to the subsequent reactions (so that two successive elastic reactions in the thin detector could not occur).

To simulate ion track structures with high accuracy, the electromagnetic model G4Em-StandardPhysics included in QGSP_BIC_HP was replaced by the model G4EmStandardPhysics_option4 in the physics list, and the production threshold for delta electrons was set at 100 eV (i.e., the lowest energy of the electrons that can be tracked in the model G4EmStandardPhysics_option4).[30] Valentin et al. in Ref. 31 implemented the "MuElec" (now named "MicroElec") processes in Geant4 to take into account the generation and transport of low-energy electrons down to 16.7 eV in silicon, so more detailed ion track structures could be simulated. Since the simulations with the "MuElec" processes were too time-consuming, they were not employed in our work. Considering that most secondary particles move at an angle to the axis of the cylinder detector, the radial energy deposition distribution should not be significantly affected by using the 100 or 16.7 eV electron production threshold.

For every incident particle, its interaction type in the silicon detector, i.e., the elastic reaction, nonelastic reaction, direct ionization (for protons), or no interaction (for neutrons), and the energy deposited in every shell were recorded. Note that the deposited energy of a proton elastic/nonelastic event contains the energy deposited by the proton's ionization process before the nuclear reaction takes place in the defined detector.

III. RESULTS AND DISCUSSIONS

The proton and neutron elastic and nonelastic cross sections at 17 energies over the 1-1000 MeV range extracted from Geant4 are shown in Fig. 3. With the above cross section biasing technique, 1 mb means that about 415 reaction events occur in the defined detector for 10^6 incident particles. Except a few nonelastic event numbers at 1 and 2 MeV, other nonelastic ones and all the elastic ones are between $\sim 10^4$ and $\sim 10^5$, which are enough for the statistics of the related energy deposition characteristics. The proton ionization event numbers at all energies are also enough for the statistics of the proton track structure. A good agreement can be seen between the proton LET derived from the simulated proton ionization events and that from the SRIM code over the 1-1000 MeV range in Fig. 4, which verifies the reliability of the electromagnetic model in Geant4 simulations to some extent.

A. Cumulative deposited energy

1. Proton and neutron elastic events

For the proton elastic events, a Si nucleus gets only a very small portion of the incident proton energy. Taking the case at 1000 MeV as an example, the recoil energy of ^{28}Si is between

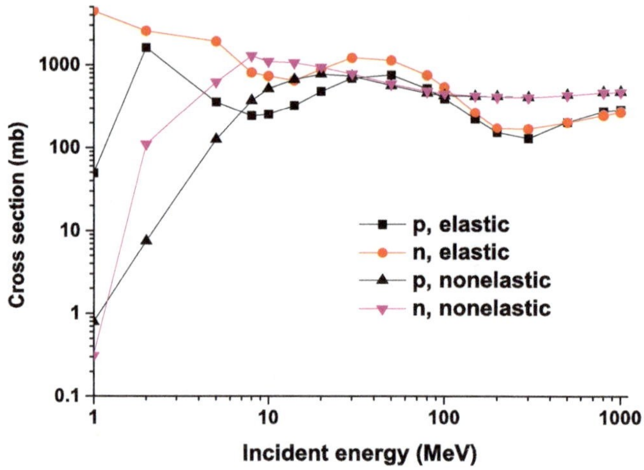

FIG. 3. Proton and neutron elastic and nonelastic cross sections at 17 energies over the 1-1000 MeV range extracted from Geant4.

2.1 eV and 10.35 MeV in our simulations. Note that 2.1 eV is so small that the displacement and thus the ionization of ^{28}Si cannot occur, and it just becomes a small amount of heat. In addition, this energy can be neglected in the energy deposition statistics. The average recoil energy of ^{28}Si from the proton elastic events over the 1-1000 MeV range is shown in Fig. 5. We can see that it rises with increasing incident energy, and reaches its maximum (about 0.51 MeV) at 14 MeV, and then drops. From the energy-range relation in Fig. 6, it is easy to know that the ranges of most recoiled Si nuclei are less than 1 μm, and the ranges of all scattered protons are so large (even 1 MeV protons' ranges are up to 16 μm) that they can deposit energy outside the defined detector. Over the 5-1000 MeV range, the scattered protons' LETs are so low that the recoiled Si nuclei contribute most of the deposited energy in the detector, and thus the average cumulative deposited energy per elastic event grows rapidly within 1 μm or less and then reaches "saturation", as shown in Fig. 7(a). The saturation value rises with incident energy, and reaches a maximum (about 0.35 MeV) at 14 MeV, and then drops. This trend is consistent with that of the average recoil energy of Si nuclei. Since some recoiled Si nuclei go out of the defined detector in the simulations, the saturation energy is some lower than the corresponding average recoil energy of Si nuclei. From 100 to 1000 MeV, the saturation value keeps unchanged basically (drops by only 15 keV).

For the neutron elastic events, the cumulative deposited energy over the 1-1000 MeV range is shown in Fig. 7(b). Clearly, its trend with incident energy is similar to that for the proton elastic

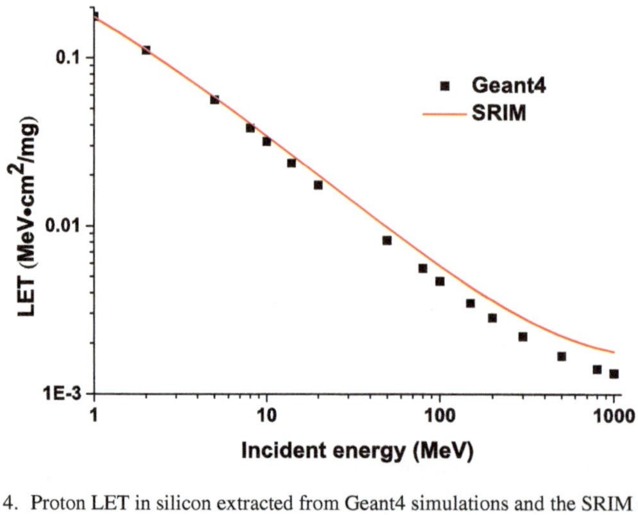

FIG. 4. Proton LET in silicon extracted from Geant4 simulations and the SRIM code.

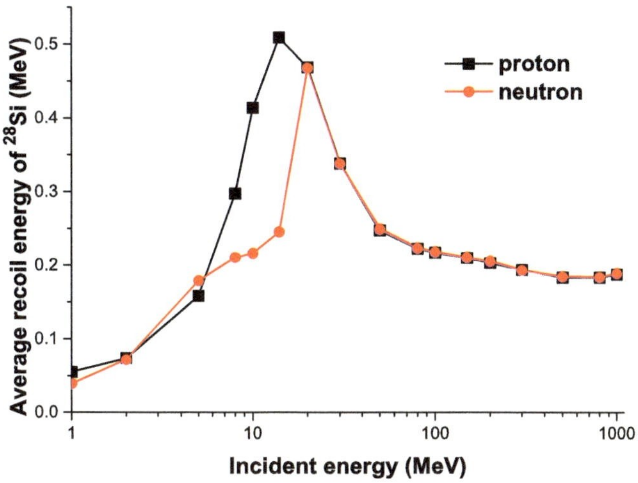

FIG. 5. Average recoil energy of ^{28}Si from the proton and neutron elastic events.

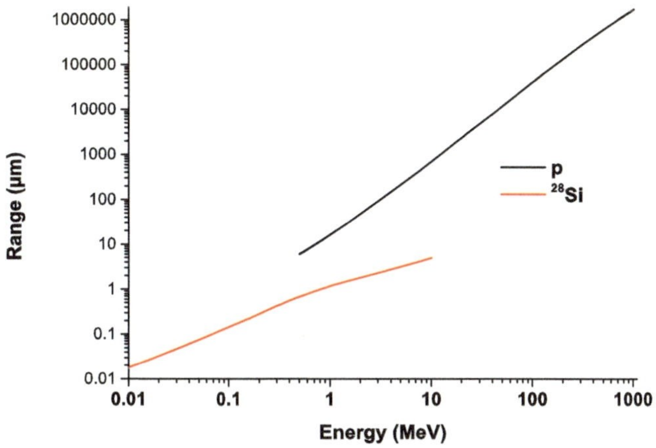

FIG. 6. Range of protons and ^{28}Si in silicon from the SRIM code.

events, and an evident difference is that it reaches a maximum (about 0.33 MeV) at 20 MeV instead of 14 MeV. At and above 20 MeV, both the average recoil energy of ^{28}Si and the cumulative deposited energy are almost the same as the proton elastic events, as shown in Figs. 5 and 7. Over this energy

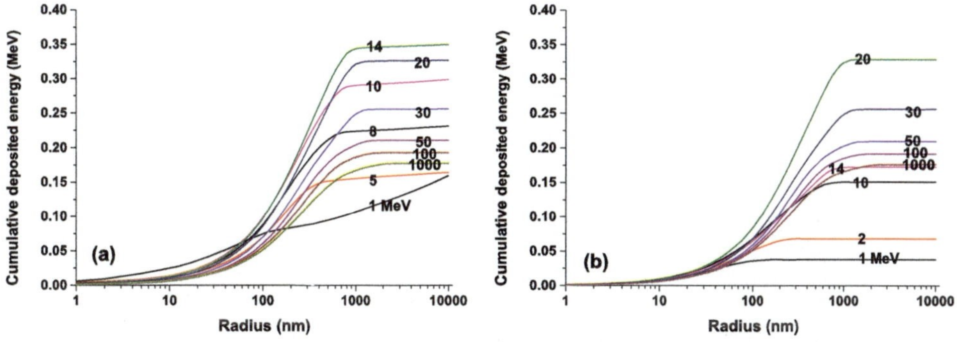

FIG. 7. Cumulative deposited energy for the proton (a) and neutron (b) elastic events at different energies over the 1-1000 MeV range.

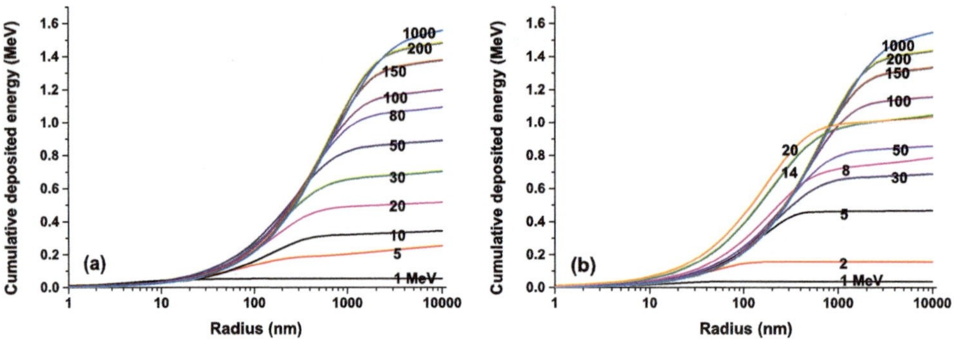

FIG. 8. Cumulative deposited energy for the proton (a) and neutron (b) nonelastic events at different energies over the 1-1000 MeV range.

range, the scattered protons' LETs are too low to create big differences in the cumulative deposited energy.

2. Proton and neutron nonelastic events

For the proton nonelastic events, the cumulative deposited energy within 10 μm rises from 0.08 to 1.56 MeV with incident energy basically as shown in Fig. 8(a). At 1 and 2 MeV, almost all the deposited energy distributes within 100 nm, because only the proton capture reactions as nonelastic reactions can take place and the ranges of the produced P nuclei are below 100 nm. At and above 5 MeV, the produced light products such as protons and alpha particles can deposit energy beyond 10 μm. From the rising trend of the cumulative deposited energy around 10 μm, it is clear that the energy deposited beyond 10 μm increases with incident energy since the reaction products' energies and ranges increase. From 150 to 1000 MeV, the cumulative deposited energy within 10 μm rises by only 180 keV.

For the neutron nonelastic events, the cumulative deposited energy within 10 μm rises from 0.04 to 1.05 MeV over the 1-14 MeV range, and drops sharply from 1.03 to 0.69 MeV over the 20-30 MeV range, and then rises from 0.69 to 1.54 MeV over the 30-1000 MeV range, as shown in Fig. 8(b). It is noted that a neutron nonelastic event at 14 MeV deposits more energy within 10 μm than one at 30 and even 50 MeV on average. At 1 and 2 MeV, almost all the deposited energy distributes within 100 and 200 nm respectively, because only the neutron capture reactions as nonelastic reactions can take place. At and above 5 MeV, the deposited energy distributes beyond 10 μm due to the production of light products such as protons and alpha particles. At and above 30 MeV, the cumulative deposited energy is almost the same as that for the proton nonelastic events.

B. Equivalent LET and critical radius

To get the overall characteristics of the energy deposition from proton/neutron reactions in silicon, two quantities are calculated: 1) the equivalent LET, the average energy deposited in the whole detector per elastic/nonelastic/total reaction event over the target thickness and the silicon density; 2) the critical radius, within which the deposited energy per event is 50% or 90% of the energy deposited in the whole detector.

1. Equivalent LET

The equivalent LET (LET_{eqv}), which represents the average magnitude of the deposited energy in the defined detector, is shown in Fig. 9. With this quantity, it is convenient to compare the amount of the deposited energy with heavy ions. LET_{eqv} for the proton/neutron elastic events shows a trend quite similar to the average recoil energy of Si nuclei shown in Fig. 5. Below 20 MeV, LET_{eqv} for the proton elastic events is higher than that for the neutron ones, since the scattered protons can deposit energy while neutrons cannot. At and above 20 MeV, they share almost the same values between

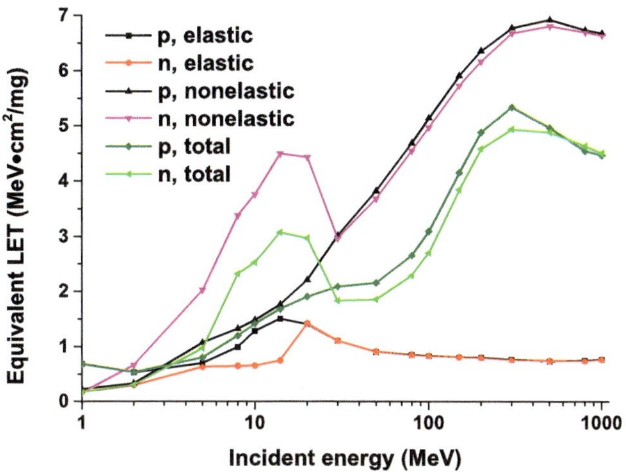

FIG. 9. Equivalent LET of the proton and neutron elastic, nonelastic and total reaction events.

0.73 and 1.40 MeV·cm^2/mg, as the energy deposited by the scattered protons in the whole detector becomes smaller with incident energy.

LET_{eqv} for the proton nonelastic events rises from 0.23 to 6.92 MeV·cm^2/mg with incident energy over the 1-500 MeV range, and then drops slightly over the 500-1000 MeV range. An evident difference between LET_{eqv} for the proton nonelastic events and that for the neutron ones is that the latter has a bump around 14 MeV. Over the 2-30 MeV range, LET_{eqv} for the proton nonelastic events is lower than that for the neutron ones, as some reaction channels are hard to open due to the coulomb barrier between protons and Si nuclei. At and above 30 MeV, they share almost the same values between 3.0 and 6.5 MeV·cm^2/mg, as the coulomb barrier is ignorable at high energy. And LET_{eqv} for the proton nonelastic events is slightly higher than that for the neutron ones, which should result from the charge of protons. For protons with energy above 5 MeV and neutrons with energy above 2 MeV, LET_{eqv} for the nonelastic events is higher than that for the elastic ones.

LET_{eqv} for the proton/neutron total reaction (including elastic and nonelastic) events varies with incident energy similarly to that for the nonelastic events. The bump around 14 MeV for the neutron total reaction events is "inherited" from the nonelastic ones. Over the 30-300 MeV range, LET_{eqv} for the proton total reaction events is close to that for the neutron ones. And the difference between them is bigger than the difference between LET_{eqv} for the proton nonelastic events and that for the neutron ones, which should result from that the proton elastic cross section is less than the neutron one while the proton and neutron nonelastic cross sections are close to each other as shown in Fig. 3. Above 300 MeV, LET_{eqv} for the proton/neutron total reaction events falls due to the rise of the proton/neutron elastic cross section.

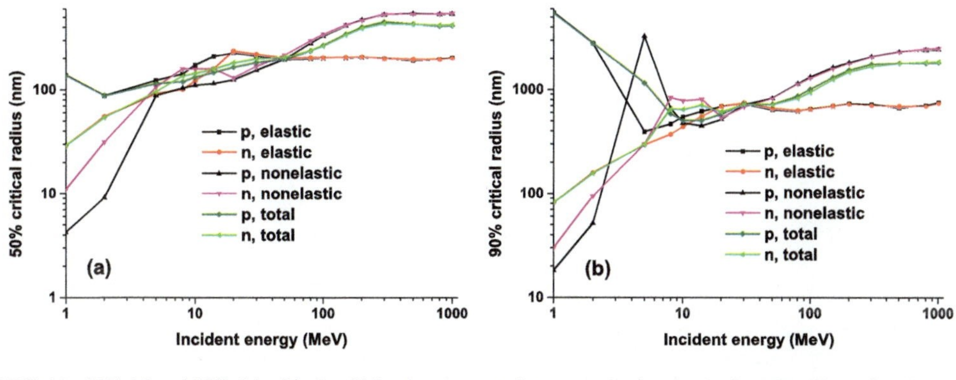

FIG. 10. 50% (a) and 90% (b) critical radii for the proton and neutron elastic, nonelastic and total reaction events.

2. Critical radius

The 50% and 90% critical radii, which reflect the radial range and structure of the energy deposition, are shown in Fig. 10. At 5 MeV, most proton nonelastic events that occur are the inelastic ones which deposit energy just like the elastic ones, so the 90% critical radius for them is up to 3286 nm and a peak turns up. It is noted that 20 MeV is a critical energy. At and above 20 MeV, the 50%/90% critical radius for the proton elastic events is almost the same as that for the neutron ones, i.e., the proton and neutron elastic events have quite similar energy deposition structure; likewise, the proton and neutron nonelastic/total reaction ones do, too. Over this energy range, the 50% critical radius for elastic events (192-232 nm) shows little change, and the 90% critical radius for them (616-734 nm) does, too; the 50% and 90% critical radii for nonelastic and thus total reaction events increase with incident energy. At and above 50 MeV, the 50% critical radius for nonelastic events (196-542 nm) is larger than that for elastic ones (192-204 nm), and the 90% critical radius for nonelastic events (817-2483 nm) is also larger than that for elastic ones (616-753 nm).

C. Cumulative event number

Fig.11 shows the cumulative event number versus the deposited energy, i.e., the number of the simulated proton/neutron total reaction events that deposit energy in the whole detector above a certain value at different incident energies. From the cumulative event number, the details about the event-by-event deposited energy distribution can be seen, and some important information as shown in Figs. 12–15 can be extracted.

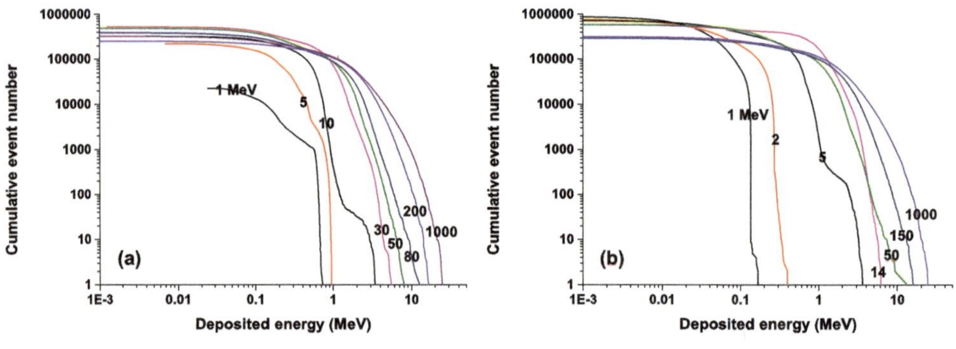

FIG. 11. Cumulative event number versus the deposited energy for the proton (a) and neutron (b) total reaction events at different energies over the 1-1000 MeV range.

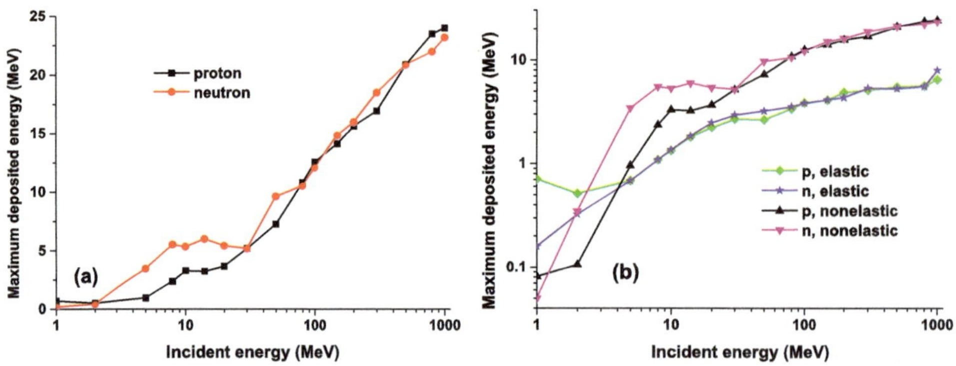

FIG. 12. Maximum deposited energy for the proton and neutron total reaction (a), elastic and nonelastic (b) events. It is taken as the average of the top five highest deposited energies to show certain regularity. Note that the ordinate scale in (a) is linear while that in (b) is logarithmic.

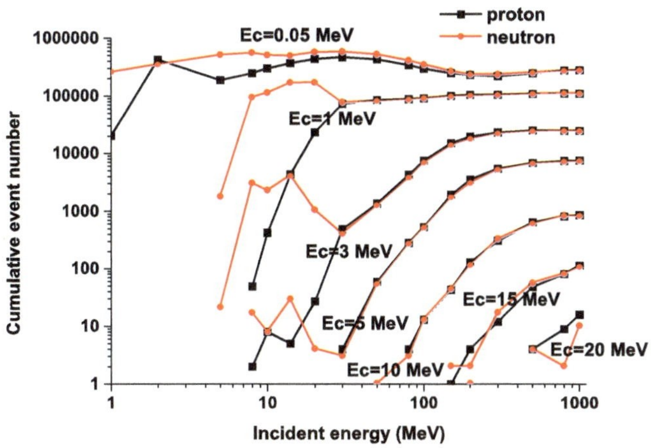

FIG. 13. Number of the events that deposit energy in the whole detector above some critical values (Ec = 0.05, 1, 3, 5, 10, 15 and 20 MeV) for the proton and neutron reaction events.

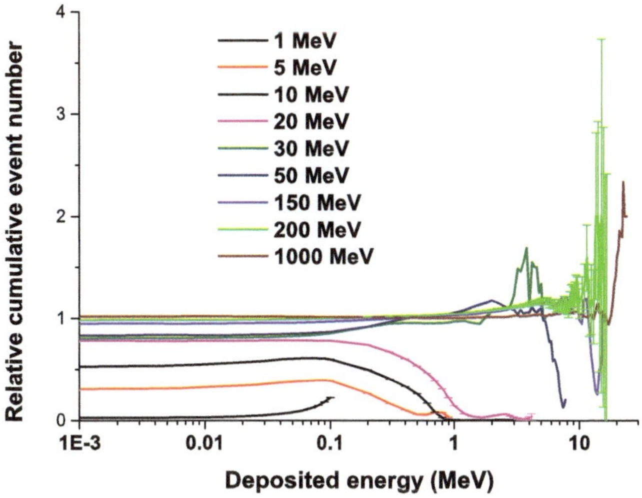

FIG. 14. Relative cumulative event number (the ratio of the cumulative event number for proton reactions to that for neutron ones) versus the deposited energy at different incident energies. Only the error bars at 1, 5, 10, 20, and 200 MeV are shown just for the sake of clarity.

1. Maximum deposited energy

The maximum deposited energy (E_m) for incident protons/neutrons at a certain energy determines that if they can induce SEEs in devices with a certain critical charge to some extent. In Fig. 12, it can be seen that E_m for the proton/neutron total reaction, elastic and nonelastic events increases with incident energy basically. And notably, E_m for the neutron nonelastic and thus total reaction events over the 10-30 MeV range does not follow this trend. In Fig. 12(b), it is clear that E_m for the proton elastic events is close to that for the neutron ones at and above 5 MeV, and E_m for the proton nonelastic events is close to that for the neutron ones at and above 30 MeV. At and above 5 MeV, E_m for the proton/neutron nonelastic events is higher than that for the elastic ones, and the difference between them increases with incident energy; at 1000 MeV, the former is higher than 23 MeV, while the latter is lower than 8 MeV. In addition, over the 5-20 MeV range, E_m for the neutron total reaction events is greater than that for the proton ones.

Based on the fact that the maximum LET for the secondary ion products generated by proton reactions with silicon is about 13 MeV·cm^2/mg, it is generally considered that protons cannot induce SEEs in the devices with the threshold LET above 15 MeV·cm^2/mg. In Ref. 32,

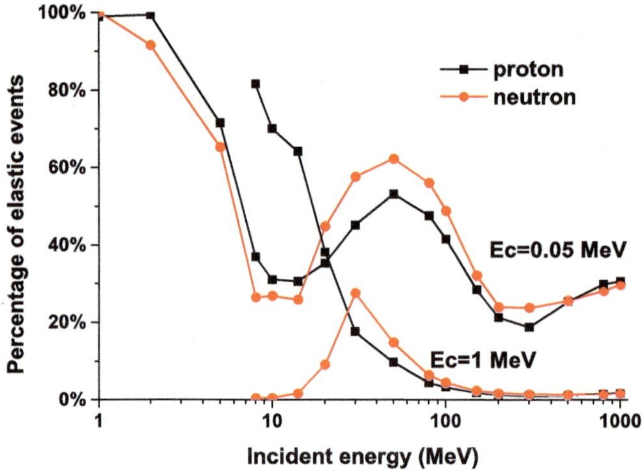

FIG. 15. Percentage of elastic events in all the reaction events that deposit energy above Ec in the whole detector.

the proton single event latchups (SELs) in two SRAMs with the threshold LET 16.3 MeV·cm^2/mg were considered to be induced by the high-energy proton nuclear reactions with high-Z materials. In our simulations, for protons at 500 MeV, the maximum deposited energy in our defined detector is up to 20.87 MeV; a heavy ion with the LET ~90 MeV·cm^2/mg, i.e., five to six times the above SEL threshold LET, produces the same energy deposition effect. Clearly, it is a geometric effect, i.e., some heavy products that exit at an angle of around 90 degrees lead to these large energy deposition events in the defined flat detector. For a 1-μm silicon cube detector, another similar simulation shows that two in 137668 proton reaction events at 500 MeV can deposit energy above 3.47 MeV (i.e., the energy deposited in the detector by a heavy ion with the LET 15 MeV·cm^2/mg) in it, and the corresponding equivalent LETs are 15.11 and 16.05 MeV·cm^2/mg, respectively. Therefore, for devices with flat sensitive volumes, it is probable that some of the high-energy proton nuclear reactions with silicon besides those with high-Z materials can also induce SEEs when the threshold LET is above 15 MeV·cm^2/mg.

2. Effects of critical deposited energy

Considering the SEE mechanism, the number of the simulated events that deposit energy above a critical value Ec in the defined detector represents the SEE cross section of a device with a critical charge to a large extent, so the effects of the critical charge on the proton/neutron SEE cross sections can be investigated from Fig. 13. First, let's talk about the proton reaction events. 1) The cumulative event number decreases with the increase of the critical deposited energy Ec. When Ec increases from 0.05 to 20 MeV, it decreases by four to five orders of magnitude. 2) The threshold energy E_{th}, where the events that deposit energy above Ec begin to occur, increases with Ec. If Ec is 1, 5, 10, 15 and 20 MeV, E_{th} is 8, 30, 80, 150 and 500 MeV, respectively. If Ec is 0.05 MeV, 1 MeV protons can deposit energy above Ec; if Ec is 20 MeV, only protons with energy above 500 MeV in our simulations can. 3) The incipient saturation energy (Es), where the curve begins to saturate, increases with Ec. If Ec is 1, 3, 5 and 10 MeV, Es is about 30, 300, 500, and 800 MeV, respectively; if Ec is above 15 MeV, Es is above 1000 MeV. 4) The shape of the cumulative event number curve is determined by Ec. If Ec is between 1 and 10 MeV, it is just like the Weibull curve; if Ec is low to 0.05 MeV, it becomes flat over the 2-1000 MeV range, and a bump turns up around 30 MeV; if Ec is up to 20 MeV, it is close to a straight line over the 500-1000 MeV range. The above discusses help much to understand the proton SEL cross sections of the five SRAMs over the 20-500 MeV range measured in Ref. 32, especially the significant variations of the proton threshold energy and the saturation cross section between different device technologies.

Then, let's talk about the neutron reaction events. The above four conclusions about the proton events also hold true for the neutron ones, but there are still some differences between them.

1) In cases like Ec is between 1 and 10 MeV, the neutron cumulative event number curve has the smaller threshold energy. 2) The neutron cumulative event number is larger than the proton one below 30 MeV. 3) In cases like Ec is between 1 and 5 MeV, there is a distinct bump around 14 MeV. If Ec is 1 MeV, the neutron cumulative event number at 14 MeV is even higher than the saturation value at 1000 MeV; if Ec is between 3 and 5 MeV, it is lower. Miller et al. predicted a relative neutron single event upset (SEU) cross section curve (Fig. 16 in Ref. 6) for a 90 nm SRAM. For this device, we can see that the SEU cross section at 14 MeV is higher than that at 200 MeV if the threshold LET is between 0.1 and 3 MeV·cm^2/mg, and it is lower if the threshold LET is between 4 and 6 MeV·cm^2/mg. Clearly, our discusses have strong implications on the SEE study.

3. Relative cumulative event number

The incipient equivalent energy, above which the proton and neutron SEE cross sections can be seen as equal, is investigated from Fig. 14, which shows the relative cumulative event number R_N, i.e., the ratio of the cumulative event number for proton reactions to that for neutron ones. To a large extent, R_N represents the ratio of the proton SEE cross section to the neutron one. Between 1 and 20 MeV, R_N is below 1 and increases with incident energy, which means the neutron SEE cross sections should be greater than the proton ones and the relative difference decreases with incident energy. Between 5 and 20 MeV, R_N decreases sharply above the deposited energy 0.1 MeV, which means the neutron SEE cross sections should be much greater than the proton ones for some devices with relatively large critical charge. At 30 MeV, R_N increases from 0.8 to 1.2 over the deposited energy range 0.001-3 MeV, and it is between 1 and 1.7 above the deposited energy 3 MeV. Above 30 MeV, the higher the incident energy is, the closer to 1 R_N is over the wider deposited energy range, and thus the more equivalent protons and neutrons are. The big deviation of R_N from 1 at the high deposited energy end should result from that the cumulative event numbers are too small to get good statistics. So roughly 30 MeV can be considered the incipient equivalent energy, which can also be seen from Fig.13. The device simulations in Ref. 3 by D. Lambert also showed that this is true for the devices with 65, 90, 130 and 180 nm technology nodes. From Fig. 14, it is easy to know that 20 MeV can also be seen as the incipient equivalent energy just for some devices with low critical charge, which should be the case in Ref. 27.

4. Percentage of elastic events

The role of the elastic scattering mechanism in the proton/neutron SEEs is discussed through Fig.15. Generally speaking, the greater Ec is, the smaller the percentage of elastic events in all the reaction events that deposit energy above Ec in the whole detector is. If Ec is 0.05 MeV, the percentage of elastic events is close to (slightly smaller than) the ratio of the elastic cross section to the total one. Over the 1-1000 MeV range, the percentage is above 18%, so the contribution of the elastic scattering to the SEE cross sections should be considered. At 1 and 2 MeV, it is noted that the percentage of elastic events is above 90%, since the elastic cross sections are much bigger than the nonelastic ones, and there is a peak at 50 MeV. If Ec is 1 MeV, the percentage of the proton elastic events is above 50% between 8 and 14 MeV, since the nonelastic reaction channels that can deposit energy above 1 MeV are open at a low probability over this energy range. If Ec is 3 MeV, the percentage of the proton/neutron elastic events is below 1% over the 1-1000 MeV range (not shown in Fig.15), so the role of the elastic scattering in the proton/neutron SEEs can be neglected. Our above conclusions are in good agreement with those in Ref. 29.

IV. SUMMARY AND CONCLUSIONS

When the proton/neutron induced nuclear reactions take place in silicon, the energy deposition information is of great significance for the SEE study. Here, this issue is investigated systematically over a wide energy range of 1-1000 MeV with Geant4 simulations. The cumulative deposited energy, equivalent LET and critical radius derived from the simulations reflect the average effects of plenty of events, while the cumulative event number reflects the event-by-event energy deposition distribution. From our simulation results, some important conclusions can be drawn.

At and above 30 MeV, protons and neutrons can be roughly considered equivalent for SEEs, i.e., the proton and neutron SEE cross sections are close to each other. Some details support this point: over this energy range, the proton and neutron nonelastic cross sections in silicon are almost equal (as shown in Fig. 3); the proton and neutron elastic/nonelastic/total reaction events have almost the same deposited energy on average in the defined detector (as shown in Fig. 9) and energy deposition structure (as shown in Figs. 7–8, 10); the event-by-event energy deposition distribution of the proton reaction events in the defined detector is close to that of the neutron ones (as shown in Figs. 13–14). With increasing incident energy, protons and neutrons become more equivalent.

Below 30 MeV, the proton SEE cross sections should be lower than the neutron ones, and the relative difference increases with decreasing incident energy, as shown in Figs.13–14. This results from the charge of protons. Over this energy range, some reaction channels are difficult to be opened due to the coulomb barrier between protons and Si nuclei while they are open for neutrons. And thus two consequences follow: the neutron nonelastic (and elastic in fact) cross section in silicon is greater than the proton one, as shown in Fig. 3; and the neutron nonelastic events can deposit more energy than the proton ones on the whole, as shown in Fig. 9. For this reason, another phenomenon that the neutron SEEs have the smaller threshold energy than the proton ones comes, as shown in Fig. 13.

It is noteworthy that there should be bumps around 14 MeV for the neutron SEE cross section curves of some devices according to Fig. 13. Similarly, this can be explained from two aspects: the relatively large neutron nonelastic cross section (as shown in Fig. 3) and the equivalent LET bump (as shown in Fig. 9) around 14 MeV. Corresponding experiments should be carried out to verify if this bump exists for some devices. Based on the unfolding technique, the SEE cross sections of 14 MeV neutrons are often employed in the low energy tail correction of the quasi-monoenergetic neutron SEEs.[33,34] If the bumps around 14 MeV exist, then the assumption that the SEE cross sections vary with energy following the Weibull fit (or the Bendel model) will introduce an error in the corrected SEE cross sections with the above technique.

The decisive effects of the critical charge on the proton/neutron SEE cross sections are well investigated. According to Fig. 13, the larger the critical charge of a device is, the higher the threshold energy and the incipient saturation energy are, and the smaller its SEE cross section is. The shape of the SEE cross section curve is also determined by the critical charge. For devices with large critical charge, the proton SEE cross sections follow the Weibull fit. For devices with low critical charge, the proton SEE cross section curves that are quite different from the Weibull curve over the 1-1000 MeV range are expected, which should be verified through corresponding experiments. The discrepancy about the incipient proton-neutron equivalent energy in Refs. 3, 27, 28 can also be explained by the critical charge. 20 MeV can be viewed as the incipient equivalent energy just for devices with low critical charge; 30 or 50 MeV can be viewed as the incipient equivalent energy for general devices.

The critical charge also plays a decisive role in the contribution of the elastic scattering to the SEE cross sections. For devices with very low critical charge, the elastic scattering contributes to the proton/neutron SEE cross sections with a percentage that is close to the ratio of the elastic cross section to the total one; and the related results show that the elastic scattering mechanism is not negligible, as shown in Fig. 15. Over the 1-5 MeV range, the proton/neutron elastic cross sections are much larger than the nonelastic ones, as shown in Fig. 3. So it is understandable that the elastic scattering mechanism can contribute more than 50% and even close to 100% of the proton/neutron SEE cross sections. For devices with large critical charge, the role of the elastic scattering in the proton/neutron SEE cross sections can be neglected. For protons at and above 5 MeV and neutrons at and above 2 MeV, the elastic events deposit less energy in the defined detector than the nonelastic ones on the whole; and the maximum deposited energy for elastic events is smaller than that for nonelastic ones, which means that for some devices nonelastic events can induce SEEs while elastic ones cannot. Continuing decreases in device dimension and operating voltage reduce the critical charge required to produce a SEE,[35] and thus the elastic scattering mechanism is playing a more and more important role in the proton/neutron SEEs for the devices of new generations.

Our simulation results, as depicted in Fig. 12, show that generally protons/neutrons at the higher energy have the greater maximum deposited energy in the defined detector, which implies that they can induce SEEs in devices with the higher critical charge or threshold LET. Our simulations also indicate that the geometric effect of the sensitive volume should be considered rather than only the LETs of the secondary products from nuclear reactions are considered in the SEE study. If the sensitive volumes are flat, it is expected that protons/neutrons only through their nuclear reactions with silicon can induce SEEs in the devices with the threshold LET above 15 MeV·cm^2/mg.

As the secondary particles from nuclear reactions can traverse a column of cells in devices at grazing angles, protons/neutrons should have the ability to induce MBUs with a large multiplicity in nano-devices. Recent studies have shown that MCUs (multiple cell upsets) of ten or more bits can be induced by neutrons of energy above 200 MeV in a 65 nm SRAM with a critical charge of 1.3 fC.[23] As shown in Figs. 7–10, at and above 50 MeV, the proton/neutron nonelastic events deposit more energy in a wider range, and thus should have a stronger ability to induce MBUs/MCUs with a larger multiplicity than the elastic ones. In addition, the ability should grow with incident energy basically, since both the magnitude and range of the energy deposition of the proton/neutron nonelastic events increase. The radial energy deposition distribution as shown in Figs. 7–8 is helpful to estimate the proton and neutron MBU multiplicity of a device with a certain feature size.

The above conclusions about the energy deposition characteristics from the proton and neutron induced nuclear reactions in silicon and their impacts on SEEs are drawn from such a simple model as shown in Fig. 2. This model is not realistic enough for the proton/neutron SEEs: a device has many cells instead of one; over-layers are not considered; other device materials, especially high-Z materials, are not considered; the SEE sensitive volume is usually not taken as such a cylinder; that some nuclear reactions occurring outside the sensitive volume can also deposit energy inside is not considered; protons/neutrons irradiate the device at random in SEE experiments but not only at a certain position. Despite these shortcomings, this model gives the reasonable aspects of the proton and neutron SEE cross sections in good agreements with the existing SEE experiments and simulations especially in Refs. 3, 6, 32, 29, which convinces us of its reliability in explaining some related issues of the proton and neutron SEEs.

Let's sum up. The average and event-by-event characteristics of the energy deposition from the 1-1000 MeV proton and neutron induced elastic and nonelastic reactions in silicon are investigated systematically in detail with Geant4 simulations. Based on the simulated energy deposition characteristics, many important issues about the proton and neutron SEEs, including the energy dependence of the SEE cross sections, the equivalence and differences of protons and neutrons in SEEs, the effects of the critical charge on SEEs, the geometric effect of the sensitive volumes, and the role of the elastic scattering mechanism in SEEs, are discussed. With the above systematic study, a comprehensive and profound understanding about the proton and neutron SEEs is reached from the physical perspective.

[1] D. Binder, E. C. Smith, and A. B. Holman, IEEE Trans. Nucl. Sci. **22**, 2675 (1975).

[2] E. Petersen, John Wiley & Sons (New Jersey, 2011) p. 1.

[3] D. Lambert, F. Desnoyers, and D. Thouvenot, in *2009 European Conference on Radiation & Its Effects on Components and Systems (RADECS)*, Bruges, Belgium, 2009, IEEE, pp. 148–154.

[4] E. Normand, IEEE Trans. Nucl. Sci. **45**, 2904 (1998).

[5] A. Akkerman, J. Barak, and Y. Lifshitz, IEEE Trans. Nucl. Sci. **49**, 1539 (2002).

[6] F. Miller, C. Weulersse, T. Carrière, N. Guibbaud, S. Morand, and R. Gaillard, IEEE Trans. Nucl. Sci. **60**, 2789 (2013).

[7] S. Serre, S. Semikh, S. Uznanski, J. L. Autran, D. Munteanu, G. Gasiot, and P. Roche, IEEE Trans. Nucl. Sci. **59**, 714 (2012).

[8] C. Inguimbert, S. Duzellier, T. Nuns, and F. Bezerra, IEEE Trans. Nucl. Sci. **54**, 2394 (2007).

[9] M. C. Ursule, C. Inguimbert, and T. Nuns, IEEE Trans. Nucl. Sci. **63**, 2159 (2016).

[10] J. F. Ziegler and W. A. Lanford, Science **206**, 776 (1979).

[11] E. Normand and T. J. Baker, IEEE Trans. Nucl. Sci. **40**, 1484 (1993).

[12] J. Barak, IEEE Trans. Nucl. Sci. **48**, 1937 (2001).

[13] A. Haran, J. Barak, L. Weissman, D. David, and E. Keren, IEEE Trans. Nucl. Sci. **58**, 848 (2011).

[14] J. Barak, IEEE Trans. Nucl. Sci. **53**, 3336 (2006).

[15] A. Akkerman, M. Murat, and J. Barak, Nucl. Instrum. Methods Phys. Res. B **269**, 1630 (2011).

[16] M. Raine, G. Hubert, M. Gaillardin, P. Paillet, and A. Bournel, IEEE Trans. Nucl. Sci. **58**, 2607 (2011).

[17] M. Raine, G. Hubert, M. Gaillardin, L. Artola, P. Paillet, S. Girard, J.-E. Sauvestre, and A. Bournel, IEEE Trans. Nucl. Sci. **58**, 840 (2011).

[18] M. Raine, A. Valentin, M. Gaillardin, and P. Paillet, IEEE Trans. Nucl. Sci. **59**, 2697 (2012).

[19] A. S. Kobayashi, A. L. Sternberg, L. W. Massengill, R. D. Schrimpf, and R. A. Weller, IEEE Trans. Nucl. Sci. **51**, 3312 (2004).
[20] G. Cellere, A. Paccagnella, M. Murat, J. Barak, A. Akkerman, R. Harboe-Sorensen, A. Virtanen, A. Visconti, and M. Bonanomi, Journal of Applied Physics **108**, 124907 (2010).
[21] C. S. Dyer, C. Comber, P. R. Truscott, C. Sanderson, C. Underwood, M. Oldfield, A. Campbell, S. Buchner, and T. Meehan, IEEE Trans. Nucl. Sci. **46**, 1486 (1999).
[22] N. Ikeda, S. Kuboyama, S. Matsuda, and T. Handa, IEEE Trans. Nucl. Sci. **52**, 2200 (2005).
[23] M. A. Clemens, B. D. Sierawski, K. M. Warren, N. A. Dodds, R. A. Weller, R. A. Reed, P. E. Dodd, M. R. Shaneyfelt, J. R. Schwank, S. A. Wender, and R. C. Baumann, IEEE Trans. Nucl. Sci. **58**, 2591 (2011).
[24] S. Buchner, A. Campbell, R. Reed, and S. Kuboyama, IEEE Trans. Nucl. Sci. **51**, 3270 (2004).
[25] T. Mérelle, F. Saigné, B. Sagnes, G. Gasiot, P. Roche, T. Carrière, M.-C. Palau, F. Wrobel, and J.-M. Palau, IEEE Trans. Nucl. Sci. **52**, 1538 (2005).
[26] S. Agostinelli, J. Allisonet, K. Amako *et al.*, "GEANT4-A simulation toolkit," Nucl. Instrum. Methods Phys. Res. A **506**, 250 (2003).
[27] C. S. Dyer, S. N. Clucas, C. Sanderson, A. D. Frydland, and R. T. Green, IEEE Trans. Nucl. Sci. **51**, 2817 (2004).
[28] JESD, "JEDEC Standard JESD89A, Measurement and reporting of alpha particle and terrestrial cosmic ray-induced soft errors in semiconductor devices measurement and reporting of alpha particles and terrestrial cosmic ray-induced soft errors in semiconductor devices," Oct. 2006, see www.jedec.org.
[29] C. Inguimbert, S. Duzellier, R. Ecoffet, and J. Bourrieau, IEEE Trans. Nucl. Sci. **44**, 2243 (1997).
[30] Geant4 User's Guide for Application Developers, available online at: http://geant4.web.cern.ch/geant4/UserDocumentation/UsersGuides/ForApplicationDeveloper/fo/BookForAppliDev.pdf.
[31] A. Valentin, M. Raine, M. Gaillardin, and P. Paillet, Nucl. Instr. Meth. B **52**, 124 (2012).
[32] J. R. Schwank, M. R. Shaneyfelt, J. Baggio, P. E. Dodd, J. A. Felix, V. Ferlet-Cavrois, P. Paillet, D. Lambert, F. W. Sexton, G. L. Hash, and E. Blackmore, IEEE Trans. Nucl. Sci. **52**, 2622 (2005).
[33] D. Lambert, J. Baggio, G. Hubert, P. Paillet, S. Girard, V. Ferlet-Cavrois, O. Flament, F. Saigné, J. Boch, B. Sagnes, N. Buard, and T. Carrière, IEEE Trans. Nucl. Sci. **53**, 1890 (2006).
[34] T. Granlund, B. Granbom, and N. Olsson, IEEE Trans. Nucl. Sci. **51**, 2922 (2004).
[35] M. P. King, R. A. Reed, R. A. Weller, M. H. Mendenhall, R.D. Schrimpf, B. D. Sierawski, A. L. Sternberg, B. Narasimham, J. K. Wang, E. Pitta, B. Bartz, D. Reed, C. Monzel, R. C. Baumann, X. Deng, J. A. Pellish, M. D. Berg, C. M. Seidleck, E. C. Auden, S. L. Weeden-Wright, N. J. Gaspard, C. X. Zhang, and D. M. Fleetwood, IEEE Trans. Nucl. Sci. **60**, 4122 (2013).

Large amplitude ion-acoustic solitons in warm negative ion plasmas with superthermal electrons

Kishan Kumar[a] and M. K. Mishra[b]
Department of Physics, University of Rajasthan, Jaipur 302004, India

Large amplitude ion-acoustic solitons are investigated in a plasma consisting of warm adiabatic positive and negative-ions and hot superthermal electrons having kappa distributions. Using Pseudo-potential method an energy integral equation is derived for the system. The latter is analysed to examine the existence regions of the solitary waves. It is found that negative ion concentration (α), spectral index (k) and ionic temperature ratio (σ_1 or σ_2) significantly influence the characteristic of the solitons. Our numerical analysis shows that the system also supports rarefactive solitons for some selected set of plasma parameters. It is also found that large amplitude ion-acoustic compressive and rarefactive solitons exist simultaneously for the same values of plasma parameters. Further an increase in the superthermality (i.e. decreasing the value of spectral index k) leads to shrinking the existing domain of the large amplitude ion-acoustic solitons. The amplitude of the compressive/rarefactive solitons increases with the increase in negative ion concentration (α). Whereas, on increasing ionic temperature ratio (σ_1 or σ_2) the amplitude of the compressive/rarefactive soliton decreases. The effect of negative-ion concentration (α), temperature ratio of two ion species (σ_1 and σ_2), Mach number (M) and spectral index (k) on the characteristics of solitons are discussed in detail. The results of the present investigation may be helpful to understand the nonlinear ion-acoustic solitary waves in space plasma and laboratory plasmas, where two distinct groups of ions and non-Boltzmann distribution electrons are present.

I. INTRODUCTION

In a dispersive and nonlinear medium solitary waves evolve as solitons due to exact balance of nonlinearity and dispersion. Washimi and Taniuti[1] were the first to show that the propagation of ion-acoustic solitary waves in a plasma can be governed by the Korteweg-de Vries (KdV) equation. The first experimental observation of ion-acoustic solitons was made by Ikezi et al.[2] Later on there has been a great deal of interest in the theoretical[3–8] and experimental[9–15] study of KdV and modified KdV (m-KdV) ion-acoustic solitons in multicomponent plasmas consisting of positive ions, electrons and negative ions. Dauxois and Peyrard[16] have discussed different classes of solitons in detail. Optical solitons are explained by Kivshar and Agarwal.[16] Recently, theoretical investigations on discrete solitons are also studied by several authors.[17] Dark and bright solitons are studied by Balakrishnan et. al[18] and Roy et. al.[19] During the last few years, negative ion plasmas attracted attention due to their importance in Laboratory experiments and space plasma observations. Space plasma observations show that considerable number of negative ions exit in the Earth's ionosphere[20–22] and Cometary comac.[23] Negative ion plasmas are also found in neutral beam sources,[24] plasma processing reactors[25] and in low temperature laboratory experiments.[26–29] Due to the presence of negative ions, the number

[a]Email: kishanjindal08@gmail.com
[b]mishramukesh105@yahoo.com

of electrons decreases to satisfy the charge neutrality condition. The decrease in number of electrons reduces the shielding effect. Due to this most of the phenomena are affected by the presence of negative ions and by the deficiency of electrons.

Space plasma observations by satellite mission shows a high energy tail in the distribution functions due to the presence of superthermal electrons. The electrons distribution obeys a power law and departs from being Maxwellian.[30–34] At a much higher velocity than the electron thermal speed,[35–38] the presence of excess super thermal electrons populations is due to velocity space diffusion. The superthermal plasma behaviour was also observed in various experimental plasma, such as laser-matter interactions or in plasma turbulence.[39] Such behaviour is effectively modelled by a kappa (or generalized Lorentzian) distribution function.[40–43] The kappa distribution was first suggested by Vasyliunas[40] to model space plasmas and was latter adopted by several researchers. The distribution of superthermal particles is characterized by the spectral index k, which is called kappa distribution function. The smaller value of k indicates more superthermal particles in the tail of distribution function and the kappa distribution approaches the Maxwellian as $k \rightarrow \infty$.

Tagare and Reddy[44] have studied the effect of ionic temperature on ion-acoustic solitons in negative ion plasma with nonthermal electrons, using reductive perturbation method. Later on Gill et al.[45] have also studied small amplitude ion-acoustic solitons and double layers in a plasma consisting of positive and negative ions with nonthermal electrons. Using KdV equation Malik and Stroth[46] have investigated the effect of ionic temperature on the propagation characteristics of fast compressive solitons and slow compressive solitons in a negative-ion containing inhomogeneous and magnetized plasma having non isothermal two temperature electrons. El-Labany et al.[47] have studied small amplitude nonlinear ion-acoustic solitary waves in a warm magneto plasma with positive-negative ions and nonthermal electrons. Mehdipoor[48] has derived KdV and m-KdV equations for solitary waves in negative ion plasmas with non-Maxwellian electrons. Recently, Mishra and Jain[49] have studied ion-acoustic waves in magnetized negative ion plasma with nonthermal electrons. Ion acoustic Gardner solitons has been studied by Rehman and Mishra.[50] All these studies are limited to small amplitude case.

Arbitrary amplitude ion-acoustic solitons in a plasma consisting of cold ions and superthermal electrons have studied by Saini et al.[51] They found that only positive ion-acoustic solitons exit in the plasma. Sabry et al.[52] have studied fully nonlinear ion-acoustic solitary waves in a plasma with cold positive-negative ions and nonthermal electrons following Cairn's[53] distribution. They found that both positive and negative potential exit together for the same values of parameters. Recently, El-Labany et al.[54] have studied nonlinear wave propagation of large amplitude ion-acoustic solitary waves in cold negative ion plasmas with superthermal electrons having Kappa distribution. They found that only compressive solitons exist in the system. In all these papers the ionic species are considered to be cold. Whereas, in realistic situation ions always possesses some finite temperature. In addition to this these studies have not reported about the existence of rarefactive solitons.

Large amplitude ion-acoustic solitons in a plasma consisting of warm positive and negative ions with nonthermal electrons following Schamel distribution have been studied by Gill et al.[55] It was predicted that only compressive solitons exit in the plasma. However, their results[55] do not carry over to the cold ionic species case with vanishing ionic temperature ratio (i.e., taking $\sigma_1 = 0$ or $\sigma_2 = 0$; the expressions of n_1, n_2 and $V(\phi)$ become infinite). Later on, Kaur and Gill[56] also studied large amplitude ion-acoustic solitons in warm multispecies plasma with non-Maxwellian electrons having Cairns[53] distribution. In this case also their results[56] do not carry over to the cold ionic species case with vanishing ionic temperature ratio (i.e., taking $\sigma_1 = 0$ or $\sigma_2 = 0$; the expressions of n_1, n_2 and $V(\phi)$ become infinite). It may be noted here that during the derivation of the expressions for n_1 and n_2; these authors[55,56] have used an assumption. This leads to error, and results of the cold ionic species case are then not recoverable in the limit of vanishing ionic temperature ratio. Therefore, it is necessary to reconsider the large amplitude ion-acoustic solitons in a negative ion plasma having finite ion temperature along with superthermal electrons and to find out the possibility of the existence of rarefactive solitons in the plasma.

The purpose of the paper is to study the large amplitude ion-acoustic solitons in a multicomponent collision-less plasma consisting of warm adiabatic positive and negative ions and hot superthermal electrons. Our aim is to determine the range of permitted Mach number values for the existence of large amplitude ion-acoustic solitons in a negative ion plasmas with excess superthermal electrons and also to demonstrate their dependence on superthermality through the function k. Using pseudo-potential method, an energy integral equation for the system has been derived. The conditions for the large amplitude ion-acoustic solitons have been discussed. Our analysis is more general than the earlier studies of large amplitude ion-acoustic solitons and our results reduce to those obtained by Saini et al.[51] in the limit of $\alpha = 0$, $\sigma_1 = 0$, $\sigma_2 = 0$ and $\varepsilon_z = 1$ (i.e. cold positive ion species plasma with superthermal electrons). It is found that the system also supports rarefactive ion-acoustic solitons for the selected set of plasma parameters. On the other hand, El-Labany et al.[54] have reported that only compressive large amplitude solitons exit in the cold negative ion plasmas with superthermal electrons. It is also investigated that compressive and rarefactive solitons exist together for the same values of plasma parameters. This result agrees with previous investigations.[52,54] Further an increase in the superthermality (i.e. decreasing the value of spectral index k) leads to shrinking the existing domain of the large amplitude ion-acoustic solitons. It is also found that the amplitude of the compressive/rarefactive solitons increases with increase in negative ion concentration (α). Whereas, on increasing ionic temperature ratio (σ_1 *or* σ_2), the amplitude of the compressive/rarefactive solitons decreases. The effect of the various plasma parameters on the characteristic of the solitons are discussed in detail.

The paper is organized in the following pattern; In Sec. II, the basic set of equation is given and Sagdeev potential for the system is derived. Applying boundary conditions to Sagdeev potential, the necessary conditions are obtained in Sec. III for the existence of large amplitude ion acoustic solitons. In Sec. IV, numerical analysis and the results obtained for solitons are discussed in detail. Final conclusions of the present investigations are contained in Sec. V.

II. BASIC EQUATIONS AND DERIVATION OF THE SAGDEEV POTENTIAL

We consider a collisionless, unmagnetized, plasma consisting of warm adiabatic positive and negative ions, as well as hot superthermal electrons. The nonlinear behaviour of ion-acoustic waves is described by the following set of normalized equations:

$$\frac{\partial n_p}{\partial t} + \frac{\partial (n_p, u_p)}{\partial x} = 0, \tag{1}$$

$$\frac{\partial n_p}{\partial t} + u_p \frac{\partial u_p}{\partial x} = -\frac{1}{\beta}\frac{\partial \phi}{\partial x} - \frac{3\sigma_1}{\beta z_1} n_p \frac{\partial n_p}{\partial x}, \tag{2}$$

$$\frac{\partial n_n}{\partial t} + \frac{\partial (n_n, u_n)}{\partial x} = 0, \tag{3}$$

$$\frac{\partial n_n}{\partial t} + u_n \frac{\partial u_n}{\partial x} = \frac{\varepsilon_z}{\mu\beta}\frac{\partial \phi}{\partial x} - \frac{3\sigma_2}{\mu\beta z_1} n_n \frac{\partial n_n}{\partial x}, \tag{4}$$

$$\frac{\partial^2 \phi}{\partial x^2} = n_e + \frac{\alpha\varepsilon_z n_n}{(1-\alpha\varepsilon_z)} - \frac{n_p}{(1-\alpha\varepsilon_z)}, \tag{5}$$

$$\beta = \frac{\mu + \alpha{\varepsilon_z}^2}{\mu(1-\alpha\varepsilon_z)}, \varepsilon_z = \frac{z_2}{z_1}, \sigma_1 = \frac{T_{ip}}{T_e}, \sigma_2 = \frac{T_{in}}{T_e}, \alpha = \frac{{n_n}^{(0)}}{{n_p}^{(0)}} and \mu = \frac{M_n}{M_p}$$

In the above equations; n_p, u_p and n_n, u_n are the density and the fluid velocity of the positive and negative ionspecies respectively. The subscripts e, p and n stand for electrons, positive ions and negative ions respectively. Here T_e is the electron temperature and $\mu = M_n/M_p$ is the mass ratio of negative to positive ion species, α is the equilibrium density ratio of negative ion species to positive ion species, ε_z is the charge multiplicity ratio of negative ion species to positive ion species. All the densities n_p, n_n and n_e are normalized with respect to their corresponding equilibrium values, namely ${n_p}^{(0)}$, ${n_n}^{(0)}$ and ${n_e}^{(0)}$ respectively. Which are the unperturbed density of positive ions, negative ions and electrons. The velocities u_p and u_n are normalized with respect to the ion sound speed c_s. The electrostatic wave potential ϕ is normalized by thermal potential T_e/e and the space and time are normalized by Debye length $\lambda_{De} = \left(T_e/4\pi n_e^{(0)}e^2\right)^{1/2}$ and inverse of the ion plasma frequency in the mixture $\omega^{-1}{}_{pi}$ respectively.

Where

$$c_s = \left(\frac{T_e \beta z_1}{M_p}\right)^{1/2} \quad \text{and} \quad \omega^{-1}{}_{pi} = \left(\frac{4\Pi n_e^{(0)} e^2 z_1 \beta}{M_p}\right)^{-1/2}$$

The electrons follow kappa distribution, which is given by

$$n_e = \left(1 - \frac{\phi}{k - \frac{3}{2}}\right)^{-k+\frac{1}{2}} \tag{6}$$

Where k is the spectral index.

To find out the Sagdeev's pseudopotential from the Eqs. (1)–(5); we assume that all the dependent variables depend on a single independent variable $\xi = (x - Mt)$, where $M = v/c_s$ is the Mach number. Using the above transformation in Eqs. (1)–(5), the fluid equations become

$$-M\frac{\partial n_p}{\partial \xi} + \frac{\partial \left(n_p u_p\right)}{\partial \xi} = 0 \tag{7}$$

$$-M\frac{\partial u_p}{\partial \xi} + u_p\frac{\partial u_p}{\partial \xi} = -\frac{1}{\beta}\frac{\partial \phi}{\partial \xi} - \frac{3\sigma_1}{\beta z_1} n_p \frac{\partial n_p}{\partial \xi} \tag{8}$$

$$-M\frac{\partial n_n}{\partial \xi} + \frac{\partial \left(n_n u_n\right)}{\partial \xi} = 0 \tag{9}$$

$$-M\frac{\partial u_n}{\partial \xi} + u_n\frac{\partial u_n}{\partial \xi} = \frac{\varepsilon_z}{\mu\beta}\frac{\partial \phi}{\partial \xi} - \frac{3\sigma_2}{\mu\beta z_1} n_n \frac{\partial n_n}{\partial \xi} \tag{10}$$

$$\frac{d^2\phi}{d\xi^2} = n_e + \frac{\alpha\varepsilon_z}{1 - \alpha\varepsilon_z} n_n - \frac{1}{1 - \alpha\varepsilon_z} n_p \tag{11}$$

Integrating Eqs. (7)–(10) and using the necessary boundary conditions $n_p = 1, n_n = 1, u_p = 0, u_n = 0, \phi = 0$, as the plasma is assumed to be unperturbed at $|\xi| \to \infty$;. We get the quadratic equations.

$$\left(\frac{3\sigma_1}{\beta z_1}\right) n_p^4 - \left(M^2 - \frac{2\phi}{\beta} + \frac{3\sigma_1}{\beta Z_1}\right) n_p^2 + M^2 = 0 \tag{12}$$

$$\left(\frac{3\sigma_2}{\mu\beta z_1}\right) n_n^4 - \left(M^2 + \frac{2\phi\varepsilon_z}{\mu\beta} + \frac{3\sigma_2}{\mu\beta Z_1}\right) n_n^2 + M^2 = 0 \tag{13}$$

From equations (12) and (13); n_p and n_n are given by

$$n_p = \frac{\sqrt{2}M}{\left\{\left(M^2 - \frac{2\phi}{\beta} + \frac{3\sigma_1}{\beta Z_1}\right) + \sqrt{\left(M^2 - \frac{2\phi}{\beta} + \frac{3\sigma_1}{\beta Z_1}\right)^2 - \frac{12\sigma_1 M^2}{\beta Z_1}}\right\}^{1/2}} \tag{14}$$

$$n_n = \frac{\sqrt{2}M}{\left\{\left(M^2 + \frac{2\phi\varepsilon_z}{\mu\beta} + \frac{3\sigma_2}{\mu\beta Z_1}\right) + \sqrt{\left(M^2 + \frac{2\phi\varepsilon_z}{\mu\beta} + \frac{3\sigma_2}{\mu\beta Z_1}\right)^2 - \frac{12\sigma_2 M^2}{\mu\beta Z_1}}\right\}^{1/2}} \tag{15}$$

Our expression of positive ion density (n_p) reduces to those obtained by Saini et al.[46] in the limit of $\alpha = 0$, $\sigma_1 = 0$, $\sigma_2 = 0$ and $\varepsilon_z = 1$(i.e. cold positive ion species plasma with superthermal electrons). Inserting (14)–(15) into (11) multiplying both sides by $\frac{d\phi}{d\xi}$ and integrating once, we obtain

$$\frac{1}{2}\left(\frac{d\phi}{d\xi}\right)^2 + V(\phi) = 0 \tag{16}$$

Where V(ϕ) is the Sagdeev potential which is given by

$$V(\phi) = 1 + \frac{1}{(1-\alpha\varepsilon_z)}\left[\beta M^2(1+\alpha\mu) + \frac{1}{Z_1}(\sigma_1 + \sigma_2\alpha)\right]$$

$$-\frac{\sqrt{2}M\beta}{(1-\alpha\varepsilon_z)}\left[\frac{2\sigma_1 M^2}{\beta Z_1}\left\{\frac{1}{\left(\left(M^2 - \frac{2\phi}{\beta} + \frac{3\sigma_1}{\beta Z_1}\right) + \sqrt{\left(M^2 - \frac{2\phi}{\beta} + \frac{3\sigma_1}{\beta Z_1}\right)^2 - \frac{12\sigma_1 M^2}{\beta Z_1}}\right)^{3/2}}\right\} + \frac{1}{2}\left\{\left(M^2 - \frac{2\phi}{\beta} + \frac{3\sigma_1}{\beta Z_1}\right) + \sqrt{\left(M^2 - \frac{2\phi}{\beta} + \frac{3\sigma_1}{\beta Z_1}\right)^2 - \frac{12\sigma_1 M^2}{\beta Z_1}}\right\}^{\frac{1}{2}}\right]$$

$$-\frac{\alpha\sqrt{2}M\beta\mu}{(1-\alpha\varepsilon_z)}\left[\frac{2\sigma_2 M^2}{\mu\beta Z_1}\left\{\frac{1}{\left(\left(M^2 + \frac{2\phi\varepsilon_z}{\mu\beta Z_1} + \frac{3\sigma_2}{\mu\beta Z_1}\right) + \sqrt{\left(M^2 + \frac{2\phi\varepsilon_z}{\mu\beta Z_1} + \frac{3\sigma_2}{\mu\beta Z_1}\right)^2 - \frac{12\sigma_2 M^2}{\mu\beta Z_1}}\right)^{3/2}}\right\} + \frac{1}{2}\left\{\left(M^2 + \frac{2\phi\varepsilon_z}{\mu\beta Z_1} + \frac{3\sigma_2}{\mu\beta Z_1}\right) + \sqrt{\left(M^2 + \frac{2\phi\varepsilon_z}{\mu\beta Z_1} + \frac{3\sigma_2}{\mu\beta Z_1}\right)^2 - \frac{12\sigma_2 M^2}{\mu\beta Z_1}}\right\}^{\frac{1}{2}}\right]$$

$$-\left\{1 + \frac{2\phi}{3-2k}\right\}^{\frac{3}{2}-k} \tag{17}$$

It may be noted that our expression for Sagdeev potential V(ϕ) reduces to those obtained by Saini et al.[51] in the limit of $\alpha = 0$, $\sigma_1 = 0$, $\sigma_2 = 0$ and $\varepsilon_z = 1$ (i.e. cold positive ion plasma with superthermal electrons). Equation (16) represents the energy integral for a classical particle which oscillates in a potential $V(\phi)$, with a unit mass and a velocity $d\phi/d\xi$ at position ϕ.

III. LARGE AMPLITUDE ION-ACOUSTIC SOLITONS

For the existence of large amplitude Solitons, the Sagdeev potential must satisfy the following conditions[57,58]

$$\text{(i) } V(\phi) \text{ and } \frac{dV(\phi)}{d\phi} = 0, \qquad \text{at } \phi = 0. \tag{18}$$

$$\text{(ii) } V(\phi)\big|_{\phi=\phi_m} = 0 \text{ and } \frac{\partial V}{\partial\phi}\big|_{\phi=\phi_m} < 0, \qquad \text{for } \phi_m < 0$$
$$\text{and } V(\phi)\big|_{\phi=\phi_m} = 0 \text{ and } \frac{\partial V}{\partial\phi}\big|_{\phi=\phi_m} > 0, \qquad \text{for } \phi_m > 0 \tag{19}$$

$$\text{(iii) } V(\phi) < 0, \qquad \text{for } 0 < |\phi| < |\phi_m| \tag{20}$$

From Eqs. (20), the Sagdeev potential $V(\phi)< 0$, when the potential ϕ lies between zero and (ϕ_m), where (ϕ_m) is the amplitude of the soliton. For the rarefactive solitons ($\phi_m < \phi < 0$) and for the compressive solitons ($0 < \phi < \phi_m$). The appearance of localized soliton solution is expressed by condition (18). Furthermore, for the soliton amplitude to be zero as $|\xi| \to \infty$, we require $\frac{d^2V}{d\phi^2} < 0$ at $\phi = 0$. The latter guarantees double roots. Therefore, a particle starting at $\phi = 0$ will be roll back to its initial position making a single transit from the point of reflection, which happens at $\phi = \phi_m$. Here ϕ_m is known as the amplitude of the soliton.

Now from Eqs. (17), the value of Sagdeev potential $V(\phi)$ at $\phi = \phi_m$ must be zero. The minimum and maximum of Mach number for formation of solitons are estimated by physical parameters of this plasma. The lower limit of Mach number ($M_{\min}$) can be obtained by the condition $\frac{d^2V}{d\phi^2} < 0$.

This gives,

$$\frac{1}{(1-\alpha\varepsilon_z)}\left[\frac{z_1}{M^2\beta z_1-3\sigma_1}+\frac{\varepsilon_z{}^2\alpha z_1}{M^2\beta\mu z_1-3\sigma_2}\right]<\frac{(1-2k)}{(3-2k)} \tag{21}$$

This gives the lower limit of Mach number M_{min} and for given fixed values of α, σ_1, σ_2, ε_z and k; the system may support soliton solution for the values of Mach number $M > M_{min}$. The above Eqs. (21) which gives M_{min} also reduces to those obtained by Saini et al.[51] in the limit of $\alpha = 0$, $\sigma_1 = 0$, $\sigma_2 = 0$ and $\varepsilon_z = 1$(i.e. cold positive ion species plasma with superthermal electrons).

On the other hand n_p should be real, which dictates that the value of square root in denominator of Eqs. (14) must have only positive values. It is possible only when the following condition is satisfied

$$\left[\left(M-\sqrt{\frac{3\sigma_1}{\beta z_1}}\right)^2-\frac{2\phi}{\beta}\right]>0 \tag{22}$$

Therefore, the upper limit on the positive potential side is given by

$$\phi_c=\frac{\beta}{2}\left(M-\sqrt{\frac{3\sigma_1}{\beta z_1}}\right)^2 \tag{23}$$

ϕ_c is the critical value of the potential for which the value of n_p remains positive and beyond this the value of n_p become negative . This implies that the amplitude ϕ_m can not exceed the critical value of the potential ϕ_c, which puts a limit on the maximum value of the Mach number. Now substituting this critical value of potential $\phi_c=\frac{\beta}{2}\left(M-\sqrt{\frac{3\sigma_1}{\beta z_1}}\right)^2$ into Equ. (17), we obtain

$$1+\frac{1}{(1-\alpha\varepsilon_z)}\left[\beta M^2(1+\alpha\mu)+\frac{1}{Z_1}(\sigma_1+\sigma_2\alpha)\right]-\frac{M\sqrt{M}}{(1-\alpha\varepsilon_z)}\left\{\frac{4\sigma_1}{z_1}\left(\frac{\beta z_1}{3\sigma_1}\right)^{\frac{3}{4}}+\frac{4\sigma_2\alpha}{z_1}\left(\frac{\mu\beta z_1}{3\sigma_2}\right)^{\frac{3}{4}}\right\}$$
$$-\left\{1+\frac{\beta}{(3-2k)}\left(M-\sqrt{\frac{3\sigma_1}{\beta z_1}}\right)^2\right\}^{\left(\frac{3}{2}-k\right)}\geq 0 \tag{24}$$

The above Equ. (24) gives the upper limit of Mach number M_{max} and for given fixed values of α, σ_1, σ_2, ε_z and k; the system may support soliton solution for the values of Mach number $M < M_{max}$. It may be noted that the above Equ. (24) also reduces to those obtained by Saini et al.[51] in the limit of $\alpha = 0$, $\sigma_1 = 0$, $\sigma_2 = 0$ and $\varepsilon_z = 1$ (i.e. cold positive ion species plasma with superthermal electrons). The range of Mach number M are determined by equation (21) and (24).

IV. DISCUSSION

For the existence of ion-acoustic solitons, the value of the sagdeev potential $V(\phi)$ must be negative between ϕ =0 and $\phi = \phi_m$ (ϕ_m represent the amplitude of the solitons). To investigate the existence of large amplitude ion-acoustic solitons, all the criteria mentioned in Eqs. (18), (19) and (20) should be satisfied. The acceptable range of the Mach number may be found by solving the transcendental equations (21) and (24) for a selected set of the plasma parameters. It may be noted that from Equ. (17) that the Sagdeev potential is a function of various plasmas parameters i.e. negative ion concentration (α), positive ion temperature ratio (σ_1), negative ion temperature ratio (σ_2), spectral index (k) , mass ratio (μ) and Mach number (M).

From Fig. (1), it may be noted that the existence region of large amplitude ion-acoustic solitary wave is confined between the minimum Mach number [due to Equ. (21)] and maximum Mach number [due to Equ. (24)]. It is also investigated that the maximum and minimum Mach number increases as the spectral index (k) increases. It is seen that as the superthermality increases (i.e. the spectral index (k) decreases) the existence region for the large amplitude ion acoustic solitary wave decreases. Fig. (2) shows the variation of that maximum and minimum Mach number with negative ion

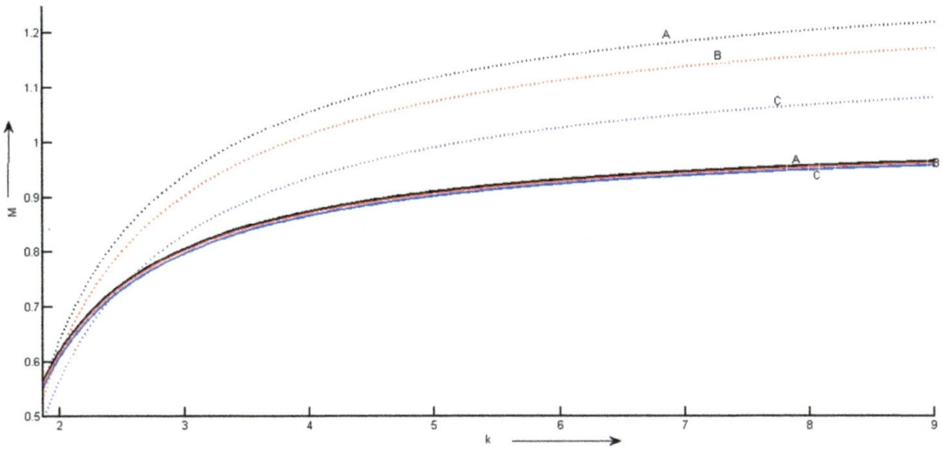

FIG. 1. The Maximum (dotted line) and minimum (solid line) Mach number M versus spectral index k for value of $\sigma_1 = 0.017$, $\sigma_2 = 0.001$, $\mu = 0.1$ having different values of negative ion concentration α. The curves A, B and C represent for $\alpha = 0.00099, 0.0099, 0.0299$, respectively. The existence region for the solitary pulses lie between the maximum and minimum Mach number curves.

concentration α. It is also seen that for fixed value of α, if we increases μ in that case $M_{\min}$ decreases and $M_{\max}$ increases.

A. Compressive solitons

Fig. (3) shows the variation of sagdeev potential $V(\phi)$ with potential ϕ of the large amplitude ion-acoustic compressive solitons for M=1.2, $\sigma_1 = 0.001$, $\sigma_2 = 0.01$, k=5 and $\alpha = 0.00099$ having the different value of $\mu = 0.02$ (solid line), 0.04 (dashed line), 0.8 (dotted line). From the figure it can be seen that the amplitude of compressive soliton decreases with increase in mass ratio μ. This figure also shows that well depth of the Sagdeev potential decreases as the mass ratio is increased. Fig. (4) shows the variation of the sagdeev potential $V(\phi)$ with potential ϕ of the large amplitude ion-acoustic compressive solitons for $\mu = 0.1$, $\sigma_1 = 0.001$, $\sigma_2 = 0.01$, k=5 and $\alpha = 0.00099$ having different value of M = 1.16 (solid line), M = 1.18 (dashed line) and M = 1.2 (dotted line). This figure clearly shows that the amplitude of the compressive soliton increases with increases in Mach

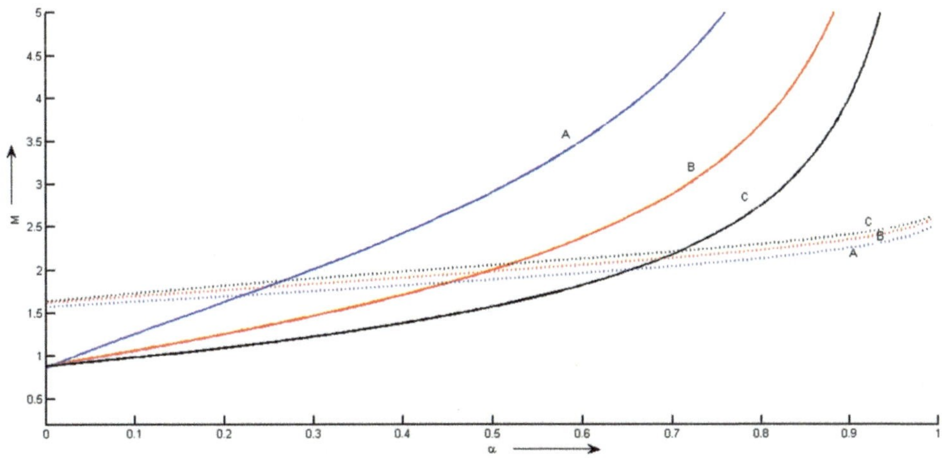

FIG. 2. The Maximum (dotted line) and minimum (solid line) Mach number M versus negative ion concentration α for k=5, $\sigma_1 = 0.01$, $\sigma_2 = 0.001$ having different values of mass ratio. The curves A, B and C represent for $\mu = 0.1, 0.3, 0.8$, respectively. The existence region for the solitary pulses lie between the maximum and minimum Mach number curves.

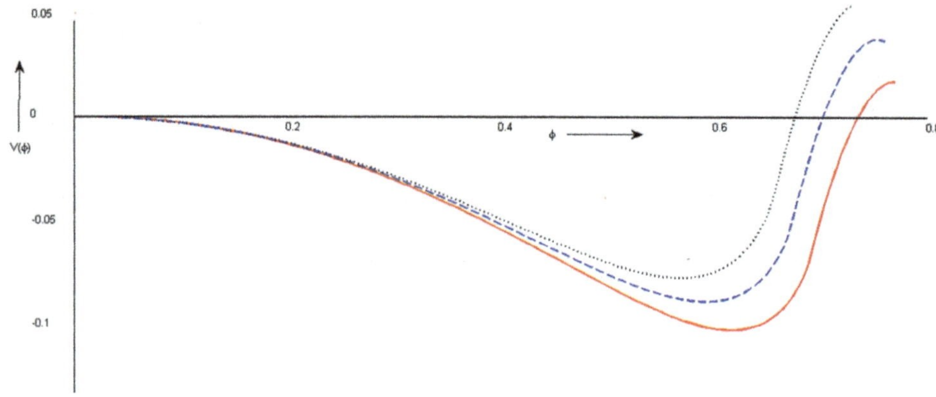

FIG. 3. Variation of the Sagdeev potential $V(\phi)$ with potential ϕ of the large amplitude ion-acoustic compressive solitons for M=1.2, $\sigma_1 = 0.001$, $\sigma_2 = 0.01$, k=5, and $\alpha = 0.00099$ having different values of μ. Where $\mu = 0.02$ (solid line), $\mu = 0.04$ (dashed line), $\mu = 0.8$ (dotted line).

number M. But, in this case well depth of the Sagdeev potential increases as the Mach number is increased. The variation of amplitude of compressive solitons with temperature ratio of positive ion to non-thermal electron species (σ_1) and negative ion to non-thermal electron species (σ_2) have been shown in Fig. (5). From the figure, it is found that an increases in the temperature ratio of (σ_1) or (σ_2), decreases the amplitude for a given set of parameters. The decreases in amplitude on increasing ion temperatures (σ_1 or σ_2) implies that the dispersion is enhanced due to increases in ion thermal motion. It is also found that well depth of the Sagdeev potential decreases slightly in comparison to the previous case. In Fig. (6), we have shown the variation of Sagdeev potential $V(\phi)$ with potential ϕ for three different values of spectral index (k). It is found that an increase in superthermality (i. e. decreasing the value of spectral index k) increases the amplitude of the compressive solitons. The well depth of the Sagdeev potential also increases.

B. Rarefactive solitons

El-Labany et al.[54] have reported that only compressive large amplitude solitary waves exist in the cold negative ion plasmas with superthermal electrons. However, we in our investigation have found that rarefactive large amplitude solitary waves also exist in negative ion plasma having superthermal electrons. We have shown these large amplitude rarefactive solitons in Fig. (7)–(9). Fig. (7) shows $V(\phi)$ as a function of ϕ of the large amplitude ion acoustic rarefactive solitons for M=1.2, $\sigma_1 = 0.1$,

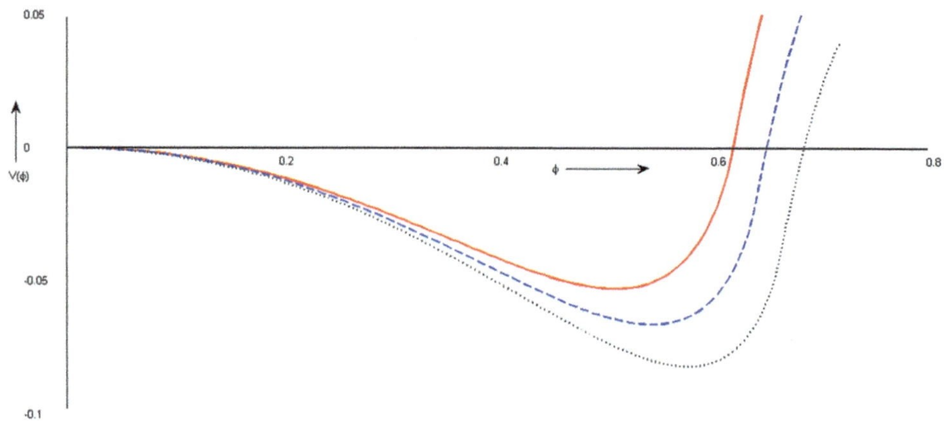

FIG. 4. Variation of the Sagdeev potential $V(\phi)$with potential ϕ of the large amplitude ion acoustic compressive solitons for $\mu = 0.1$,$\sigma_1 = 0.001$, $\sigma_2 = 0.01$, k=5, and $\alpha = 0.00099$ having different value of M. Where M=1.16 (solid line), M=1.18 (dashed line), M=1.2 (dotted line).

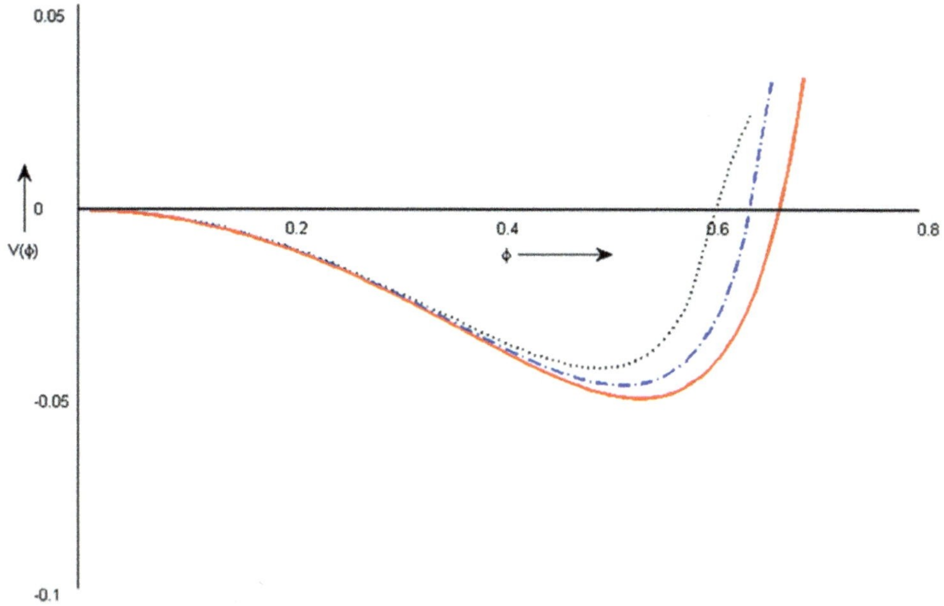

FIG. 5. Variation of the Sagdeev potential $V(\phi)$ with potential ϕ of the large amplitude ion acoustic compressive solitons for $\mu = 0.1$, M = 1.2 k = 7, and $\alpha = 0.00099$ having different value of σ_1 and σ_2. Where $(\sigma_1, \sigma_2) = (0, 0)$ (solid line),(σ_1, σ_2) = (0.002, 0.01) (dashed line), (σ_1, σ_2) =(0.005, 0.03) (dotted line).

$\sigma_2 = 0.001$, $\mu = 0.1$, k=5 and three different values of negative-ion concentration $\alpha = 0.2549$ (solid line), $\alpha = 0.2749$ (dashed line) , $\alpha = 0.2949$ (dotted line). From the figure it can be seen that the amplitude of the rarefactive solitons increases with increase in the negative ion concentration (α) and well depth of the Sagdeev potential also increases. It may be noted that for the given set of fixed parameter the system support rarefactive solitons only for large values of negative ion concentration. In Fig. (8), we have shown the variation of Sagdeev potential $V(\phi)$ with potential ϕ of the large amplitude rarefactive solitons for M=1.2, $\sigma_1 = 0.1$, $\sigma_2 = 0.01$, $\alpha = 0.2549$, k=5 and for three different value of mass ratio (μ). $\mu = 0.3$ (Solid line), $\mu = 0.46$ (dashed line), $\mu = 0.6$ (dotted line). It is found that an increase in mass ratio decreases the amplitude of the rarefactive solitons and well depth of the Sagdeev potential also decreases. The variation of amplitude of rarefactive solitons with temperature ratio (σ_1 and σ_2) have been shown in Fig. (9). In this figure we have shown the variation of $V(\phi)$ with

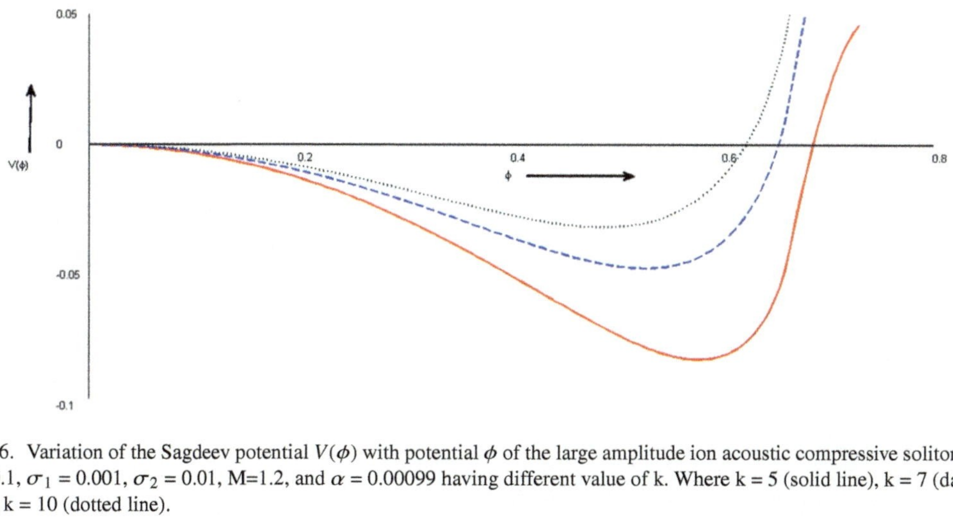

FIG. 6. Variation of the Sagdeev potential $V(\phi)$ with potential ϕ of the large amplitude ion acoustic compressive solitons for $\mu = 0.1$, $\sigma_1 = 0.001$, $\sigma_2 = 0.01$, M=1.2, and $\alpha = 0.00099$ having different value of k. Where k = 5 (solid line), k = 7 (dashed line), k = 10 (dotted line).

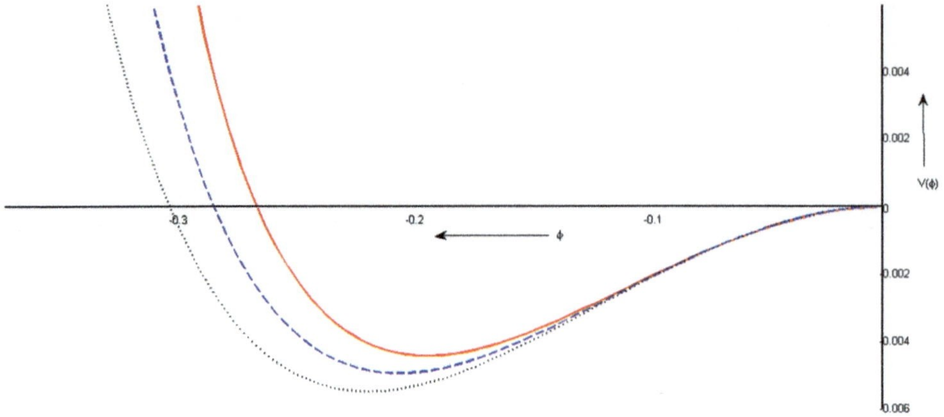

FIG. 7. Variation of the Sagdeev potential $V(\phi)$ with potential ϕ of the large amplitude ion acoustic rarefactive solitons for $\mu = 0.1$, $\sigma_1 = 0.1$, $\sigma_2 = 0.001$, k = 5, and M = 1.2 having different value of α. Where $\alpha = 0.2549$ (solid line), $\alpha = 0.2749$ (dashed line), $\alpha = 0.2949$ (dotted line).

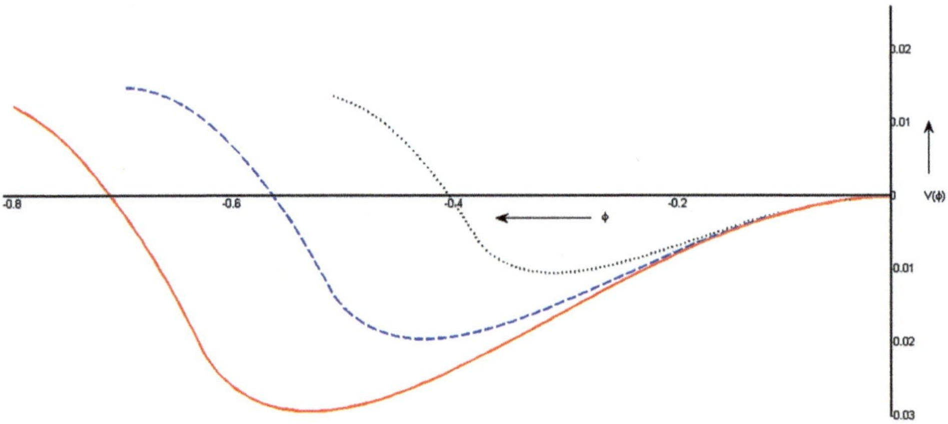

FIG. 8. Variation of the Sagdeev potential $V(\phi)$ with potential ϕ of the large amplitude ion-acoustic rarefactive solitons for M=1.2 , $\sigma_1 = 0.1$, $\sigma_2 = 0.01$, k=5, and $\alpha = 0.2549$ having different values of μ. Where $\mu = 0.3$ (solid line), $\mu = 0.46$ (dashed line), $\mu = 0.6$ (dotted line).

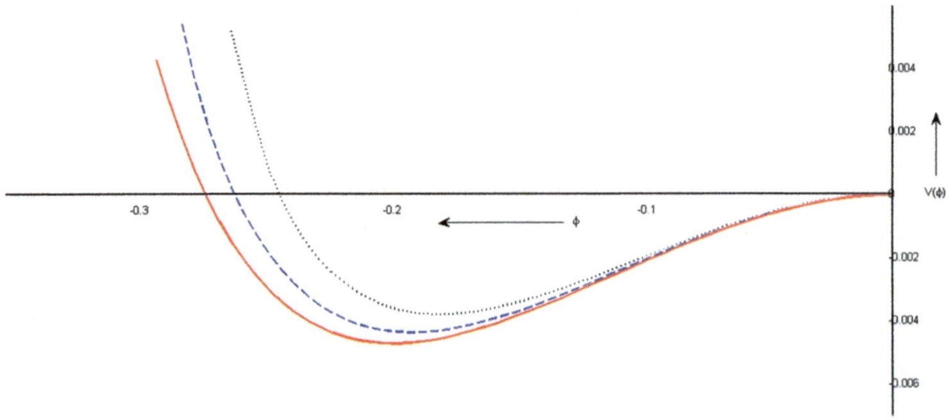

FIG. 9. Variation of the Sagdeev potential $V(\phi)$ with potential ϕ of the large amplitude ion acoustic rarefactive solitons for $\mu = 0.1$, M = 1.2, k = 5, and $\alpha = 0.2549$ having different value of σ_1 and σ_2. Where $(\sigma_1,\sigma_2) = (0, 0)$ (solid line), $(\sigma_1,\sigma_2) = (0.01, 0.002)$ (dashed line), $(\sigma_1,\sigma_2) = (0.1, 0.005)$ (dotted line).

potential ϕ for $\mu = 0.1$, M=1.2, k=5, $\alpha = 0.2549$ and for three different values of (σ_1,σ_2); $(\sigma_1,\sigma_2) = (0, 0)$ (solid line), $(\sigma_1,\sigma_2) = (0.01, 0.002)$ (dashed line), $(\sigma_1,\sigma_2) = (0.1, 0.005)$ (dotted line). From the figure, it is found that an increase in temperature ratio of either species (i.e. positive or negative) decreases the amplitude of the rarefactive solitons.

C. Coexistence of compressive and rarefactive solitons

During numerical investigations, we have also found that for the selected set of parameters, compressive and rarefactive solitons exist simultaneously. We have shown these in Fig. (10) –(12).

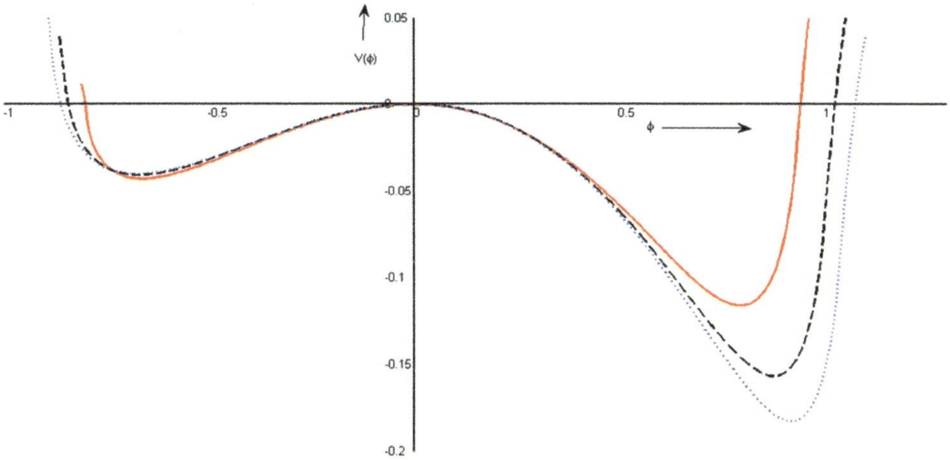

FIG. 10. Variation of the Sagdeev potential $V(\phi)$ with potential ϕ of the large amplitude ion acoustic compressive and rarefactive solitons for $\mu = 0.8$, M = 1.2 , $\sigma_1 = 0.001$, $\sigma_2 = 0.00001$ and k = 12 having different values of α. Where $\alpha = 0.15$ (solid line), $\alpha = 0.18$ (dashed line), α= 0.1949 (dotted line).

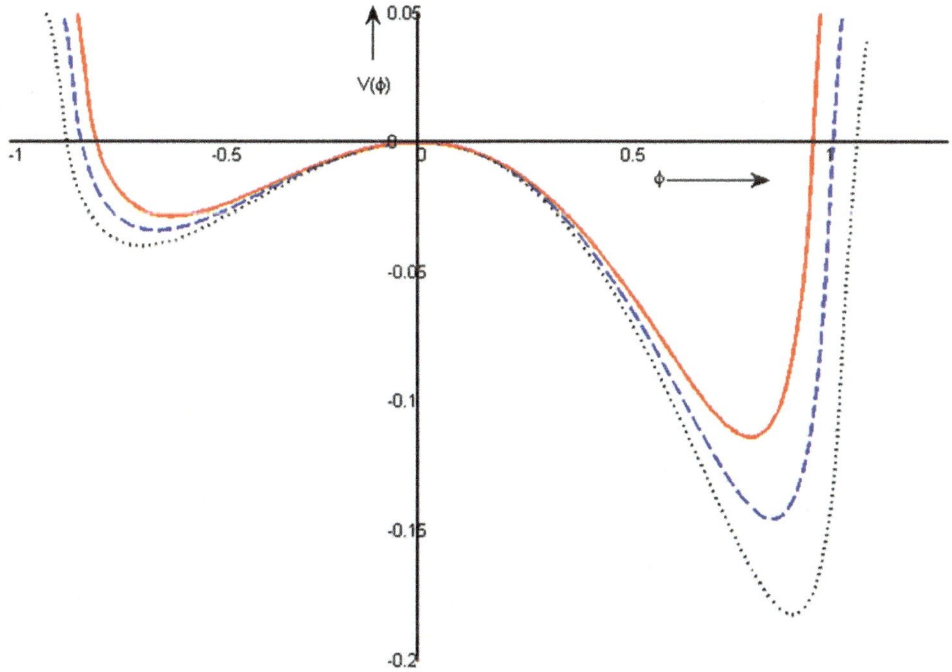

FIG. 11. Variation of the Sagdeev potential $V(\phi)$ with potential ϕ of the large amplitude ion acoustic compressive and rarefactive solitons for $\mu = 0.8$, k = 12, $\sigma_1 = 0.001$, $\sigma_2 = 0.0001$ and $\alpha = 0.1949$ having different values of M. Where M= 1.16 (solid line), M= 1.18 (dashed line), M = 1.2 (dotted line).

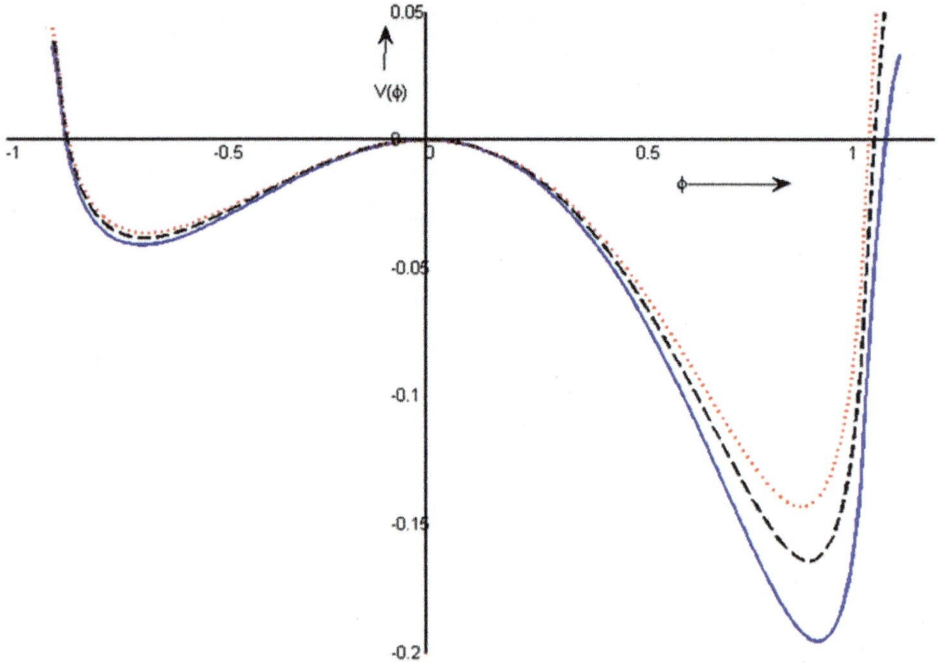

FIG. 12. Variation of the Sagdeev potential $V(\phi)$ with potential ϕ of the large amplitude ion acoustic compressive and rarefactive solitons for $\mu = 0.8$, M = 1.2, $\sigma_1 = 0.001$, $\sigma_2 = 0.0001$ and $\alpha = 0.1949$ having different values of k. Where k = 11 (solid line), k = 14 (dashed line), k = 18 (dotted line).

In Fig. (10), we have shown the variation of Sagdeev potential $V(\phi)$ with potential ϕ of the large amplitude ion-acoustic compressive and rarefactive solitons for M = 1.2, $\mu = 0.8$, $\sigma_1 = 0.001$, $\sigma_2 = 0.00001$, k = 12 and three different values of negative ion-concentration (α), 0.15 (solid line), 0.18 (dashed line), 0.1949 (dotted line). It is found that there exist both compressive and rarefactive solitons for the selected set of plasma parameters. It is also found that on increasing the value of α the amplitude of the compressive/rarefactive solitons increases. It is also observed that the amplitude of the compressive soliton is large in comparison to rarefactive soliton. It is also seen that the well depth of the Sagdeev potential is large in the case of compressive soliton, whereas it is small in rarefactive soliton. Fig. (11) shows Sagdeev potential $V(\phi)$ as a function of ϕ for $\mu = 0.8$, k=12, $\sigma_1 = 0.001$, $\sigma_2 = 0.0001$, $\alpha = 0.1949$ and three different value of Mach number (M), M = 1.16 (solid line), M= 1.18 (dashed line) , M = 1.2 (dotted line). In this case also the system supports both compressive and rarefactive solitons simultaneously. This figure shows that on increasing the value of Mach number (M), the amplitude of compressive/rarefactive solitons increases. It is also found that the amplitude of the compressive solitons is large in comparison to the rarefactive solitons. This figure also shows that the well depth of the Sagdeev potential in compressive solitons is large in comparison to rarefactive solitons. In Fig. (12) we have shown the variation of Sagdeev potential $V(\phi)$ with potential ϕ for $\mu = 0.8$, M=1.2, σ_1 =0.001, $\sigma_2 = 0.0001$, $\alpha = 0.1949$ and three different values of spectral index k, k = 11 (solid line), k = 14 (dashed line), k = 18 (dotted line). It is found that for these set of parameters also both compressive and rarefactive solitons exist simultaneously. It is found that on increasing superthermality (i. e. decreasing the value of k) the amplitude of compressive/rarefactive solitons increases. It can be seen from the graph that amplitude of the compressive solitons is large in comparison to rarefactive solitons. It also show that well depth of the Sagdeev potential is large in compressive soliton case in comparison to rarefactive solitons.

V. CONCLUSIONS

We have studied the large amplitude ion acoustic solitary waves in a collisionless, unmagnetized plasma consisting of warm adiabatic positive and negative ions with superthermal electrons having kappa distribution. It is found that the system also supports large amplitude rarefactive ion acoustic solitons for the selected set of plasma parameters. Whereas, El-Labany et al.[54] have reported that in negative ion plasma having superthermal electrons the system supports only compressive ion acoustic solitons. It is also investigated that large amplitude ion-acoustic compressive and rarefactive solitons exist together for the same values of plasma parameters. Our investigations also reveal that an increase in the superthermality (i.e. decreasing the value of spectral index k) leads to shrinking the existence domain of large amplitude ion acoustic solitons. The amplitude of the rarefactive/compressive solitons increases with the increase in negative ion concentration (α).Whereas, on increasing ionic temperature ratio (σ_1 or σ_2), the amplitude of the compressive/rarefactive soliton decreases. The results of the present investigation may be helpful to understand the nonlinear ion-acoustic solitary waves in space plasma and laboratory plasmas, where two distinct groups of ions and non-Boltzmann distribution electrons are present.

[1] H. Washimi and T. Taniuti, Phys. Rev. Lett. **17**, 996 (1966).
[2] H. Ikezi, R. J. Taylor, and D. R. Baker, Phys. Rev. Lett. **25**, 11 (1970).
[3] G. C. Das and S. G. Tagare, Plasma Phys. **17**, 1025 (1975).
[4] S. Watanabe, Journal of the Physical Society of Japan **53**, 950 (1984).
[5] B. C. Kalita and M. K. Kalita, Phys. Fluids B **2**, 674 (1990).
[6] M. K. Mishra and R. S. Chhabra, Phys. Plasmas **3**, 12 (1996).
[7] D. K. Singh and H. K. Malik, Physics of Plasmas **13**, 082104 (2006).
[8] M. K. Mishra, R. S. Tiwari, and J. K. Chawla, Phys. Plasmas **19**, 062303 (2012).
[9] Y. Nakamure and I. Tsukabayashi, Phys. Rev. Lett. **52**, 2356 (1984).
[10] Y. Nakamura, J. Plasma Phys. **38**, 461 (1987).
[11] J. L. Cooney, M. T. Gavin, and K. E. Lonngren, Phys. Fluids **B3**, 2758 (1991).
[12] J. E. Williams, J. L. Cooney, D. W. Dossey, and K. E. Lonngren, Phys. Rev. Lett. A **45**, 5897 (1992).
[13] Y. Nakamura, H. Bailung, and K. E. Lonngren, Physics of Plasmas **6**, 9 (1999).
[14] N. C. Adhikary, M. K. Deka, and H. Balunga, Physics of Plasma **16**, 063701 (2009).
[15] H. Bailung, S. K. Sharma, and Y. Nakamura, Phys. Rev. Lett. **107**, 255005 (2011).
[16] T. Dauxois and M. Peyrard, "*Physics of solitons*",(cambridge Univ. Press, Cambridge, 2006); Y. S. Kivshar and G. P. Agarwal, "*Optical solitons: from Fibres to Photonic Crystals*" (Academic press, London, 2002).
[17] S. Flach and A. V. Gorbach, Phys. Rep. **467**, 1 (2008); Y. V. Khartasov, B. A. Malomed, and L. Torner, Rev. Mod. Phys. **83**, 247 (2011); P. G. Kevrekidis, IMA J. Appl. Maths. **76**, 389 (2011).
[18] R. Balakrishnan, I. I. Satija, and C. W. Clark, Phys. Rev. Lett. **103**, 230403 (2009); R. Balakrishnan, Phys. Rev. B **39**, 174 (1989).
[19] R. Roy, P. Giri, B. Das, K. Choudhary, A. Ghosal, and A. K. Bandyopadhyay, AIP Advances **4**, 087101 (2014).
[20] H. Massey, *Negative ions*, 3^{rd}ed, (Cambridge University press, Cambridge, 1976), p. 663.
[21] W. Swider, in *ionospheric modelling*, edited by J. N. Korenkov (Birkhauser, Basel, 1988), p. 403.
[22] Yu. I. Portnyagin, O. F. Klyuev, A. A. Shidlovsky, A. N. Evdokimov, T. W. Buzdigar, P. G. Matukhin, S. G. Pasynkov, K. N. Shamshev, V. V. Sokolov, and N. D. Semkin, Adv. Space Res. **11**, 89 (1991).
[23] P. H. Chaizy, H. Reme, J. A. Sauvaud, C. D'uston, R. P. Lin, D. F. Larson, D. L. Mitchell, K. A. Anderson, C. W. Carlson, A. Korth, and D. A. Mendis, Nature (London) **349**, 393 (1991).
[24] M. Bacal and G. W. Hamilton, Phys. Rev. Lett. **42**, 1538 (1979).
[25] R. A. Gottscho and C. E. Gaebe, IEEE Trans. Plasma Sci. **14**, 92 (1986).
[26] J. Jacquinot, B. D. Mcvey, and J. E. Scharer, Phys. Rev. Lett. **39**, 88 (1977).
[27] Y. Nakamara, T. Odagiri, and I. Tsukabayashi, Plasma Phys. Controlled Fusion **39**, 115004 (2001).
[28] A. Weingarten, R. Arad, Y. Maron, and A. Fruchtman, Phys. Rev. Lett. **87**, 115004 (2001).
[29] R. Ichiki, S. Yoshimura, T. Watanabe, N. Nakamura, and Y. Kawai, Phys. Plasmas **9**, 4481 (2002).
[30] V. Pierrard and J. Lemaire, J. Geophys. Res. **101**, 7923, doi:10.1029/95ja03802 (1996).
[31] S. P. Christon, D. G. Mitchell, D. J. Williams, L. A. Frank, C. Y. Huang, and T. E. Eastman, J. Geophys. Res. **93**, 2562, doi: 10.1029/ja093ia04p02562 (1988).
[32] M. Maksimovic, V. Pierrard, and P. Riley, Geophys. Res. Lett. **24**, 1151, doi:10.1029/97gl00992 (1997).
[33] S. M. Krimigis, J. F. Carbary, E. P. Keath, T. P. Armstrong, L. J. Lanzerotti, and G. Gloeckler, J. Geophys. Res. **88**, 8871, doi:10.1029/ja088ia11p08871 (1983).
[34] V. Pierrard, H. Lamy, and J. Lemaire, J. Geophys. Res. **109**, A02118 (1996).
[35] M. P. Leubner, Phys. Plasmas **11**, 1308 (2004).
[36] V. Formisano, G. Moreno, F. Palmiotto, and P. C. Hedgecock, J. Geophys. Res. **78**, 3714, doi:10.1029/ja078i019p03714 (1973).
[37] E. Marsch, K. H. Muhlhauser, R. Schwenn, H. Rosenbauer, W. Phillip, and F. M. Neubauer, J. Geophys. Res. **87**, 52, doi: 10.1029/ja087ia01p00052 (1982).
[38] M. Lazar, R. Schlickeiser, S. Poedts, and R. C. Tautz, Mod. Not. R. Astron. Soc. **390**, 168 (2008).
[39] S. Magni, H. E. Roman, R. Barni, C. Riccardi, T. Pierre, and D. Guyomarc'h, Phy. Rev. E **72**, 0264403 (2005).
[40] V. M. Vasyliunas, J. Geophys. Res. **73**, 2839, doi:10.1029/ja073i009p02839 (1968).
[41] M. A. Hellberg, R. L. Mace, R. J. Armstrong, and G. Karlstad, J. Plasmas Phys. **64**, 433 (2000).
[42] A. Hasegawa, K. Mima, and M. Duong-Van, Phys. Rev. Lett. **54**, 2608 (1985).
[43] T. K. Baluku and M. A. Hellberg, Phys. Plasmas **15**, 123705 (2008).

[44] S. G. Tagare and R. Virupakshi Reddy, Plasma Physics and Controlled Fusion **29**, 671 (1987).
[45] T. S. Gill, P. Bala, H. Kaur, N. S. Saini, S. Bansal, and J. Kaur, Eur. Phys. J. D **31**, 91 (2004).
[46] H. K. Malik and U. Stroth, Plasma Sources Sci. Technol. **17**, 035005 (2008).
[47] S. K. El-Labany, R. Sabry, W. F. El-Taibany, and E. A. Elghmaz, Physics of Plasmas **17**, 042301 (2010).
[48] M. Mehdipoor, Astrophys. Space Sci. **348**, 115 (2013).
[49] M. K. Mishra and S. K. Jain, J. Plasma Phys. **79**, 893 (2013).
[50] M. A. Rehman and M. K. Mishra, Phys. Plasma **23**, 012302 (2016).
[51] N. S. Saini, I. Kourakis, and M. A. Hellberg, Physics of Plasmas **16**, 062903 (2009).
[52] R. Sabry, W. M. Moslem, and P. K. Shukla, Physics of Plasmas **16**, 032302 (2009).
[53] R. A. Cairns, A. A. Mamun, R. Bingham, R. Bostrom, R. O. Dendy, C. M. C. Nairn, and P. K. Shukla, Geophys. Rev. Lett. **22**, 2709, doi:10.1029/95gl02781 (1995).
[54] S. K. El-Labany, R. Sabry, E. F. El-Shamy, and D. M. Khedr, J. Plasma Physics **79**, 613 (2013).
[55] T. S. Gill, H. Kaur, and N. P. Singh, Physics of Plasmas **10**, 3927 (2003).
[56] H. Kaur and T. S. Gill, Journal of Physics: Conference Series **208**, 012080 (2010).
[57] R. Z. Sagdeev, Rev. Mod. Phys. **51**, 11 (1979).
[58] S. K. Jain and M. K. Mishra, Astro. Phys. Space Sci. **345**, 395 (2013).

22

Effect of ionic substitutions on the magnetic properties of strontium hexaferrite: A first principles study

Vivek Dixit,[1] Seong-Gon Kim,[1,a] Jihoon Park,[2] and Yang-Ki Hong[3]
[1]*Department of Physics and Astronomy, Mississippi State University, Mississippi State, MS 39762, USA and Center for Computational Sciences, Mississippi State University, Mississippi State, MS 39762, USA*
[2]*Korea Institute of Materials Science, Changwon, Gyeongnam 51508, Republic of Korea*
[3]*Department of Electrical and Computer Engineering and MINT Center, The University of Alabama, Tuscaloosa, AL 35487, USA*

We investigated the effect of substitution of various ions at the Fe sites on magnetic properties of strontium hexaferrite ($SrFe_{12}O_{19}$) using first principles method based on density functional theory. The site occupancies of substituted atoms were estimated by calculating the substitution energies of different configurations. The formation probabilities of configurations were used to calculate magnetic properties of substituted strontium hexaferrite. A total of 21 elements (M) were screened for their possible substitution in strontium hexaferrite, $SrFe_{12-x}M_xO_{19}$ with $x = 0.5$. In each case the site preference of the substituted atom and the magnetic properties were calculated. We found that Bi, Sb, Sn, and Sc can effectively increase the magnetization and P, Co, Al, Ga, and Ti can enhance the anisotropy field when substituted into strontium hexaferrite.

I. INTRODUCTION

M-type strontium hexaferrite, $SrFe_{12}O_{19}$(SFO) is one of the most commonly used materials for hard magnets. It is used in electric motors, microwave devices, recording media, magneto-optic media, telecommunication, and electronic industry etc. SFO has large saturation magnetization, high coercivity, high Curie temperature, large uniaxial magnetic anisotropy, and excellent chemical stability.[1–5] Ferrite magnets are produced in large quantities as low-cost moderate-performance permanent magnets. Even a minor improvement in the magnetic properties of SFO has a major technological importancc.

Several investigations on the substituted strontium hexaferrite have been carried out to further improve its magnetic and electrical properties. Ghazaiel et al have studied the substitution of Al, Bi, Cr, and Mn in $BaFe_{12-x}M_xO_{19}$ for $x = 1.0$.[6] Other recent investigations include: Ga,[7–13] In,[8,14–16] Er-Ni,[5] Co-Ti,[17] Nd-Co,[18] Mg-Ti,[19] Co-Zr,[20] Bi-Cr,[21] La,[22,23] Sm,[24] Pr,[25] Nd,[26] Sn-Mg,[4] and Zn-Sn.[27–29] However, there have been limited first-principles studies on M-type hexaferrite compared to experimental studies. Fang et al investigated the electronic structure of SFO using density-functional theory (DFT).[30] Park et al have calculated the exchange interaction of SFO from the differences of the total energy of different collinear spin configurations.[31] Magnetism in La-substituted SFO has been studied using DFT.[32,33] Zn-Sn substituted SFO has been studied by Liyanage et al.[29] The site occupancies and magnetic properties of Al, In, and Ga-substituted SFO have been investigated by Dixit et al.[34,35]

The objective of this study is to investigate the changes induced by the substitution of various ions on the magnetization and magnetic anisotropy of SFO. In this work, we used first-principles

[a]Corresponding author: sk162@msstate.edu

total-energy calculations to study the site occupation and magnetic properties of M-substituted strontium hexaferrite $SrFe_{12-x}M_xO_{19}$ (SFMO) with $x = 0.5$. We determined the structure of various configurations with substituted M atoms at different Fe sites and computed their formation probabilities, which was further used to compute the weighted average of the magnetic properties.

II. METHODS

SFO has a hexagonal magnetoplumbite crystal structure that belongs to $P6_3/mmc$ space group. Fig. 1(a) shows a unit cell of SFO which contains 64 atoms of two formula units. Magnetism in SFO arises from Fe^{3+} ions occupying five crystallographically inequivalent sites in the unit cell, three octahedral sites ($2a$, $12k$, and $4f_2$), one tetrahedral site ($4f_1$), and one trigonal bipyramidal site ($2b$). SFO is a ferrimagnetic material that has 16 Fe^{3+} ions with spins in the majority direction ($2a$, $2b$, and $12k$ sites) and 8 Fe^{3+} ions with spins in the minority direction ($4f_1$ and $4f_2$ sites) as indicated in Fig. 1(b). In this work, DFT calculations were performed with projector augmented wave (PAW) potential as implemented in the Vienna Ab initio Simulation Package (VASP).[36,37] All calculations were spin polarized according to the ground state ferrimagnetic ordering of Fe spins.[30,38] Electronic wave functions were expanded in a plane-wave basis with a energy cutoff of 520 eV. Reciprocal space was sampled with a $7 \times 7 \times 1$ Monkhorst-Pack k-point mesh[39] with a electron smearing of 0.2 eV applied through the Methfessel-Paxton method.[40] Relaxation of the electronic degrees of freedom was performed until the change in free energy and the band structure energy was less than 10^{-7} eV. Geometric optimization was performed to relax the positions of ions and cell shape until the change in total energies between two ionic steps was less than 10^{-4} eV. Ionic positions, cell volume, and cell shape were allowed to change during the relaxation calculations. Exchange-correlation effect was described using the Perdew-Burke-Ernzerhof (PBE)[41] generalized gradient approximation. To improve the description of localized Fe $3d$ electrons, we employed the GGA+U

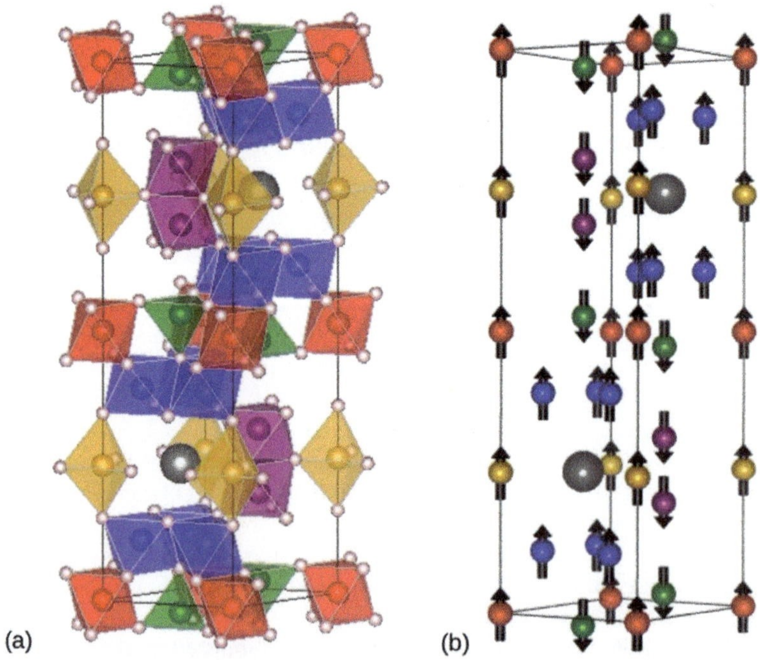

FIG. 1. (a) A unit cell of strontium hexaferrite containing two formula units. Two large grey spheres are Sr atoms and small pink spheres are O atoms. Colored spheres enclosed by polyhedra formed by O atoms represent Fe^{3+} ions in different inequivalent sites: $2a$ (red), $2b$ (yellow), $12k$ (blue), $4f_1$ (green), and $4f_2$ (purple). (b) The direction of local magnetic moment on Fe centers are shown by arrows. Magnetic moments of the Fe^{3+} ions at $2a$, $2b$ and $12k$ point upwards, while downwards at $4f_1$ and $4f_2$. For clarity, O atoms are not shown.

method in the simplified rotationally invariant approach described by Dudarev et al.[42] This method requires an effective U parameter (U_{eff}) to be equal to the difference between the Hubbard parameter U and the exchange parameter J. For Fe atoms U_{eff} was set to 3.7 eV, based on previous study.[29] The U_{eff} for all other elements was set to be zero. The Magnetocrystalline anisotropy energy was calculated by first performing an accurate collinear calculation in the ground state, and then the spin-orbit coupling calculations were performed in two different spin orientations within non-collinear setup.

III. RESULTS AND DISCUSSION

Magnetic properties of SFO can be tuned by the substitution of impurity atoms at the Fe sites. Each substitution is unique and modifies the magnetic properties in a distinct manner. Once we understand how different substitutions change the magnetic properties of SFO, we can choose a suitable combination of elements to achieve desired effect on the magnetic properties of the material. We investigated 21 elements, which can attain the oxidation state of 3+ like the Fe in SFO. There are five different types of Fe sites in SFO, and therefore each foreign element can be substituted in five different ways. We label these configurations according to their substitution site. To investigate the likelihood of formation of each configuration, the substitution energy was calculated using the following expression:

$$E^i_{sub} = E^i(\text{SFMO}) - [E(\text{SFO}) - E(\text{Fe}) + E(\text{M})] \quad (1)$$

E^i(SFMO) is the total energy per unit cell of substituted SFO in configuration [i], and E(SFO) is the total energy per unit cell of SFO. E(Fe) is the cohesive energy of Fe in its ground state bcc structure and E(M) is the cohesive energy of element M in its most stable crystal structure. A negative value of E_{sub} implies that the unit cell of the substituted SFO has lower energy than pure SFO and therefore such a substitution at Fe sites is thermodynamically favorable. Out of 21 elements that were investigated, 12 elements have one or more configurations with a negative substitution energy. A high negative value of E_{sub} of elements like Sc, Al, and Ti indicates that these elements can be easily substituted. For example, Al has $E_{sub} = -6.04$ eV, and it has been shown experimentally that eight Fe atoms of $SrFe_{12}O_{19}$ can be substituted by Al atoms.[1] The substitution energies for the element considered in this study are plotted in Fig. 2 in the ascending order of their lowest substitution energies. Substitutions up to Sb show negative E_{sub}'s. For the subsequent analysis, we will consider the elements with a negative E_{sub} only.

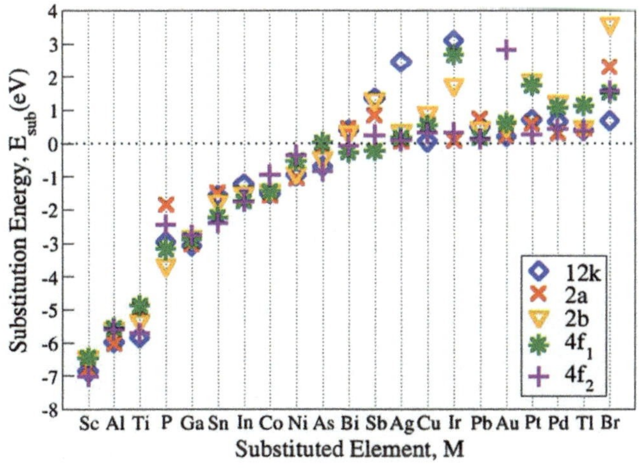

FIG. 2. Substitution energies per unit cell, E_{sub}, of element M in $SrFe_{11.5}M_{0.5}O_{19}$. The elements are listed in the ascending order of their lowest substitution energies.

To understand how magnetic properties of SFO change due to substitution, it is imperative to study the site preference of the substituted atoms. Here, we consider the $x = 0.5$ case which corresponds to the substitution of one M element in the unit cell. There are five inequivalent Fe sites in SFO unit cell. Therefore, one M atom can be substituted in five different ways, resulting in five different configurations, viz. [12k], [2a], [2b], [4f_1], and [4f_2]. In the first order approximation, we assume that the doped atoms are uniformly distributed in the sample and there is no concentration gradient. The E^i_{sub} for each configuration with different M atoms were calculated, which was then used to calculate the relative substitution energy, ΔE. The ΔE is the difference between E^i_{sub} of a particular configuration and the most stable configuration. Therefore, the configuration with $\Delta E = 0.0$ corresponds to the configuration with the lowest E^i_{sub} and thermodynamically most preferable site for the substituted atom. For example, Co has $\Delta E = 0.0$ for the [2a] configuration. It means that at 0 K $SrFe_{23.5}Co_{0.5}O_{19}$ system has the lowest energy when Co atom substitutes the Fe atom at the 2a site. The relative substitutional energies for all 12 elements are presented in Table I.

In order to understand the site preference of substituted M atoms at elevated temperatures, we computed the formation probability of configuration [i] using the Maxwell-Boltzmann statistical distribution[43]

$$P_i = \frac{g_i \exp(-\Delta G_i/k_B T)}{\sum_j g_j \exp(-\Delta G_j/k_B T)} \tag{2}$$

where g_i is the multiplicity of configuration [i] (number of equivalent configurations), and

$$\Delta G_i = \Delta E^i_{\text{sub}} + P\Delta V_i - T\Delta S_i \tag{3}$$

ΔG_i is the change of the free energy of configuration [i] relative to that of the ground state configuration. ΔE^i_{sub}, ΔV_i, and ΔS_i are the substitution energy change, volume change, entropy change for configuration [i]; P and k_B are pressure and Boltzmann constant. Formation probability, P_{1000}, at 1000 K (typical annealing temperature of SFO) is required to calculate magnetic properties of substituted SFO (Eq. 7). This is due to the fact that during annealing process substituted atoms get chance and sufficient energy to overcome the activation barrier due to bonds with neighbouring atoms and occupy the thermodynamically most favorable sites. Table I shows the values of P_{1000} for different configurations. Previously, Sc substituted barium hexaferrite was studied by Somogyvári et al.[44] On the basis of Mössbauer spectroscopy and neutron diffraction the authors concluded that substituted Sc atoms occupy the 4f_2, 12k and 2a sites. This result is in agreement with site occupancies of Sc shown in Table I. Our prediction, that Sn atoms most likely occupy the 4f_2 and 12k sites, is also consistent with an earlier experimental study.[45] Similarly, results of experimental studies by Choi and Albanese,[46,47] which used Mössbauer spectroscopy, show that substituted Al atoms most likely

TABLE I. Relative substitution energies and formation probabilities for different configurations of $SrFe_{11.5}M_{0.5}O_{19}$. M is the substituted element, ΔE is the change in substitution energy in eV compared to the most stable configuration, P_{1000} is the formation probability at 1000 K.

	[12k]		[2a]		[2b]		[4f_1]		[4f_2]	
M	ΔE	P_{1000}	ΔE	P_{1000}	ΔE	P_{1000}	ΔE	P_{1000}	ΔE	P_{1000}
Sc	0.158	0.32	0.234	0.03	0.580	0.00	0.566	0.00	0.000	0.66
Ti	0.000	0.96	0.840	0.00	0.498	0.00	0.966	0.00	0.116	0.08
Co	0.061	0.65	0.000	0.22	0.151	0.04	0.144	0.08	0.616	0.00
Ni	0.140	0.54	0.000	0.46	0.471	0.00	0.515	0.00	0.676	0.00
Al	0.033	0.80	0.000	0.20	0.437	0.00	0.468	0.00	0.412	0.00
In	0.014	0.53	0.030	0.07	0.268	0.01	0.007	0.19	0.000	0.21
P	0.707	0.00	1.808	0.00	0.000	0.99	0.503	0.00	1.212	0.00
Bi	0.673	0.00	0.714	0.00	0.594	0.00	0.000	0.90	0.198	0.09
Sb	1.577	0.00	1.056	0.00	1.531	0.00	0.000	0.99	0.464	0.00
Ga	0.000	0.85	0.029	0.10	0.245	0.01	0.186	0.03	0.311	0.01
As	0.135	0.38	0.710	0.00	0.375	0.00	0.849	0.00	0.000	0.61
Sn	0.176	0.28	0.915	0.00	0.640	0.00	0.908	0.00	0.000	0.72

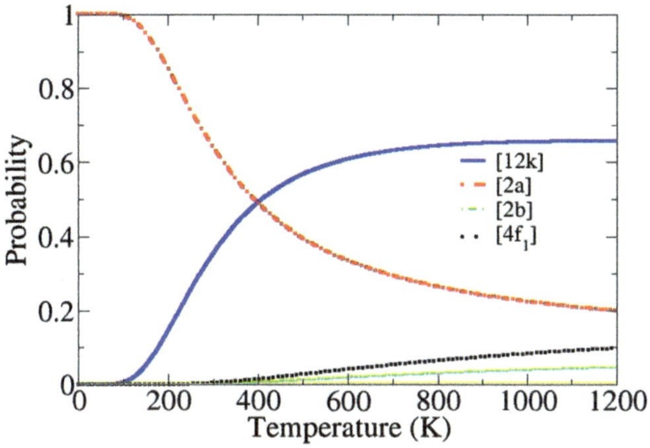

FIG. 3. Temperature dependence of the formation probability of different configurations of $SrFe_{11.5}Co_{0.5}O_{19}$. The configurations with negligible probability are not shown.

occupy the $12k$ site followed by the $2a$ site, match with results shown here. The substituted elements Ti, P, Bi, and Sb mainly occupy the $12k$, $2b$, $4f_1$, and $4f_1$ sites, respectively. In the case of As, Sn, and Sc substitutions the most probable site is the $4f_2$ followed by the $12k$. Co and Ni atoms prefer to occupy the $12k$ and $2a$ sites. The formation probabilities of different configurations of $SrFe_{11.5}M_{0.5}O_{19}$ for M = Co, and Sb, calculated using Eq. 2, are shown in Fig. 3, and Fig. 4. Fig. 3 shows the change in the formation probability of different configurations of $SrFe_{11.5}Co_{0.5}O_{19}$ as temperature rises. We note that the site occupation probability of $2a$ site falls, while occupancy of the $12k$ site rises with the increase in temperature. This is due to higher multiplicity of site $12k$ than that of $2a$ (P_i is proportional to g_i as given in Eq. 2). In the case of $SrFe_{11.5}Sb_{0.5}O_{19}$ as shown in Fig. 4, the formation probability of all configurations are constant throughout the temperature range we considered; this implies that the preferred site of the substituted Sb atoms is $4f_1$ site, and it does not change with temperature.

The knowledge of site preference of the substituted atoms helps to predict the effect of substitution on the magnetic properties of SFO. The magnetic moment of the 12 Fe atoms at the majority spin sites ($12k$, $2a$, and $2b$) points in the direction parallel to the c-axis and those of the other 8 Fe atoms at the minority spin sites point in the opposite direction. Therefore, if the substituted non-magnetic foreign element prefers the minority sites, it will reduce the negative contribution to the total magnetic moment and results in the enhancement of the total magnetic moment of the unit cell.

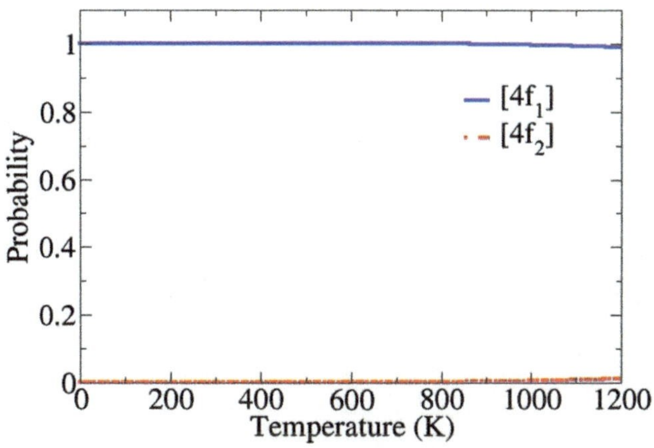

FIG. 4. Temperature dependence of the formation probability of different configurations of $SrFe_{11.5}Sb_{0.5}O_{19}$. The configurations with negligible probability are not shown.

TABLE II. Total magnetic moments (m_{tot}) and magnetocrystalline anisotropy energies (E_a) for different configurations of $SrFe_{11.5}M_{0.5}O_{19}$. All quantities are per unit cell. m_{tot} is in μ_B and E_a is in meV.

	[12*k*]		[2*a*]		[2*b*]		[4*f*1]		[4*f*2]	
M	m_{tot}	E_a	m_{tot}	E_a	m_{tot}	E_a	m_{tot}	E_a	m_{tot}	E_a
Sc	35	0.779	35	0.870	35	0.659	45	0.820	45	0.729
Ti	36	0.799	36	0.850	36	0.249	44	0.170	44	0.699
Co	35	0.900	35	0.789	39	1.560	41	0.479	45	1.590
Ni	36	2.810	36	1.840	38	4.460	42	1.039	42	0.209
Al	35	0.800	35	0.950	35	0.670	45	0.860	45	0.830
In	35	0.760	35	0.880	35	0.510	45	0.820	45	0.730
P	33	0.680	37	1.589	37	0.929	43	0.200	43	0.229
Bi	35	0.650	35	0.799	35	0.000	45	0.840	45	0.299
Sb	36	0.600	33	0.969	33	0.630	43	0.150	43	0.339
Ga	35	0.780	35	0.890	35	0.669	45	0.810	45	0.840
As	35	0.820	35	0.710	35	0.370	45	0.680	43	0.490
Sn	36	0.830	36	0.900	34	0.660	44	0.180	44	0.660

Magnetization is related to the total magnetic moment (m_{tot}), thus magnetization of SFO can be tuned by changing the substituted element. m_{tot} was determined for each configuration and listed in Table II.

To investigate the effect of substitution on the magnetic anisotropy of $SrFe_{11.5}M_{0.5}O_{19}$, we computed the magnetocrystalline anisotropy energy, magnetic anisotropy constant, and anisotropy field. The magnetocrystalline anisotropy energy, E_a, in the present case, is defined as the difference between the two total energies where the spin quantization axes are aligned along two different directions:[48]

$$E_a = E_{(100)} - E_{(001)} \tag{4}$$

where, $E_{(100)}$ is the total energy with spin quantization axis in the magnetically hard plane and $E_{(001)}$ is the total energy with spin quantization axis in the magnetically easy axis. The total energies in Eq. (4) are computed by the non-self-consistent calculations, where the spin densities are kept constant.[49] E_a for different configurations of $SrFe_{12-x}M_xO_{19}$ are shown in Table II.

The uniaxial magnetic anisotropy constant, K_1, can be computed as:[50,51]

$$K_1 = \frac{E_a}{V \sin^2 \theta} \tag{5}$$

where V is the equilibrium volume of the unit cell, and θ is the angle between the two spin quantization axis orientations (90° in the present case). The anisotropy field, H_a, which is related to the coercivity can be expressed as:[52]

$$H_a = \frac{2K_1}{M_s} \tag{6}$$

where K_1 is the magnetocrystalline anisotropy constant and M_s is the saturation magnetization.

Lattice parameters (a, c), volume (V), E_a, m_{tot}, and P_{1000} of the five configurations are then used to calculate structure and magnetic properties of M-substituted SFO from their weighted averages:

$$\langle Q \rangle = \sum_i P_{1000}(i) Q_i \tag{7}$$

where $P_{1000}(i)$ and Q_i are the formation probability at 1000 K and a physical quantity Q of the configuration [i]. We note that SFMO considered in the present work loses its magnetic properties at a typical annealing temperature (1000 K or higher) that is near or above its Curie temperature and the values in Table IV refer to their ground state properties. We use the formation probability at the high temperature only to compute the weighted averages, since the crystalline configurations of SFMO will be distributed according to these values during the annealing process.

TABLE III. Lattice parameters and volume of $SrFe_{11.5}M_{0.5}O_{19}$ unit cell.

M	a (Å)	c (Å)	V (Å^3)
Fe	5.93	23.21	707.29
Sc	5.94	23.41	717.12
Ti	5.94	23.40	712.00
Co	5.92	23.16	703.27
Ni	5.93	23.17	704.14
Al	5.91	23.14	702.50
In	5.95	23.30	716.52
P	5.97	23.94	707.38
Bi	5.95	23.53	724.99
Sb	5.95	23.36	717.48
Ga	5.93	23.20	707.13
As	5.92	23.37	710.23
Sn	5.94	23.41	716.86

Lattice parameters a and c, and volume V of hexagonal unit cell of $SrFe_{12-x}M_xO_{19}$ ($x = 0.5$) are shown in Table III. Lattice parameters and volumes are weighted averages using the results from all five configurations and their formation probabilities (Eq. 7). The maximum increase in volume was found to be 2.5% for Bi-substituted SFO. Al-substituted SFO showed a maximum decrease in volume of 0.62%. The substitution of the foreign atoms does not cause any significant change of the size of the unit cell; this is due to the fact that their ionic radii are close to that of Fe^{3+} ion. Increase of the unit cell volume upon the substitution of Sc, In, Bi, and Sn is in agreement with the previous experimental results (Refs. 44, 14, 53 and 45). Similarly, reduction of the unit cell volume upon substitution of Al, Ga, and Co was also observed by Albanese,[14] and Mallick.[54]

The magnetic properties of all 12 elements are presented in Table IV. The calculated magnetic properties are: volume magnetization (M_s), mass magnetization (σ), E_a, K_1, and H_a. M = Fe represents the case of unsubstituted SFO. Fig. 5 and Fig. 6 show the percentage change in the M_s and the H_a, respectively, compared to the unsubstituted SFO. We note that M_s increases when Sc, Sn, Bi, and Sb are substituted. Increase in H_a was observed when SFO was substituted by Ti, Co, Al, P, and Ga atoms. We found four elements that give rise to increase of M_s and significant decrease in the H_a. These elements are: Sc, Bi, Sb, and Sn. Their substitution has increased M_s by 3.5%. 8.7%, 5.0%, and 3.1%, respectively, while the reduction in H_a was found to be −15.3%, −17.3%, −83.5%, and −20.3%, respectively, compared with unsubstituted SFO.

TABLE IV. Magnetic properties of substituted SFO ($SrFe_{11.5}M_{0.5}O_{19}$). M_s, σ, E_a, K_1, and H_a are the volume magnetization, mass magnetization, magnetocrystalline anisotropy energy, uniaxial magnetic anisotropy constant, and anisotropy field, respectively. Relative difference with respect to the pure SFO (M=Fe) values are given in the parentheses.

M	M_s (emu/cm^3)		σ (emu/g)		E_a (meV)		K_1 (kJ/m^3)		H_a (kOe)	
Fe	524.219	—	110.19	—	0.850	—	193.000	—	7.350	—
Sc	542.509	(+3.5%)	109.97	(−0.2%)	0.749	(−11.9%)	167.317	(−13.3%)	6.224	(−15.3%)
Ti	496.003	(−5.4%)	96.71	(−12.2%)	0.792	(−6.9%)	178.106	(−7.7%)	7.469	(+1.6%)
Co	465.368	(−11.2%)	91.35	(−17.1%)	0.866	(+1.9%)	197.222	(+2.2%)	8.604	(+17.1%)
Ni	474.144	(−9.6%)	94.37	(−14.3%)	0.666	(−21.7%)	151.471	(−21.5%)	6.405	(−12.9%)
Al	462.050	(−11.9%)	93.19	(−15.4%)	0.828	(−2.6%)	188.907	(−2.1%)	8.193	(+11.5%)
In	509.312	(−2.9%)	99.61	(−9.6%)	0.773	(−9.1%)	172.780	(−10.5%)	6.863	(−6.6%)
P	480.232	(− 8.4%)	98.21	(−10.9%)	0.923	(+8.6%)	209.077	(+8.3%)	8.647	(+17.6%)
Bi	569.879	(+8.7%)	110.18	(−0.0%)	0.790	(−7.1%)	174.495	(−9.6%)	6.077	(−17.3%)
Sb	550.251	(+5.0%)	109.57	(−0.6%)	0.151	(−82.3%)	33.688	(−82.5%)	1.214	(−83.5%)
Ga	464.271	(−11.4%)	92.40	(−16.1%)	0.792	(−6.9%)	179.356	(−7.1%)	7.739	(+5.3%)
As	516.173	(−1.5%)	104.01	(−5.6%)	0.616	(−27.5%)	138.965	(−28.0%)	5.337	(−27.4%)
Sn	540.248	(+3.1%)	106.67	(−3.2%)	0.708	(−16.8%)	158.126	(−18.1%)	5.857	(−20.3%)

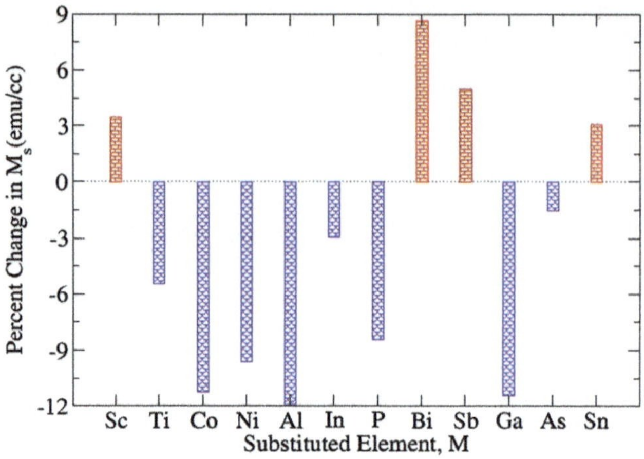

FIG. 5. Percent change in M_s of $SrFe_{11.5}M_{0.5}O_{19}$ compared to unsubstituted SFO.

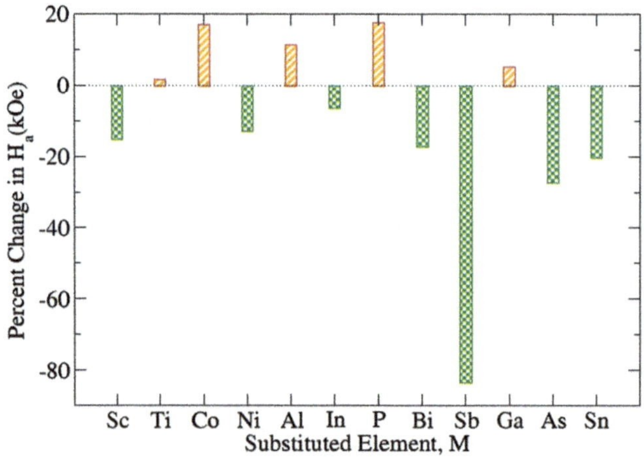

FIG. 6. Percent change in H_a of $SrFe_{11.5}M_{0.5}O_{19}$ compared to unsubstituted SFO.

Such substitutions can be useful for magnetic storage applications. Co substitution leads to an increase in H_a (hence coercivity) and a decrease in σ, this agrees with experimental studies by Mallick.[54] Reduction in σ as well as K_1 upon the substitution of Al, Ga and Sc is consistent with the experimental work of Albanese et al.[14] Similar to our results, Marino-Castellanos et al[45] showed that Sn substitution in SFO resulted in slight increase in M_s (for $x \sim 0.5$) and decrease in K_1.

IV. CONCLUSIONS

We investigated a total of 21 elements for substitution in SFO. To check the possibility of substitution, the E_{sub} was calculated for all the elements. Out of these 21 elements, 12 elements have one or more configurations with a negative substitution energy. Elements with a negative E_{sub} were further investigated for their site preference in SFO. Site occupation probabilities at elevated temperatures was estimated using Boltzmann distribution which was then used to calculate the magnetic properties of substituted SFO. Elements that show an increased M_s are Bi, Sb, Sn, and Sc. On the other hand, an enhancement in H_a has been found in the case of P, Co, Al, Ga, and Ti substitution. Elements that show increase in M_s with considerable decrease in H_a are Sc, Bi, Sb, and Sn. These substitutions are important for high density magnetic recording and magneto-optical storage applications.

ACKNOWLEDGMENTS

This research was supported by the Creative Materials Discovery Program through the National Research Foundation of Korea (NRF) funded by the Ministry of Science, ICT, and Future Planning (2016M3D1A1919181). It was also partly supported by the Center for Computational Sciences (CCS) at Mississippi State University. Computer time allocation has been provided by the High Performance Computing Collaboratory (HPC^2) at Mississippi State University.

[1] H. Wang, B. Yao, Y. Xu, Q. He, G. Wen, S. Long, J. Fan, G. Li, L. Shan, B. Liu, L. Jiang, and L. Gao, Journal of Alloys and Compounds **537**, 43 (2012).
[2] R. C. Pullar, Progress in Materials Science **57**, 1191 (2012).
[3] Z. Pang, X. Zhang, B. Ding, D. Bao, and B. Han, Journal of Alloys and Compounds **492**, 691 (2010).
[4] A. Davoodi and B. Hashemi, Journal of Alloys and Compounds **509**, 5893 (2011).
[5] M. N. Ashiq, M. J. Iqbal, M. N. ul Haq, P. H. Gomez, and A. M. Qureshi, Journal of Magnetism and Magnetic Materials **324**, 15 (2012).
[6] T. B. Ghzaiel, W. Dhaoui, A. Pasko, and F. Mazaleyrat, Journal of Alloys and Compounds **671**, 245 (2016).
[7] G. K. Thompson and B. J. Evans, Journal of Applied Physics **75**, 6643 (1994).
[8] T. M. Clark, B. J. Evans, G. K. Thompson, and S. Freeman, Journal of Applied Physics **85**, 5229 (1999).
[9] G. Albanese, M. Carbucicchio, and A. Deriu, Physica Status Solidi (A) **23**, 351 (1974).
[10] G. Albanese, G. Asti, and P. Batti, Il Nuovo Cimento B Series **10**(58), 467 (1968).
[11] I. Bsoul and S. Mahmood, Journal of Alloys and Compounds **489**, 110 (2010).
[12] M. Awawdeh, I. Bsoul, and S. Mahmood, Journal of Alloys and Compounds **585**, 465 (2014).
[13] M. J. Iqbal, M. N. Ashiq, P. Hernandez-Gomez, and J. M. Munoz, Journal of Magnetism and Magnetic Materials **320**, 881 (2008).
[14] G. Albanese and A. Deriu, Ceramurgia International **5**, 3 (1979).
[15] C. N. Chinnasamy, T. Sakai, S. Sivasubramanian, A. F. Yang, C. Vittoria, and V. G. Harris, Journal of Applied Physics **103** (2008).
[16] A. Tsuzuki, K. Kani, K. Watari, S. Kawakami, and Y. Torii, Journal of Materials Science Letters **11**, 893 (1992).
[17] W. Zhang, Y. Bai, X. Han, L. Wang, X. Lu, and L. Qiao, Journal of Alloys and Compounds **546**, 234 (2013).
[18] Z. Zhou, Z. Wang, X. Wang, Q. Li, M. Jin, and J. Xu, Journal of Superconductivity and Novel Magnetism **28**, 1773 (2015).
[19] V. V. Soman, V. Nanoti, and D. Kulkarni, Ceramics International **39**, 5713 (2013).
[20] R. Mudsainiyan, S. Chawla, and S. Meena, Journal of Alloys and Compounds **615**, 875 (2014).
[21] S. Shakoor, M. N. Ashiq, M. A. Malana, A. Mahmood, M. F. Warsi, M. N. ul Haq, and N. Karamat, Journal of Magnetism and Magnetic Materials **362**, 110 (2014).
[22] D. Seifert, J. Töpfer, F. Langenhorst, J.-M. L. Breton, H. Chiron, and L. Lechevallier, Journal of Magnetism and Magnetic Materials **321**, 4045 (2009).
[23] J. Wang, C. Ponton, R. Grössinger, and I. Harris, Journal of Alloys and Compounds **369**, 170 (2004).
[24] J. Wang, C. Ponton, and I. Harris, Journal of Magnetism and Magnetic Materials **234**, 233 (2001).
[25] J. Wang, C. Ponton, and I. Harris, Journal of Alloys and Compounds **403**, 104 (2005).
[26] J. Wang, C. Ponton, and I. Harris, IEEE Transactions on Magnetics **38**, 2928 (2002).
[27] A. Ghasemi, V. Šepelák, X. Liu, and A. Morisako, Journal of Applied Physics **107**, 09A734 (2010).
[28] A. Ghasemi and V. Šepelák, Journal of Magnetism and Magnetic Materials **323**, 1727 (2011).
[29] L. S. Liyanage, S. Kim, Y.-K. Hong, J.-H. Park, S. C. Erwin, and S.-G. Kim, Journal of Magnetism and Magnetic Materials **348**, 75 (2013).
[30] C. Fang, F. Kools, R. Metselaar, R. Groot, and Others, Journal of Physics: Condensed Matter **15**, 6229 (2003).
[31] J. Park, Y.-K. Hong, S.-G. Kim, S. Kim, L. S. Liyanage, J. Lee, W. Lee, G. S. Abo, K.-H. Hur, and S.-Y. An, Journal of Magnetism and Magnetic Materials **355**, 1 (2014).
[32] M. Küpferling, P. Novák, K. Knížek, M. W. Pieper, R. Grössinger, G. Wiesinger, and M. Reissner, Journal of Applied Physics **97**, 10F309 (2005).
[33] P. Novák, K. Knížek, M. Küpferling, R. Grössinger, and M. W. Pieper, The European Physical Journal B - Condensed Matter and Complex Systems **43**, 509 (2005).
[34] V. Dixit, C. N. Nandadasa, S.-G. Kim, S. Kim, J. Park, Y.-K. Hong, L. S. I. Liyanage, and A. Moitra, Journal of Applied Physics **117**, 243904 (2015).
[35] V. Dixit, C. N. Nandadasa, S.-G. Kim, S. Kim, J. Park, and Y.-K. Hong, Journal of Applied Physics **118**, 203908 (2015).
[36] G. Kresse and J. Furthmüller, Phys. Rev. B **54**, 11169 (1996).
[37] G. Kresse and D. Joubert, Phys. Rev. B **59**, 1758 (1999).
[38] E. F. Gorter, Proc. IEEE **104B**, 2555 (1957).
[39] H. Monkhorst and J. Pack, Phys. Rev. B **13**, 5188 (1976).
[40] M. Methfessel and A. T. Paxton, Phys. Rev. B **40**, 3616 (1989).
[41] J. P. Perdew, K. Burke, and M. Ernzerhof, Phys. Rev. Lett. **77**, 3865 (1996).
[42] S. L. Dudarev, G. A. Botton, S. Y. Savrasov, C. J. Humphreys, and A. P. Sutton, Phys. Rev. B **57**, 1505 (1998).
[43] X. B. Liu, Y. Ma, Z. Altounian, Q. Zhang, and J. Ping Liu, Journal of Applied Physics **115**, 17A702 (2014).
[44] Z. Somogyvári, E. Sváb, K. Krezhov, L. Kiss, D. Kaptás, I. Vincze, E. Beregi, and F. Bourée, Journal of Magnetism and Magnetic Materials **304**, e775 (2006), proceedings of the 17th International Symposium on Soft Magnetic Materials.
[45] P. Marino-Castellanos, J. Somarriba-Jarque, and J. Anglada-Rivera, Physica B: Condensed Matter **362**, 95 (2005).
[46] D. H. Choi, S. W. Lee, S. Y. An, S.-I. Park, I.-B. Shim, and C. S. Kim, IEEE Transactions on Magnetics **39**, 2884 (2003).
[47] G. Albanese, Journal of Magnetism and Magnetic Materials **147**, 421 (1995).

[48] P. Ravindran, A. Delin, P. James, B. Johansson, J. M. Wills, R. Ahuja, and O. Eriksson, Phys. Rev. B **59**, 15680 (1999).
[49] G. H. O. Daalderop, P. J. Kelly, and M. F. H. Schuurmans, Phys. Rev. B **41**, 11919 (1990).
[50] F. Muñoz, A. Romero, J. Mejía-López, and J. Morán-López, Journal of Nanoparticle Research **15**, 1524 (2013).
[51] J. Smit and H. Wijn, *Ferrites* (Philips Technical Library, Eindhoven, 1959).
[52] C. Kittel, Rev. Mod. Phys. **21**, 541 (1949).
[53] I. Auwal, H. Güngünes, A. Baykal, S. Güner, S. E. Shirsath, and M. Sertkol, Ceramics International **42**, 8627 (2016).
[54] K. K. Mallick, P. Shepherd, and R. J. Green, Journal of Magnetism and Magnetic Materials **312**, 418 (2007).

23

On the coupling effects of piezoelectricity and flexoelectricity in piezoelectric nanostructures

Liwen He,[1,a] Jia Lou,[1,a] Aibing Zhang,[1] Huaping Wu,[2] Jianke Du,[1] and Ji Wang[1]

[1] *Piezoelectric Device Laboratory, Department of Mechanics and Engineering Science, Ningbo University, Ningbo, Zhejiang 315211, China*

[2] *Key Laboratory of E&M (Zhejiang University of Technology), Ministry of Education & Zhejiang Province, Hangzhou 310014, China*

Flexoelectricity is a novel kind of electromechanical coupling phenomenon that is prevalent in all solid dielectrics and usually of vital importance in nanostructures and soft materials. Although the fundamental theory of flexoelectric solids and related beam or plate theories were extensively studied in recent years, the coupling effect of flexoelectricity and piezoelectricity in piezoelectric nanostructures has not been completely clarified yet. In the present work, a geometrically nonlinear piezoelectric plate model is established with a focus on the coupling effect. The constitutive equations for piezoelectric plates are derived under both the electrically short-circuit and open-circuit conditions. It is found that due to the coupling between flexoelectricity and piezoelectricity, stretching-bending coupling stiffness arises in the homogeneous plate and its specific value relies on the applied electrical boundary conditions. The effects of the flexoelectric-piezoelectric coupling on the effective mechanical behavior and the electromechanical behavior of nanobeams and nanoplates are also discussed. The developed model and presented results are expected to benefit the design and analysis of piezoelectric and flexoelectric devices and systems.

I. INTRODUCTION

The last decade has witnessed the rapid and significant development of piezoelectric field-effect transistor, piezoelectric-gated diodes, piezoelectric resonators and energy harvesters for future wirelessly self-powered nanosystems.[1] It is evident that electromechanical coupling of dielectrics, including piezoelectric materials, plays a vital role in the performance of these devices. Piezoelectricity, as a conventional electromechanical coupling effect, refers to electrical polarization in response to a uniform strain and only exists for non-centrosymmetric crystals. Recently, flexoelectricity, which is a relatively new electromechanical coupling effect referred to polarization due to a non-uniform strain or a strain gradient, has attracted a surge of scientific interest. Since flexoelectricity is induced by breaking the inversion symmetry via the application of a strain gradient,[2] it is a universal property allowed by all dielectrics including centrosymmetric crystals. In general, such flexoelectric effect is not important relative to the piezoelectric effect for macro-scale piezoelectric materials. However, due to large strain gradients typically exhibited in nano-scale structures, the flexoelectricity becomes manifest and may significantly influence the electromechanical behaviors of nano-scale piezoelectrics.[3–6] A thorough and compressive review of the fundamental mechanisms, its physical consequence in nanostructures and potential applications of flexoelectricity in solids has been conducted by Nguyen et al.,[7] Yudin and Tagantsev,[8] Zubko et al.[9] and also many distinguished researchers in a newly published book.[10] For example, flexoelectricity is found to be responsible for modifying

[a]Corresponding authors. E-mail: physi_mechanism@163.com (Liwen He), jiajia_smile@163.com (Jia Lou).

some physical characteristics of ferroelectrics, including the shifts of domain configurations and the ferroelectric hysteresis loop,[11] the vanishing of switchable spontaneous polarization[12] and the formation of a non-switchable dead layer.[13] The flexoelectricity induced electric polarization may enable effective electromechanical coupling responses in nonpiezoelectric dielectric materials. In fact, it has already been demonstrated that apparent piezoelectric composite materials can be created without using any piezoelectric constituents by employing the flexoelectric effect of its constituents.[14,15]

Recently, lots of efforts have been made to increase our understanding of flexoelectricity via both theoretical and experimental approaches. For instance, Ma and Cross[16–19] conducted a series of experiments to quantitatively measuring the flexoelectric coefficient of ferroelectric ceramics, which were found to be proportional to their dielectric permittivity. On the theoretical modeling aspects, Sharma's group[2] and Shen's group[20,21] established the theoretical framework of flexoelectricity by extending the linear piezoelectricity developed by Touplin. Based on these pioneering studies, the flexoelectric effect on electromechanical coupling behaviors of piezoelectric nanostructures has been extensively investigated. For example, Majdoub et al.[3] applied the framework to the Euler-Bernoulli beam model and found that the apparent piezoelectric coefficients of both piezoelectric and nonpiezoelectric nanobeams were significantly enhanced by the flexoelectricity. The effect of the flexoelectricity on the electric potential distribution in a bending ZnO nanowire was analytically predicted by Liu et al.[5] Based on the modified Euler-Bernoulli beams and Timoshenko beam models with the incorporation of the flexoelectric effect, the size-dependent electroelastic responses and dynamic behavior of slender piezoelectric nanobeams were investigated by Yan and Jiang.[22,23] Tagantsev and Yudin studied the flexoelectric effect in finite samples[24–26] including thin plates with a focus on the effects of surface polarization effects and surface piezoelectric effect. Besides these works on thin plates, a Kirchhoff-Love type plate model was also proposed to study the electroelastic responses of piezoelectric nanoplates by Zhang et al.[27] and Liang et al.[28,29] However, almost all of these works did not consider large deformation of the nanobeams and nanoplates. In fact, large deformation is very prevalent in nanostructures due to their high flexibility. Moreover, except a recent study on one dimensional analysis of piezoelectric thin films,[30] the coupling of the piezoelectric effect and the flexoelectric effect in nanostructures and its consequence on the electromechanical behaviors of nanostructures have not been completely clarified yet. Motivated by such a gap, a geometrically nonlinear plate model will be established in the present paper, which adopts the von Karman's nonlinear strain and accounts for the coupling of piezoelectric effect and flexoelectric effect.

The rest of the present paper is organized as follows. In Section 2, a general three dimensional theory of flexoelectricity is firstly presented based on which a new von Karman plate model accounting for the flexoelectricity of piezoelectric nanoplates is developed. In this model, both the in-plane displacement and the transverse displacement of the plate are taken into account rather than only considering the transverse deflection as most previous works assumed. During the model development, the electrical field distributions through the thickness of the plate are firstly determined under both the open-circuit and closed-circuit conditions, then the constitutive equations for the piezoelectric nanoplate are obtained analytically under the two different conditions, and finally the governing equations and associated boundary conditions are also derived via the standard variational method. It is found that due to the coupling between flexoelectricity and piezoelectricity, stretching-bending coupling stiffness arises in the homogeneous nanoplate. Furthermore, simple analytical solutions are derived in Section 3 for the deflection of a piezoelectric and flexoelectric nanobeam subject to an axial compressive load on its geometrical mid-plane, the polarization of a piezoelectric and flexoelectric nanoplate under a uniform bending moment and the deflection of a piezoelectric and flexoelectric nanoplate subject to a voltage in order to demonstrate the effects of the flexoelectric-piezoelectric coupling on the effective mechanical behavior and the electromechanical behavior of the piezoelectric nanoplate. Finally, some conclusions are drawn in Section 4.

II. THEORETICAL MODELING

A. Formulation for 3D piezoelectric solids with flexoelectricity

The electrical enthalpy density of a three dimensional piezoelectric solid with flexoelectricity is assumed to be[31,32]

$$H = -\frac{1}{2}a_{kl}E_kE_l + \frac{1}{2}c_{ijkl}\varepsilon_{ij}\varepsilon_{kl} - e_{ijk}\varepsilon_{ij}E_k - \frac{1}{2}\mu_{klij}\left(E_k\varepsilon_{ij,l} - \varepsilon_{ij}E_{k,l}\right) - \frac{1}{2}b_{ijkl}E_{i,j}E_{k,l}, \tag{1}$$

where ε_{ij} and E_k are the Cartesian components of the strain tensor and electric field, a_{kl}, c_{ijkl} and e_{ijk} are the dielectric constants, the elastic modulus and the piezoelectric constants respectively, μ_{klij} is the flexoelectric coefficient which reflects the flexoelectric response of the electromechanical system, and b_{ijkl} is the electrical field gradient modulus.

According to the thermodynamic theory, we have:

$$\begin{aligned}
\sigma_{ij} &= \frac{\partial H}{\partial \varepsilon_{ij}} = c_{ijkl}\varepsilon_{kl} - e_{ijk}E_k + \frac{1}{2}\mu_{klij}E_{k,l},\\
\tau_{ijl} &= \frac{\partial H}{\partial \varepsilon_{ij,l}} = -\frac{1}{2}\mu_{klij}E_k,\\
D_i &= -\frac{\partial H}{\partial E_i} = a_{ij}E_j + e_{jki}\varepsilon_{jk} + \frac{1}{2}\mu_{ijkl}\varepsilon_{kl,j},\\
Q_{ij} &= -\frac{\partial H}{\partial E_{i,j}} = -\frac{1}{2}\mu_{ijkl}\varepsilon_{kl} + b_{ijkl}E_{k,l},
\end{aligned} \tag{2}$$

where σ_{ij} is the classical Cauchy stress, τ_{ijl} is a higher order stress that is work conjugated to the strain gradient $\varepsilon_{ij,l}$, and D_i and Q_{ij} are the electric displacement and electric quadpole respectively.

Applying the variational method to the electrical enthalpy of the system with the internal and boundary loads taken into account:

$$\delta G = \delta\left(\int_V H\mathrm{d}V - \int_V b_iu_i\mathrm{d}V + \int_V \rho_e\varphi\mathrm{d}V - \int_{S_\sigma} \bar{t}_iu_i\mathrm{d}S + \int_{S_\omega} \bar{\omega}\varphi\mathrm{d}S\right), \tag{3}$$

the following governing equations and associated boundary conditions are obtained:[31]

$$\begin{cases}\left(\sigma_{ij} - \tau_{ijk,k}\right)_{,j} + b_i = 0,\\ \left(D_i - Q_{ij,j}\right)_{,i} = \rho_e,\end{cases} \tag{4}$$

$$\begin{cases}\left(\sigma_{ij} - \tau_{ijk,k}\right)n_j + (\Delta_l n_l)n_jn_k\tau_{ijk} - \Delta_j\left(n_k\tau_{ijk}\right) = \bar{t}_i, \text{ or } u_i = \bar{u}_i,\\ \tau_{ijk}n_jn_k = 0, \text{ or } \Delta u_i = \Delta\bar{u}_i,\\ \left(D_i - Q_{ij,j}\right)n_j = -\bar{\omega}, \text{ or } \varphi = \bar{\varphi},\\ Q_{ij}n_in_j = 0, \text{ or } \Delta\varphi = \Delta\bar{\varphi},\end{cases} \tag{5}$$

where b_l and ρ_e are the body force and body free charge density, $\bar{t}_i$ and $\bar{\omega}$ are the surface traction and surface free charge density, φ is the classical electrical potential, *i.e.* $E_i = -(\nabla\varphi)_i$, u_i are the components of the displacement vector, and the operators $\Delta \equiv n_k\partial_k$, $\Delta_j \equiv (\delta_{jk} - n_jn_k)\partial_k$ and ∂_k denotes the partial derivative with respect to x_k.

It should be pointed out that in above derivation, the Maxwell stress which plays a significant role in soft electroactive materials such as dielectric elastomers[33] and flexoelectric polymers[34] is ignored in the present paper as many researchers have done,[22,32] because its effect on hard ceramics is negligibly small and it is also not considered in classical piezoelectric theory.[35] The elastic strain gradient effect is also not considered in the present electrical enthalpy for it is expected not to have a notable influence on the electromechanical coupling properties. Moreover, since our main goal of the present paper is to clarify the coupling effect of bulk piezoelectricity and bulk flexoelectricity within piezoelectric nanostructures, surface effects such as surface polarization and surface piezoelectricity are also not taken into account in the following formulation and discussion, although surface effects have strong effects on effective electromechanical properties of piezoelectric nanostructures.[24,26] Actually, it can be inferred from the theoretical framework presented in the present work that surface effects can also be taken in consideration without difficulty. In the following subsections, the three dimensional theory presented above will be used to develop a geometrically nonlinear plate model accounting for both the flexoelectric and piezoelectric effects.

B. Kinematics of piezoelectric nanoplates

Firstly, we consider a piezoelectric and flexoelectric nanoplate with uniform thickness. Following the classical Kirchhoff-Love hypothesis, the deformation of the plate can be assumed as:

$$
\begin{aligned}
u_x(x,y,z,t) &= u(x,y,t) - z\frac{\partial w(x,y,t)}{\partial x},\\
u_y(x,y,z,t) &= v(x,y,t) - z\frac{\partial w(x,y,t)}{\partial y},\\
u_z(x,y,z,t) &= w(x,y,t),
\end{aligned}
\tag{6}
$$

where u and v are the in-plane components of the displacement of any point on the mid-plane (the O-xy plane) of the nanoplate in a Cartesian coordinate system O-xyz, and w is the out-of-plane component of the displacement in the thickness direction, i.e. along the z coordinate. Accounting for the von-Karman's nonlinear strain,[36] the strains and strain gradients are obtained:

$$
\begin{aligned}
\varepsilon_{xx} &= \varepsilon_{xx0} - zw_{,xx} = u_{,x} + w_{,x}^2/2 - zw_{,xx},\\
\varepsilon_{yy} &= \varepsilon_{yy0} - zw_{,yy} = v_{,y} + w_{,y}^2/2 - zw_{,yy},\\
\varepsilon_{xy} &= \varepsilon_{xy0} - zw_{,xy} = \left(u_{,y} + v_{,x}\right)/2 + w_{,x}w_{,y} - zw_{,xy},
\end{aligned}
\tag{7}
$$

$$
\begin{aligned}
&\eta_{xxx} = \varepsilon_{xx,x} = \varepsilon_{xx0,x} - zw_{,xxx},\ \eta_{xxy} = \varepsilon_{xx,y} = \varepsilon_{xx0,y} - zw_{,xxy},\ \eta_{xxz} = \varepsilon_{xx,z} = -w_{,xx},\\
&\eta_{yyy} = \varepsilon_{yy,y} = \varepsilon_{yy0,y} - zw_{,yyy},\ \eta_{yyx} = \varepsilon_{yy,x} = \varepsilon_{yy0,x} - zw_{,yyx},\ \eta_{yyz} = \varepsilon_{yy,z} = -w_{,yy},\\
&\eta_{xyx} = \varepsilon_{xy,x} = \varepsilon_{xy0,x} - zw_{,xyx},\ \eta_{xyy} = \varepsilon_{xy,y} = \varepsilon_{xy0,y} - zw_{,xyy},\ \eta_{xyz} = \varepsilon_{xy,z} = -w_{,xy}.
\end{aligned}
\tag{8}
$$

Since the lateral dimension in the O-xy plane is much larger than the thickness of the plate, the components of the strain gradient in the thickness direction (including η_{xxz}, η_{yyz} and η_{xxz}) are expected to be much larger than the corresponding components along the lateral direction (including η_{xxx}, η_{xxy}, η_{yyy}, η_{yyx}, η_{xyx} and η_{xyy}), consequently only the former components including η_{xxz}, η_{yyz} and η_{xyz}, are considered and the other components of the strain gradient as listed above are neglected here. Following the same reasoning, only the electrical field component E_z in the thickness direction and its gradient $E_{z,z}$ in the thickness direction are taken into account, and the other two components (including E_x and E_y) of the electrical field vector and the gradients of these two components are all neglected.

C. Electrical field analysis of piezoelectric nanoplates

According to the general constitutive equation (2), the constitutive equation for transversely isotropic piezoelectric nanostructures with flexoelectricity can be written in the following matrix form:

$$
\begin{Bmatrix} \sigma_{xx}\\ \sigma_{yy}\\ \sigma_{xy}\\ \tau_{xxz}\\ \tau_{yyz}\\ \tau_{xyz} \end{Bmatrix}
=
\begin{bmatrix}
c_{11} & c_{12} & 0 & 0 & 0 & 0\\
c_{12} & c_{11} & 0 & 0 & 0 & 0\\
0 & 0 & c_{66} & 0 & 0 & 0\\
0 & 0 & 0 & 0 & 0 & 0\\
0 & 0 & 0 & 0 & 0 & 0\\
0 & 0 & 0 & 0 & 0 & 0
\end{bmatrix}
\begin{Bmatrix} \varepsilon_{xx}\\ \varepsilon_{yy}\\ 2\varepsilon_{xy}\\ \eta_{xxz}\\ \eta_{yyz}\\ \eta_{xyz} \end{Bmatrix}
-
\begin{bmatrix}
e_{31} & -\mu_{31}/2\\
e_{31} & -\mu_{31}/2\\
0 & 0\\
\mu_{31}/2 & 0\\
\mu_{31}/2 & 0\\
0 & 0
\end{bmatrix}
\begin{Bmatrix} E_z\\ E_{z,z} \end{Bmatrix},
\tag{9}
$$

$$
\begin{Bmatrix} D_z\\ Q_{zz} \end{Bmatrix}
=
\begin{bmatrix}
e_{31} & e_{31} & 0 & \mu_{31}/2 & \mu_{31}/2 & 0\\
-\mu_{31}/2 & -\mu_{31}/2 & 0 & 0 & 0 & 0
\end{bmatrix}
\begin{Bmatrix} \varepsilon_{xx}\\ \varepsilon_{yy}\\ \varepsilon_{xy}\\ \eta_{xxz}\\ \eta_{yyz}\\ \eta_{xyz} \end{Bmatrix}
+
\begin{bmatrix} a_{33} & 0\\ 0 & b_{33} \end{bmatrix}
\begin{Bmatrix} E_z\\ E_{z,z} \end{Bmatrix},
\tag{10}
$$

where $c_{11} - c_{12} = 2c_{66}$.

The electrical potential is related with the electrical field through $E_z = -\partial\varphi/\partial z$.

The electrical governing equation reads:

$$(D_z - Q_{zz,z})_{,z} = 0. \tag{11}$$

Integrating Eq. (11), we have:

$$D_z - Q_{zz,z} = C(x, y). \tag{12}$$

Substituting Eq. (10) into the electrical governing equation (11), a differential equation for the electrical potential is obtained:

$$-a_{33}\varphi_{,zz} - e_{31}\left(w_{,xx} + w_{,yy}\right) + b_{33}\varphi_{,zzzz} = 0. \tag{13}$$

The general solution of the equation is easily found to be:

$$\varphi = -\frac{e_{31}}{2a_{33}}\left(w_{,xx} + w_{,yy}\right)z^2 + c_1 + c_2 z + c_3\sinh\lambda z + c_4\cosh\lambda z, \tag{14}$$

where $\lambda = \sqrt{a_{33}/b_{33}}$ has a dimension of Length^{-1} and its specific value depends on the electrical field gradient modulus b_{33}. In the limit of $b_{33} \to 0$, *i.e.* neglecting the electrical field gradient energy, λ becomes infinity.

For the studied nanoplate, two different electrical boundary conditions will be considered, which are the voltage-loaded condition (or electrically short-circuit) and the open-circuit condition respectively.

1. Voltage-loaded condition

$$Q_{zz}(\pm h/2) = 0,\ \varphi(h/2) = V,\ \varphi(-h/2) = 0. \tag{15}$$

By $Q_{zz}(\pm h/2) = 0$, it is found that:

$$c_3 = \frac{\mu_{31}}{2a_{33}\sinh(\lambda h/2)}\frac{h}{2}\left(w_{,xx} + w_{,yy}\right), \tag{16}$$

$$c_4 = -\frac{\mu_{31}}{2a_{33}\cosh(\lambda h/2)}\left(\varepsilon_{xx0} + \varepsilon_{yy0}\right) + \frac{e_{31}}{a_{33}\lambda^2\cosh(\lambda h/2)}\left(w_{,xx} + w_{,yy}\right). \tag{17}$$

Hence c_3 and c_4 can be uniquely determined only from the first two boundary conditions.

By $\varphi(h/2) = V$, $\varphi(-h/2) = 0$, c_1 and c_2 can also be determined:

$$c_1 = \frac{e_{31}}{2a_{33}}\left(\frac{h}{2}\right)^2\left(w_{,xx} + w_{,yy}\right) - c_4\cosh(\lambda h/2) + \frac{V}{2}, \tag{18}$$

$$c_2 = -\frac{2c_3}{h}\sinh(\lambda h/2) + \frac{V}{h}. \tag{19}$$

Substituting Eqs. (16)(17)(18) and (19) into Eq. (14), the electrical potential is determined:

$$\begin{aligned}\varphi = &\left(\frac{V}{2} + \frac{V}{h}z\right) + \frac{e_{31}}{2a_{33}}\left(\frac{h^2}{4} - z^2\right)\left(w_{,xx} + w_{,yy}\right) \\ &+ c_3\left[\sinh\lambda z - \sinh(\lambda h/2)\tfrac{2z}{h}\right] + c_4\left[\cosh\lambda z - \cosh(\lambda h/2)\right]\end{aligned}, \tag{20}$$

and the electrical filed component E_z and its gradient $E_{z,z}$ are consequently derived:

$$E_z = -\frac{V}{h} + \frac{e_{31}}{a_{33}}z\left(w_{,xx} + w_{,yy}\right) + \frac{\mu_{31}}{2a_{33}}\left(w_{,xx} + w_{,yy}\right) - c_3\lambda\cosh\lambda z - c_4\lambda\sinh\lambda z, \tag{21}$$

$$E_{z,z} = \frac{e_{31}}{a_{33}}\left(w_{,xx} + w_{,yy}\right) - c_3\lambda^2\sinh\lambda z - c_4\lambda^2\cosh\lambda z. \tag{22}$$

It is noted that hyperbolic sinusoidal functions arise in the expressions of φ, E_z and $E_{z,z}$. This is due to the boundary layer effect near the top and bottom electrodes, which will almost not affect the distribution of these electrical quantities far away from the top and bottom electrodes, *i.e.*,

$$E_z = -\frac{V}{h} + \frac{e_{31}}{a_{33}} z \left(w_{,xx} + w_{,yy}\right) + \frac{\mu_{31}}{2a_{33}} \left(w_{,xx} + w_{,yy}\right), \ z \notin \text{the two boundary layers.} \tag{23}$$

It should be pointed out that the boundary layer associated terms should be included in the derivation process for it will directly influence the stress resultants such as the axial force and bending moment. Tagantsev and Yudin[24] also reported that the polarization within the very thin boundary layers plays an crucial role in the bulk flexoelectricity effect. The thickness of the boundary layers is of the magnitude of nanometers for ceramic materials,[37] and this specific value depends on the selection of piezoelectric materials. In the present paper, since the deformation of the plate is assumed to follow the Kirchhoff-Love kinematic relations as given in Eq. (6), the thickness of the boundary layers only depends on the ratio of the electrical field gradient modulus b_{33} to the dielectric constant a_{33} through $\lambda = \sqrt{a_{33}/b_{33}}$. In a general case, the thickness of the boundary layers not only depends on these electrical properties of the materials, but relies on the elastic properties (including the usual elastic modulus and the strain gradient modulus) and electromechanical coupling properties (including the piezoelectric and flexoelectric properties) of the selected materials.[37]

2. Open-circuit condition

$$Q_{zz}\left(\pm h/2\right) = 0, \ \left(D_z - Q_{zz,z}\right)\left(\pm h/2\right) = 0. \tag{24}$$

Similar to the voltage-loaded condition, c_3 and c_4 can be uniquely determined only from the first two BCs in Eq. (24), hence their expressions are the same as those shown in Eqs. (16) and (17). From the last two boundary conditions of Eq. (24) and Eq. (12), it can be inferred that $D_z - Q_{zz,z} = 0$ for any z. By solving this equation, it is found that:

$$c_2 = \frac{e_{31}}{a_{33}} \left(\varepsilon_{xx0} + \varepsilon_{yy0}\right) - \frac{\mu_{31}}{a_{33}} \left(w_{,xx} + w_{,yy}\right). \tag{25}$$

Therefore, the potential under the open-circuit condition takes a different form from that under the voltage-loaded or short-circuit conditions as shown below:

$$\begin{aligned}\varphi = & -\frac{e_{31}}{2a_{33}} z^2 \left(w_{,xx} + w_{,yy}\right) + c_1 + \left[\frac{e_{31}}{a_{33}} \left(\varepsilon_{xx0} + \varepsilon_{yy0}\right) - \frac{\mu_{31}}{a_{33}} \left(w_{,xx} + w_{,yy}\right)\right] z \\ & + c_3 \sinh\lambda z + c_4 \cosh\lambda z\end{aligned}, \tag{26}$$

and the corresponding electrical filed component E_z and its gradient $E_{z,z}$ are respectively:

$$E_z = -\frac{e_{31}}{a_{33}} \left(\varepsilon_{xx} + \varepsilon_{yy}\right) + \frac{\mu_{31}}{a_{33}} \left(w_{,xx} + w_{,yy}\right) - c_3 \lambda \cosh\lambda z - c_4 \lambda \sinh\lambda z, \tag{27}$$

$$E_{z,z} = \frac{e_{31}}{a_{33}} \left(w_{,xx} + w_{,yy}\right) - c_3 \lambda^2 \sinh\lambda z - c_4 \lambda^2 \cosh\lambda z. \tag{28}$$

Similarly, we have

$$E_z = -\frac{e_{31}}{a_{33}} \left(\varepsilon_{xx} + \varepsilon_{yy}\right) + \frac{\mu_{31}}{a_{33}} \left(w_{,xx} + w_{,yy}\right), \ z \notin \text{the two boundary layers,} \tag{29}$$

which is also different from the counterpart under the voltage-loaded or short-circuit condition.

By substituting the expressions for the electrical field and electrical field gradient, *i.e.* Eqs. (21) and (22) under the voltage-loaded condition or Eqs. (27) and (28) under the open-circuit condition, into the constitutive equation (9), a simplified constitutive equation can be obtained for the nanoplate and will be used to analyze the effective mechanical behavior of the nanoplate.

D. Constitutive equations for piezoelectric nanoplates with flexoelectricity

The variation of the strain energy of the piezoelectric nanoplate reads:

$$\begin{aligned}\delta U &= \int_V \left(\sigma_{xx}\delta\varepsilon_{xx} + \sigma_{yy}\delta\varepsilon_{yy} + 2\sigma_{xy}\delta\varepsilon_{xy} + \tau_{xxz}\delta\eta_{xxz} + \tau_{yyz}\delta\eta_{yyz} + \tau_{xyz}\delta\eta_{xyz}\right)\mathrm{d}V \\ &= \int_S \left(N_{xx}\delta\varepsilon_{xx0} + N_{yy}\delta\varepsilon_{yy0} + 2N_{xy}\delta\varepsilon_{xy0} + M_{xx}\delta\kappa_{xx} + M_{yy}\delta\kappa_{yy} + 2M_{xy}\delta\kappa_{xy}\right. \\ &\quad \left. + P_{xxz}\delta\kappa_{xx} + P_{yyz}\delta\kappa_{yy}\right)\mathrm{d}A\end{aligned}, \tag{30}$$

where

$$N_{\alpha\beta} = \int_h \sigma_{\alpha\beta} \mathrm{d}z, \quad M_{\alpha\beta} = \int_h \sigma_{\alpha\beta} z \mathrm{d}z, \quad P_{\alpha\beta z} = \int_h \tau_{\alpha\beta z} \mathrm{d}z, \tag{31}$$

and $\kappa_{\alpha\beta} = -w_{,\alpha\beta}, \quad \alpha,\beta = 1,2$.

1. Voltage-loaded condition

By substituting the expressions of the stress components and higher order stress components (Eq. (9)) into Eq. (31), conducting the integrations and defining a moment $\tilde{M}_{\alpha\beta} = M_{\alpha\beta} + P_{\alpha\beta z}$, we have the following constitutive relation for the voltage-loaded piezoelectric plate:

$$\begin{Bmatrix} N_{xx} \\ N_{yy} \\ N_{xy} \\ \tilde{M}_{xx} \\ \tilde{M}_{yy} \\ \tilde{M}_{xy} \end{Bmatrix} = \begin{bmatrix} A_{11} & A_{12} & 0 & B_{11} & B_{12} & 0 \\ A_{12} & A_{11} & 0 & B_{12} & B_{11} & 0 \\ 0 & 0 & A_{66} & 0 & 0 & 0 \\ B_{11} & B_{12} & 0 & D_{11} & D_{12} & 0 \\ B_{12} & B_{11} & 0 & D_{12} & D_{11} & 0 \\ 0 & 0 & 0 & 0 & 0 & D_{66} \end{bmatrix} \begin{Bmatrix} \varepsilon_{xx0} \\ \varepsilon_{yy0} \\ 2\varepsilon_{xy0} \\ \kappa_{xx} \\ \kappa_{yy} \\ 2\kappa_{xy} \end{Bmatrix} - \begin{Bmatrix} N^e \\ N^e \\ 0 \\ M^e \\ M^e \\ 0 \end{Bmatrix}, \tag{32}$$

where $N^e = -e_{31}V, \ M^e = -\mu_{31}V/2$,

$$A_{11} = c_{11}h + \frac{\mu_{31}^2 \lambda}{2a_{33}} \tanh(\lambda h/2), \ A_{12} = c_{12}h + \frac{\mu_{31}^2 \lambda}{2a_{33}} \tanh(\lambda h/2), \ A_{66} = c_{66}h,$$

$$B_{11} = B_{12} = \frac{e_{31}\mu_{31}}{a_{33}} \left[\frac{1}{\lambda} \coth(\lambda h/2) - \frac{h}{2} \right],$$

$$D_{11} = \left(c_{11} + \frac{e_{31}^2}{a_{33}} \right) I_2 - \frac{\mu_{31}^2 h}{4a_{33}} + \frac{\mu_{31}^2 \lambda h^2}{8a_{33}} \coth(\lambda h/2) - \frac{e_{31}^2 h}{a_{33}\lambda^2} + \frac{2e_{31}^2}{a_{33}\lambda^3} \tanh(\lambda h/2),$$

$$D_{12} = \left(c_{12} + \frac{e_{31}^2}{a_{33}} \right) I_2 - \frac{\mu_{31}^2 h}{4a_{33}} + \frac{\mu_{31}^2 \lambda h^2}{8a_{33}} \coth(\lambda h/2) - \frac{e_{31}^2 h}{a_{33}\lambda^2} + \frac{2e_{31}^2}{a_{33}\lambda^3} \tanh(\lambda h/2),$$

$$D_{66} = c_{66} I_2, \ I_2 = \frac{1}{12} h^3.$$

It is evident from Eq. (32) that due to the inverse flexoelectric effect, an electrical bending moment M^e will be induced when applying a voltage between the top and bottom electrodes. The bending moment is proportional to both the applied voltage and the flexoelectric coefficient, and will drive the plate to deflect without the application of any mechanical loads. It is noted that Tagantsev and Yudin[24] also arrived the same conclusion.

It is also noted that although the plate is made of homogeneous materials, stretch-bending coupling terms $B_{\alpha\beta}$ arise. They are proportional to both the piezoelectric coefficient e_{31} and the flexoelectric coefficient μ_{31}, and thus originated from the piezoelectric-flexoelectric coupling effect. As a result, the effective bending stiffness of the nanoplate will be different from the original bending stiffness D_{11}, and the nanoplate will deflect transversely when a purely axial compressive or tensile load is applied on the geometrical mid-plane of the plate (without externally applied bending moment or voltage). Simple examples will be presented in the next section to explain the implications of the piezoelectric-flexoelectric coupling effect in detail. The constitutive equations can be rewritten as:

$$\begin{aligned} \mathbf{N} &= \mathbf{A}\boldsymbol{\varepsilon}_0 + \mathbf{B}\boldsymbol{\kappa} - \mathbf{N}^e \\ \tilde{\mathbf{M}} &= \mathbf{B}^T \boldsymbol{\varepsilon}_0 + \mathbf{D}\boldsymbol{\kappa} - \tilde{\mathbf{M}}^e, \end{aligned} \tag{33}$$

where

$$\mathbf{N}=\begin{Bmatrix}N_{xx}\\N_{yy}\\N_{xy}\end{Bmatrix},\ \tilde{\mathbf{M}}=\begin{Bmatrix}\tilde{M}_{xx}\\\tilde{M}_{yy}\\\tilde{M}_{xy}\end{Bmatrix},\ \boldsymbol{\varepsilon}_0=\begin{Bmatrix}\varepsilon_{xx0}\\\varepsilon_{yy0}\\2\varepsilon_{xy0}\end{Bmatrix},\ \boldsymbol{\kappa}=\begin{Bmatrix}\kappa_{xx}\\\kappa_{yy}\\2\kappa_{xy}\end{Bmatrix},\ \mathbf{N}^e=\begin{Bmatrix}N^e\\N^e\\0\end{Bmatrix},\ \tilde{\mathbf{M}}^e=\begin{Bmatrix}M^e\\M^e\\0\end{Bmatrix},$$

$$\mathbf{A}=\begin{bmatrix}A_{11}&A_{12}&0\\A_{12}&A_{11}&0\\0&0&A_{66}\end{bmatrix},\ \mathbf{B}=\begin{bmatrix}B_{11}&B_{12}&0\\B_{12}&B_{11}&0\\0&0&0\end{bmatrix},\ \mathbf{D}=\begin{bmatrix}D_{11}&D_{12}&0\\D_{12}&D_{11}&0\\0&0&D_{66}\end{bmatrix}.$$

From the first equation of Eqs. (33), we have:

$$\boldsymbol{\varepsilon}_0=\mathbf{A}^{-1}\left(\mathbf{N}+\mathbf{N}^e-\mathbf{B}\boldsymbol{\kappa}\right). \tag{34}$$

Substituting Eq. (34) into the second equation of Eqs. (33) yields:

$$\tilde{\mathbf{M}}_{eff}+\tilde{\mathbf{M}}^e_{eff}=\mathbf{D}_{eff}\boldsymbol{\kappa}, \tag{35}$$

where $\mathbf{D}_{eff}=\mathbf{D}-\mathbf{B}^T\mathbf{A}^{-1}\mathbf{B}$, $\tilde{\mathbf{M}}_{eff}=\tilde{\mathbf{M}}-\mathbf{B}^T\mathbf{A}^{-1}\mathbf{N}$ and $\tilde{\mathbf{M}}^e_{eff}=\tilde{\mathbf{M}}^e-\mathbf{B}^T\mathbf{A}^{-1}\mathbf{N}^e$. Their components could be called effective bending stiffness, effective mechanical moment and effective electrical moment, respectively. Consequently, the effective bending stiffness matrix for voltage-loaded plates reads:

$$\mathbf{D}_{eff}=\begin{bmatrix}D_{11}&D_{12}&0\\D_{12}&D_{11}&0\\0&0&D_{66}\end{bmatrix}-\frac{2B_{11}^2}{A_{11}+A_{12}}\begin{bmatrix}1&1&0\\1&1&0\\0&0&0\end{bmatrix}. \tag{36}$$

2. Open-circuit condition

By utilizing the same derivation method as shown above, a constitutive equation for the piezoelectric plate under open-circuit condition is obtained:

$$\begin{Bmatrix}N_{xx}\\N_{yy}\\N_{xy}\\\tilde{M}_{xx}\\\tilde{M}_{yy}\\\tilde{M}_{xy}\end{Bmatrix}=\begin{bmatrix}A_{11}&A_{12}&0&B_{11}&B_{12}&0\\A_{12}&A_{11}&0&B_{12}&B_{11}&0\\0&0&A_{66}&0&0&0\\B_{11}&B_{12}&0&D_{11}&D_{12}&0\\B_{12}&B_{11}&0&D_{12}&D_{11}&0\\0&0&0&0&0&D_{66}\end{bmatrix}\begin{Bmatrix}\varepsilon_{xx0}\\\varepsilon_{yy0}\\2\varepsilon_{xy0}\\\kappa_{xx}\\\kappa_{yy}\\2\kappa_{xy}\end{Bmatrix}, \tag{37}$$

where

$$A_{11}=\left(c_{11}+\frac{e_{31}^2}{a_{33}}\right)h+\frac{\mu_{31}^2\lambda}{2a_{33}}\tanh\left(\lambda h/2\right),\ A_{12}=\left(c_{12}+\frac{e_{31}^2}{a_{33}}\right)h+\frac{\mu_{31}^2\lambda}{2a_{33}}\tanh\left(\lambda h/2\right),$$

$$A_{66}=c_{66}h,\ B_{11}=B_{12}=\frac{e_{31}\mu_{31}}{a_{33}}\frac{1}{\lambda}\coth\left(\lambda h/2\right),\ D_{66}=c_{66}I_2,$$

$$D_{11}=\left(c_{11}+\frac{e_{31}^2}{a_{33}}\right)I_2+\frac{\mu_{31}^2\lambda h^2}{8a_{33}}\coth\left(\lambda h/2\right)-\frac{e_{31}^2h}{a_{33}\lambda^2}+\frac{2e_{31}^2}{a_{33}\lambda^3}\tanh\left(\lambda h/2\right),$$

$$D_{12}=\left(c_{12}+\frac{e_{31}^2}{a_{33}}\right)I_2+\frac{\mu_{31}^2\lambda h^2}{8a_{33}}\coth\left(\lambda h/2\right)-\frac{e_{31}^2h}{a_{33}\lambda^2}+\frac{2e_{31}^2}{a_{33}\lambda^3}\tanh\left(\lambda h/2\right),$$

The constitutive equations can also be transformed to a more compact form. However, under the electrically open-circuit condition, $\tilde{\mathbf{M}}^e$ and $\mathbf{N}^e$ vanish. Therefore the corresponding constitutive equation takes the following form:

$$\tilde{\mathbf{M}}_{eff}=\mathbf{D}_{eff}\boldsymbol{\kappa}, \tag{38}$$

where $\tilde{\mathbf{M}}_{eff} = \tilde{\mathbf{M}} - \mathbf{B}^T\mathbf{A}^{-1}\mathbf{N}$ and $\mathbf{D}_{eff} = \mathbf{D} - \mathbf{B}^T\mathbf{A}^{-1}\mathbf{B}$. It is noted that in this case, $\tilde{\mathbf{M}}_{eff}$ and $\mathbf{D}_{eff}$ should be calculated by using the corresponding stiffness under the open-circuit condition.

E. A new Karman plate theory for piezoelectric plates with flexoelectricity

With the definition $\tilde{M}_{\alpha\beta} = M_{\alpha\beta} + P_{\alpha\beta z}$, the variation of the strain energy given in Eq. (30) can be rewritten as:

$$\begin{aligned}\delta U &= \int_S \left(N_{xx}\delta\varepsilon_{xx0} + N_{yy}\delta\varepsilon_{yy0} + 2N_{xy}\delta\varepsilon_{xy0} + \tilde{M}_{xx}\delta\kappa_{xx} + \tilde{M}_{yy}\delta\kappa_{yy} + 2\tilde{M}_{xy}\delta\kappa_{xy}\right)\mathrm{d}A \\ &= \int_S \begin{pmatrix} N_{xx}\delta\left(u_{,x} + w_{,x}^2/2\right) + N_{yy}\delta\left(v_{,y} + w_{,y}^2/2\right) + \\ 2N_{xy}\delta\left(\left(u_{,y} + v_{,x}\right)/2 + w_{,x}w_{,y}\right) - \tilde{M}_{xx}\delta w_{,xx} - \tilde{M}_{yy}\delta w_{,yy} - 2\tilde{M}_{xy}\delta w_{,xy}\end{pmatrix}\mathrm{d}A\end{aligned}, \tag{39}$$

where Eq. (7) has been used.

Furthermore, the variation of external potential energy is:

$$\delta W = -\int_S q\delta w\mathrm{d}A - \int_\Gamma \left(\bar{N}_{nn}\delta u_n + \bar{N}_{ns}\delta u_s + \bar{Q}\delta w - \bar{M}_{nn}\delta w_{,n} - \bar{M}_{ns}\delta w_{,s}\right)\mathrm{d}s, \tag{40}$$

where q is the distributed pressure applied on the surface S of the plate, $\bar{N}_{nn}$, $\bar{N}_{ns}$, $\bar{Q}$, $\bar{M}_{nn}$ and $\bar{M}_{ns}$ the externally applied in-plane normal force, in-plane shear force, the transverse shear force, the normal bending moment and the twist moment per unit arclength at the edge Γ of the plate, and u_n, u_s, w, $w_{,n}$ and $w_{,s}$ are the corresponding displacements or rotation angles at the edge. Boundary loads associated with higher order stresses at the edge Γ are not considered here except the externally applied higher order moment $\bar{P}_{\alpha\beta z}$, which has already been absorbed in $\bar{M}_{nn}$ and $\bar{M}_{ns}$. These loads are clearly illustrated in Fig. 1.

The principle of minimum potential energy states that $\delta(U + W) = 0$. Applying the standard variational method to Eqs. (39) and (40), the equilibrium equations and associated boundary conditions can be obtained:

$$\begin{cases} N_{xx,x} + N_{xy,y} = 0, \\ N_{xy,x} + N_{yy,y} = 0, \\ \tilde{M}_{xx,xx} + 2\tilde{M}_{xy,xy} + \tilde{M}_{yy,yy} + \hat{N} + q = 0, \end{cases} \tag{41}$$

$$\begin{cases} N_{nn} = \bar{N}_n \ \text{ or } \ u_n = 0, \\ N_{ns} = \bar{N}_{ns} \ \text{ or } \ u_s = 0, \\ \tilde{M}_{nn} = \bar{M}_{nn} \ \text{ or } \ w_{,n} = 0, \\ Q = \bar{Q} + \bar{M}_{ns,s} \ \text{ or } \ w = 0, \end{cases} \tag{42}$$

where $\hat{N} = N_{xx}w_{,xx} + 2N_{xy}w_{,xy} + N_{yy}w_{,yy}$,

$$N_{nn} = N_{xx}n_x^2 + 2N_{xy}n_xn_y + N_{yy}n_y^2,\ N_{ns} = \left(N_{yy} - N_{xx}\right)n_xn_y + N_{xy}\left(n_x^2 - n_y^2\right),$$

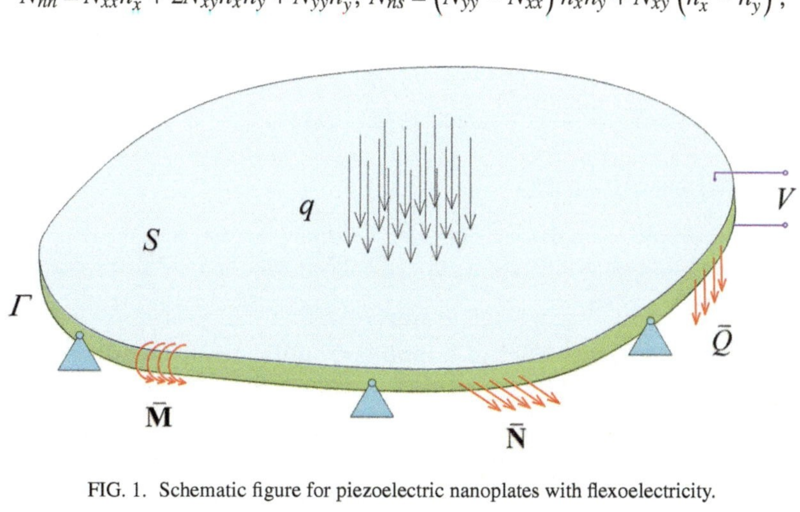

FIG. 1. Schematic figure for piezoelectric nanoplates with flexoelectricity.

$$\tilde{M}_{nn} = \tilde{M}_{xx}n_x^2 + 2\tilde{M}_{xy}n_xn_y + \tilde{M}_{yy}n_y^2,\ \tilde{M}_{ns} = \left(\tilde{M}_{yy} - \tilde{M}_{xx}\right)n_xn_y + \tilde{M}_{xy}\left(n_x^2 - n_y^2\right),$$

$$Q = \left(\tilde{M}_{xx,x} + \tilde{M}_{xy,y}\right)n_x + \left(\tilde{M}_{xy,x} + \tilde{M}_{y,y}\right)n_y + \tilde{N},$$

in which $\tilde{N} = \left(N_{xx}w_{,x} + N_{xy}w_{,y}\right)n_x + \left(N_{xy}w_{,x} + N_{yy}w_{,y}\right)n_y$. By substituting Eq. (32) or Eq. (37) into Eqs. (41) and (42), the equilibrium equations and boundary conditions in terms of the displacements for electrically short-circuit or electrically open-circuit condition can be directly obtained:

$$\begin{cases} A_{11}\left(u_{,x} + \frac{1}{2}w_{,x}^2 + v_{,y} + \frac{1}{2}w_{,y}^2\right)_{,x} + A_{66}\left[\left(u_{,y} - v_{,x}\right)_{,y} + 2w_{,x}w_{,xy}\right] - B_{11}\left(w_{,x} + w_{,y}\right)_{,xx} = 0, \\ A_{11}\left(u_{,x} + \frac{1}{2}w_{,x}^2 + v_{,y} + \frac{1}{2}w_{,y}^2\right)_{,y} + A_{66}\left[\left(v_{,x} - u_{,y}\right)_{,x} + 2w_{,y}w_{,xy}\right] - B_{11}\left(w_{,x} + w_{,y}\right)_{,yy} = 0, \\ B_{11}\Delta\left(u_{,x} + \frac{1}{2}w_{,x}^2 + v_{,y} + \frac{1}{2}w_{,y}^2\right) - D_{11}\Delta\Delta w + \hat{N} + q = 0, \end{cases} \tag{43}$$

where $\Delta = (\,)_{,xx} + (\,)_{,yy}$ is the Laplacian operator, and the expressions (7) for the strain components and the relations $A_{11} - A_{12} = 2A_{66}$, $D_{11} - D_{12} = 2D_{66}$ and $B_{11} = B_{12}$ has been utilized. These relations can be directly verified with the relation $c_{11} - c_{12} = 2c_{66}$. It is evident that by perturbing the nonlinear equilibrium equations (43) around a given initial configuration or an undeformed configuration of the nanoplate, a group of incremental linear equilibrium equations can be obtained, which can be used to analyze the linear vibration and stability of piezoelectric nanoplates with flexoelectricity (for the vibration problem, inertial terms need to be added in the equilibrium equations and the boundary conditions). The corresponding incremental equilibrium equations are not presented here due to space limitation.

It should be noted that boundary conditions associated with some higher order stresses such as τ_{yyx} and corresponding strain gradients do not arise in the above variational derivation. This is because only the strain gradients along the thickness direction are considered and boundary loads associated with higher order stresses at the edge Γ are neglected except $\bar{P}_{\alpha\beta z}$. If all the components of the strain gradient tensor are taken into account and other boundary loads associated with higher order stresses do not vanish, the "missing" boundary conditions associated with these higher order stresses and corresponding strain gradients will be derived naturally, and constitutive equations associated with these higher order stresses should also be taken into account.

It is also stressed that the electrical boundary conditions have a strong effect on the effective mechanical behavior of the considered plate as evidently shown in the constitutive equations of the plate (see the contrast between Eqs.(32) and (37)), although it is not self-evidently shown in the derived equilibrium equations (41) and boundary conditions (42). At last, it is pointed out that the derived equilibrium equations and boundary conditions as well as associated constitutive equations (32) or (37) can be easily used to analyze the static large deflection and nonlinear vibration behaviors (just need to further incorporate inertial terms) of piezoelectric nanoplates comprised of hard ceramics as long as the von Karman strain is applicable, and consequently will benefit the analysis and design of piezoelectric and flexoelectric sensors, actuators and energy harvesters and so on.

III. RESULTS AND DISCUSSIONS

In this section, simple examples are presented to show the effects of the coupling of flexoelectricity and piezoelectricity on the effective mechanical behavior and the electromechanical behavior of piezoelectric nanostructures.

A. A simply supported piezoelectric nanobeam under axial compressive loading and electrically open-circuit conditions

Firstly, let's consider a simply supported piezoelectric nanobeam under axial compressive loading and electrically open-circuit conditions as shown in Fig. 2. The equilibrium equations for the electroelastic beam can be easily obtained from the above plate theory:

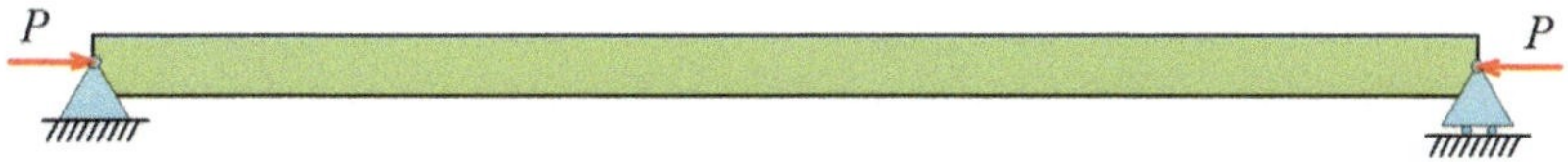

FIG. 2. Schematic figure of a simply supported thin piezoelectric and flexoelectric nanobeam under axial compressive load and electrically open-circuit condition.

$$\begin{aligned} &N_{xx,x} = 0, \\ &\tilde{M}_{xx,xx} + (N_{xx} w_{,x})_{,x} = 0, \end{aligned} \tag{44}$$

and associated boundary conditions can be derived:

$$N_{xx} = -P,\ \tilde{M}_{xx} = 0,\ w = 0,\ \text{at } x = \pm l, \tag{45}$$

where P is the applied axial compressive force.

By the first equation of Eqs. (44) and the corresponding boundary condition in Eq. (45), we have

$$N_{xx} = \text{const} = -P. \tag{46}$$

Utilizing the beam version of constitutive equation (38), we have:

$$D_{eff} w_{,xxxx} + P w_{,xx} = 0, \tag{47}$$

and the corresponding boundary conditions:

$$w_{,xx} = -B_{11}P \big/ \left(D_{eff} A_{11}\right),\ w = 0 \text{ at } x = \pm l, \tag{48}$$

where $D_{eff} = D_{11} - B_{11}^2 \big/ A_{11}$. The solution of this boundary value problem (Eqs. (47) and (48)) can be easily found to be:

$$w = \frac{B_{11}}{A_{11}} \left[\frac{\cos(kx)}{\cos(kl)} - 1 \right], \tag{49}$$

The maximum deflection at the center point of the beam in dimensionless form is

$$\frac{w_{max}}{\left(\dfrac{B_{11}}{A_{11}}\right)} = \frac{1}{\cos\left(\dfrac{\pi}{2}\sqrt{\dfrac{P}{P_0}}\right)} - 1, \tag{50}$$

where $k = \sqrt{P \big/ D_{eff}}$, $P_0 = \frac{D_{eff}\pi^2}{4l^2}$.

It can be inferred from this solution as also plotted in Fig. 3 that, due to the piezoelectric-flexoelectric coupling effect, the simply supported beam will deflect only if a little axial load is applied,

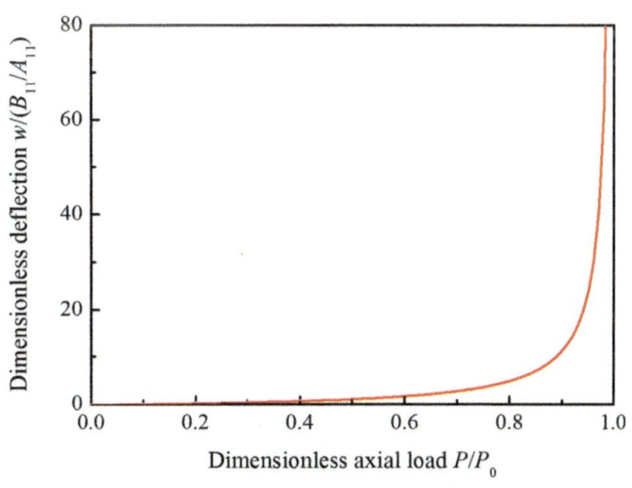

FIG. 3. The transverse deflection of the simply supported piezoelectric and flexoelectric nanobeam under axial compressive load ($P_0 = D_{eff}\pi^2/(4l^2)$).

which is completely different from the classical Euler rod buckling phenomenon.[38] Moreover, this conclusion also holds true for simply supported nanoplates with piezoelectricity and flexoelectricity coupling effects. Besides the above quantitative analysis, these results can also be qualitatively explained simply from the physical point of view. Firstly, the piezoelectric structures under a compressive load will be polarized due to piezoelectricity, and polarized charges accumulated on the top and bottom electrodes will create an electrical voltage (electrical potential difference) between the electrodes; then the generated voltage will further cause the beam to deflect because of flexoelectricity as discussed in subsection 2.4; hence the piezoelectric-flexoelectric coupling effect directly leads to the emergence of stretching-bending coupling effect in a homogeneous structure and the strength of the piezoelectric-flexoelectric coupling effect is proportional to the product of the piezoelectric coefficient and the flexoelectric coefficient as also given in Eq. (37).

It is noted that the above compression or tension induced deflection phenomenon exists as long as one end of piezoelectric nanobeams or some part of the edge of nanoplates is simply supported, but does not exist in doubly clamped nanobeams or fully clamped nanoplates. It should also be noted that when large rotation happens in the beam, the finite deformation beam model or the elastica model[38] should be utilized, for the von Karman strain is no longer suitable for the characterization of the geometrically nonlinear deformation in this case.

B. Polarization of a homogeneous piezoelectric nanoplate under uniform bending moment loading and open-circuit condition

For a homogeneous piezoelectric nanoplate under uniform bending moment and open-circuit condition, this is no need to analyze its equilibrium equation. According to the loading condition, we have $\tilde{\mathbf{M}}_{eff} = M\{1,1,0\}^{\mathrm{T}}$ and $\mathbf{N} = \{0,0,0\}^{\mathrm{T}}$. From the relations $\tilde{\mathbf{M}}_{eff} = \mathbf{D}_{eff}\boldsymbol{\kappa}$ and $\boldsymbol{\varepsilon}_0 = \mathbf{A}^{-1}(\mathbf{N} - \mathbf{B}\boldsymbol{\kappa})$ given by Eqs.(38) and (34), the curvature, the in-plane strain and the total strain of the plate can be determined to be:

$$\boldsymbol{\kappa} = \mathbf{D}_{eff}^{-1}\tilde{\mathbf{M}}_{eff},\ \boldsymbol{\varepsilon}_0 = -\mathbf{A}^{-1}\mathbf{B}\mathbf{D}_{eff}^{-1}\tilde{\mathbf{M}}_{eff},\ \text{and}\ \boldsymbol{\varepsilon} = \boldsymbol{\varepsilon}_0 + \boldsymbol{\kappa}z = \left(-\mathbf{A}^{-1}\mathbf{B} + \mathbf{I}z\right)\mathbf{D}_{eff}^{-1}\tilde{\mathbf{M}}_{eff}.$$

Utilizing Eqs. (10) (16) (17) and the expression of the electrical field (27), it is found that:

$$\begin{aligned} P_z &= D_z - \varepsilon_0 E_z \\ &= -(a_{33} - \varepsilon_0)\left\{\frac{\mu_{31}\lambda\cosh\lambda z}{2a_{33}\sinh(\lambda h/2)}\frac{h}{2}\left(w_{,xx} + w_{,yy}\right) + \frac{\lambda\sinh\lambda z}{\cosh(\lambda h/2)}\left[-\frac{\mu_{31}}{2a_{33}}\left(\varepsilon_{xx0} + \varepsilon_{yy0}\right)\right.\right. \\ &\quad \left.\left. + \frac{e_{31}}{a_{33}\lambda^2}\left(w_{,xx} + w_{,yy}\right)\right]\right\} + \frac{\varepsilon_0 e_{31}}{a_{33}}\left(\varepsilon_{xx} + \varepsilon_{yy}\right) + \left(\frac{1}{2} - \frac{\varepsilon_0}{a_{33}}\right)\mu_{31}\left(w_{,xx} + w_{,yy}\right). \end{aligned} \tag{51}$$

It is evident that the polarization driven by a uniform bending moment is not only contributed from the generated bending deformation due to the flexoelectricity, but also from the induced in-plane stretching or compression due to the piezoelectric effect. The induced in-plane stretching or compression is purely because of the existence of stretching-bending coupling which is in turn originated from the coupling of the flexoelectricity and piezoelectricity in nanostructures.

C. Direct current voltage driven bending deformation of a mechanically free homogeneous plate

For a mechanically free homogeneous plate subject to a direct current voltage, from Eq. (35) we have $\boldsymbol{\kappa} = \mathbf{D}_{eff}^{-1}\tilde{\mathbf{M}}_{eff}^{e}$, or

$$\boldsymbol{\kappa} = \left(\mathbf{D} - \mathbf{B}^T\mathbf{A}^{-1}\mathbf{B}\right)^{-1}\left(M^e\mathbf{I} - N^e\mathbf{B}^T\mathbf{A}^{-1}\right)\mathbf{I}', \tag{52}$$

where $\mathbf{I}' = \{1, 1, 0\}^T$, $N^e = -e_{31}V$, $M^e = -\mu_{31}V/2$.

It can be easily found from this equation that if the stretching-bending coupling is not taken into account, the generated curvature due to the flexoelectric effect will be $\boldsymbol{\kappa} = \mathbf{D}^{-1}(M^e\mathbf{I}')$, which is also affected by the piezoelectric coefficients through the bending stiffness $\mathbf{D}$, but is completely different from the more accurate one given in Eq. (52). Therefore, it can be concluded that the stretching-bending coupling terms induced by the coupling of the flexoelectricity and piezoelectricity directly influence the strength of the apparent converse flexoelectric effect of the considered piezoelectric plate.

IV. CONCLUSION

Based on the Kirchhoff-Love plate theory, the von Karman's nonlinear strain and the general flexoelectricity theory for piezoelectric solids, a geometrically nonlinear piezoelectric plate model is developed in the present work via the variational method, with a focus on the coupling effect between flexoelectricity and piezoelectricity and its consequence on the electromechanical behavior of piezoelectric nanostructures. The constitutive equations of the nanostructures *i.e.* the relation between the generalized forces (in-plane force, bending moment) and generalized strains (in-plane strain and bending curvature) are obtained for both the electrically short-circuit and open-circuit conditions. It is found that due to the coupling between flexoelectricity and piezoelectricity, stretching-bending coupling stiffness arises in nanostructures and its specific value relies on the applied electrical boundary conditions. As a result, the nanostructure will deflect only if a small in-plane loading is applied, which is completely different from the classical Euler rod buckling phenomenon. The polarization of a homogeneous plate subject to uniform bending moment is also studied by adopting the derived the constitutive equations under the open-circuit condition. It is found that the polarization is contributed not only from the generated bending deformation due to the flexoelectricity, but also from the induced in-plane stretching/compression due to the piezoelectricity and the stretching-bending coupling which in turn comes from the coupling of the flexoelectricity and piezoelectricity in nanostructures.

Although many assumptions have been made to drive the plate model, it can be extended without difficulty to account for the elastic strain gradient effect, the Maxwell stress[34] and surface effects such as surface polarization[26] and surface piezoelectricity.[24] The model and results presented in this work are expected to be useful for the analysis and design of piezoelectric and flexoelectric devices and systems.

ACKNOWLEDGMENTS

The present work is supported by National Natural Science Foundation of China (Grant No. 11602118, 11602117), China Postdoctoral Science Foundation funded project (Grant No. 2015M580499), and also sponsored by K.C. Wong Magna Fund in Ningbo University. The first author also thanks Prof. Deng qian and Prof. Chen changqing for helpful discussions.

[1] Z. L. Wang, Materials Today **10**, 20 (2007).
[2] R. Maranganti, N. Sharma, and P. Sharma, Physical Review B **74**, 014110 (2006).
[3] M. Majdoub, P. Sharma, and T. Cagin, Physical Review B **77**, 125424 (2008).
[4] M. Majdoub, P. Sharma, and T. Çağin, Physical Review B **78**, 121407 (2008).
[5] C. Liu, S. Hu, and S. Shen, Smart Materials and Structures **21**, 115024 (2012).
[6] Q. Deng, M. Kammoun, A. Erturk, and P. Sharma, International Journal of Solids and Structures **51**, 3218 (2014).
[7] T. D. Nguyen, S. Mao, Y. W. Yeh, P. K. Purohit, and M. C. McAlpine, Advanced Materials **25**, 946 (2013).
[8] P. Yudin and A. Tagantsev, Nanotechnology **24**, 432001 (2013).
[9] P. Zubko, G. Catalan, and A. K. Tagantsev, Annual Review of Materials Research **43**, 387 (2013).
[10] *Flexoelectricity in Solids: From Theory to Applications*; *Vol.* World Scientific Publishing Co. Pte. Ltd, 2017).
[11] D. Lee, A. Yoon, S. Jang, J.-G. Yoon, J.-S. Chung, M. Kim, J. Scott, and T. Noh, Physical Review Letters **107**, 057602 (2011).
[12] M.-W. Chu, I. Szafraniak, R. Scholz, C. Harnagea, D. Hesse, M. Alexe, and U. Gösele, Nature Materials **3**, 87 (2004).
[13] M. Majdoub, R. Maranganti, and P. Sharma, Physical Review B **79**, 115412 (2009).
[14] N. Sharma, R. Maranganti, and P. Sharma, Journal of the Mechanics and Physics of Solids **55**, 2328 (2007).
[15] J. Y. Fu, W. Zhu, N. Li, N. B. Smith, and L. E. Cross, Applied Physics Letters **91**, 182910 (2007).
[16] W. Ma and L. E. Cross, Applied Physics Letters **78**, 2920 (2001).
[17] W. Ma and L. E. Cross, Applied Physics Letters **81**, 3440 (2002).
[18] W. Ma and L. E. Cross, Applied Physics Letters **86**, 072905 (2005).
[19] W. Ma and L. E. Cross, Applied Physics Letters **88**, 232902 (2006).
[20] S. Hu and S. Shen, Science China Physics, Mechanics and Astronomy **53**, 1497 (2010).
[21] S. Shen and S. Hu, Journal of the Mechanics and Physics of Solids **58**, 665 (2010).
[22] Z. Yan and L. Jiang, Journal of Applied Physics **113**, 194102 (2013).
[23] Z. Yan and L. Jiang, Journal of Physics D: Applied Physics **46**, 355502 (2013).
[24] A. K. Tagantsev and A. S. Yurkov, Journal of Applied Physics **112**, 044103 (2012).
[25] A. S. Yurkov, Physics of the Solid State **57**, 460 (2015).
[26] A. S. Yurkov and A. K. Tagantsev, Applied Physics Letters **108**, 022904 (2016).
[27] Z. Zhang, Z. Yan, and L. Jiang, Journal of Applied Physics **116**, 014307 (2014).
[28] W. Yang, X. Liang, and S. Shen, Acta Mechanica **226**, 3097 (2015).

[29] X. Liang, W. Yang, S. Hu, and S. Shen, Journal of Physics D: Applied Physics **49**, 115307 (2016).
[30] C. Liu, H. Wu, and J. Wang, Applied Physics Letters **109**, 192901 (2016).
[31] S. Hu and S. Shen, Computers, Materials & Continua **1389**, 1 (2009).
[32] X. Liang, S. Hu, and S. Shen, Smart Materials and Structures **24**, 105012 (2015).
[33] Z. Suo, Acta Mechanica Solida Sinica **23**, 549 (2010).
[34] Q. Deng, L. Liu, and P. Sharma, Journal of the Mechanics and Physics of Solids **62**, 209 (2014).
[35] J. Yang, *An introduction to the theory of piezoelectricity*, Vol. 9 (Springer Science & Business Media, 2004).
[36] S. Timoshenko and S. Woinowsky, (McGraw-Hill, New York, 1959).
[37] A. S. Starkov and I. A. Starkov, International Journal of Solids and Structures **82**, 65 (2016).
[38] S. Timoshenko, (Tata McGraw-Hill Education, 1970).

24

Interactions between the first mode and the second Bragg gap in a cylindrical waveguide with undulated walls

Jiu-Ling Xue,[1,2] Huan Liu,[2] Ting Liu,[2] Fu-Hou Zheng,[2] Ya-Xian Fan,[1,a] and Zhi-Yong Tao[1,2,b]
[1]*Key Lab of In-fiber Integrated Optics, Ministry Education of China, Harbin Engineering University, Harbin 150001, People's Republic of China*
[2]*Physics Research Center, College of Science, Harbin Engineering University, Harbin 150001, People's Republic of China*

Bragg resonances caused by the same transverse modes can always play a major role in periodic waveguides when the period is larger than the average radius. Because of higher-order mode cutoffs, the related Bragg gaps can be identified as interactions between different spatial harmonics of the fundamental mode, and the first Bragg gaps are more intensive than the higher ones. When we alter the parameters of the periodic waveguide, especially, decrease the period, the first transverse mode can be involved in Bragg gaps. Here, we demonstrate a direct mode-stopband interaction between the first mode and the second Bragg gap, that an extraordinary passband arises in the original second Bragg gap and splits the bandgap into two. Furthermore, the extraordinary passband is mainly composed of a pure first mode, which effectively suppresses the transmission of the fundamental one. We have also investigated the influence of wall profiles on the transmission and mode purity, and have found that the defined shape factor of wall profiles is proportionally related to the width of both pass and stop bands. The results could benefit not only the understanding of wave phenomena but also the applications in mode generators, filters, and so on.

I. INTRODUCTION

Periodic structures are constituted by identical components which are arranged in an array or in a lattice. One of the most common examples of periodic structures in nature is the crystalline structure, in which the atoms form atomic lattices under inter-atomic forces. Periodic structures are of great importance since they are widely utilized in vibration isolators,[1] acoustic frequency filters[2] and so on. The initial study of periodic structures can be traced back to 1887, when Lord Rayleigh explored vibration on a periodic string.[3] Recently, more attention has been paid to determine the frequency bandgaps of periodic structures since the bandgap is an important feature of periodic structures, and it can be applied in photonic bandgap materials which can act like a kind of "semiconductors."[4] Moreover, a lot of topics has been developed in physics and technologies, such as invisibility cloak,[5] negative refraction,[6,7] and thermal management.[8]

Periodically corrugated waveguides as one special case of periodic structures have also been focused in recent decades. As early as 1974, Nayfeh had solved the problem of wave propagation in sinusoidally corrugated ducts using a multiple scales perturbation technique.[9] Later, in 1994, Bradley employed a Floquet-like theorem and characterized the waves traveling in a rectangular periodic duct as Bloch wave functions.[10] Giudice and Bernasconi studied a sinusoidally corrugated waveguide and

[a]Electronic mail: yxfan@hrbeu.edu.cn
[b]Electronic mail: zytao@hrbeu.edu.cn

derived the simple formulas of its reflected and transmitted spectral responses based on the Bragg grating theory.[11] And Hawwa compared the speed of waves propagating in the sinusoidally and rectangularly corrugated ducts with the same corrugation wavelength.[12]

The study of frequency bandgaps is also crucial for periodic corrugated waveguides. When a monochromatic wave is propagating in a periodic waveguide, both constructive and destructive interferences will arise depending on their different paths. If the incident and reflected waves have the same transverse wave front, their interference will cause Bragg resonance. Otherwise, if the interference is caused by interactions between two different transverse modes, it will result in non-Bragg resonance. Both types of resonances will lead to bandgaps, and they have been reported in recent works.[13–18] In 2013, Zhao and Tao[19] discovered that the non-Bragg bandgap is proportional to the undulation height and shape factor of the waveguide. In 2014, Sorokin derived the width of stopbands for different corrugation shapes.[20] Huang, *et al.* investigated the relation between bandgap properties and structures of periodic waveguides.[21] And, in 2015, Nielsen and Sorokin[22] demonstrated the relationship between the stopband patterns and two parameters of the corrugated waveguide, which are the impedance mismatch and the ratio of propagation times. Later in 2016, Sorokin[23] explored into the relationship between the amplitudes of the harmonics of the corrugation series and widths of the frequency band-gaps using the method of varying amplitudes. However, all the above researches were focused on the two-mode interactions, whereas the more complicated multimode interactions concerning more than two coupled modes have not been studied thoroughly. It is crucial to look into multimode interactions, since they bring unusual transmissions in the traditional stopband and might generate and transmit a single high-order mode, which has many potential uses in different fields such as integrated optics, ultrasonic testing, microfluidic devices, magnetomechanical sensors and so on. Previously, Tao *et al.* has realized the energy transmission in the Bragg gap via high-order transverse modes, confirming the possibility of indirect interaction between modes.[24] In our work, we went a step further and realized a direct interaction of the first mode and the second Bragg gaps.

Here, we will investigate interactions between the transverse mode and Bragg gap in periodically corrugated waveguides, and demonstrate that they can efficiently generate a single first mode and suppress the fundamental ones. In the following section, we will introduce the structure of the corrugated waveguide and present the appearance of the second Bragg gap. In Sec. III, we will focus on the direct mode-stopband interaction, and derive the formulas of the frequency splitting. Next, we will compare the effect of wall profiles on mode-stopband interactions in Sec. IV. In Sec. V, we will verify the generation of a single first mode based on the direct mode-stopband interactions. Finally, Sec. VI is the conclusion which summarizes our major findings.

II. THE SECOND BRAGG GAPS

The periodic waveguide to be investigated here is a cylindrical duct with corrugated walls, which is filled with an inviscid, nonconducting perfect gas, whose longitudinal sectional axonometric drawing is shown in Fig. 1. Its radial and longitudinal coordinates are denoted by r^* and z^* respectively. Λ^* is the period of the waveguide, ω^* is the circular frequency, and a^* denotes the wall corrugation amplitude. With the waveguide's mean radius R_0 and the velocity of sound v, we define the dimensionless variables as follows:

$$\begin{aligned} r &= r^*/R_0, \quad z = z^*/R_0, \quad \omega = \omega^* R_0/v, \\ \varepsilon &= a^*/R_0, \quad k = 2\pi R_0/\Lambda^*, \end{aligned} \tag{1}$$

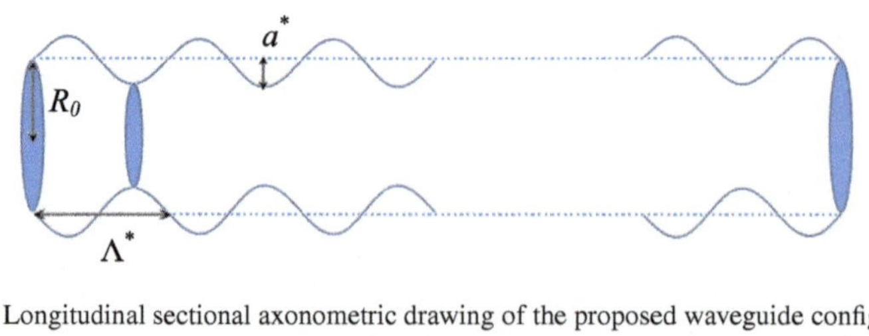

FIG. 1. Longitudinal sectional axonometric drawing of the proposed waveguide configuration.

such that ε and k represent the degree of wall fluctuation and the corrugation frequency, respectively. In this way, the dimensionless sound pressure P is governed by the Helmholtz equation,

$$\nabla^2 P + \omega^2 P = 0, \tag{2}$$

and the boundary condition holds on the wall of the waveguide, which leads to

$$\frac{\partial P}{\partial \mathbf{n}} = 0, \quad when \quad r = r_w = 1 + \varepsilon W(z), \tag{3}$$

where $W(z)$ is the wall profile and $\mathbf{n}$ is the unit outward vector normal to the wall. The homogeneous boundary condition considered here is very reasonable because the common 3 mm thick stainless steel could be treated as a rigid boundary.[25]

By expanding the sound pressure P in spatial harmonic series according to the Floquet's theorem, one gets

$$P = \sum_{n=-\infty}^{\infty} A_n \mathrm{J}_0(k_{r,n} r) \mathrm{e}^{\mathrm{i}k_{z,n} z}, \tag{4}$$

where $\mathrm{J}_0(\cdot)$ is the zero-order Bessel function, $k_{r,n} = \beta + nk$ the transverse wave number, $k_{z,n}$ the longitudinal wave number of the nth space harmonic, and β $(-k/2 \leq \beta \leq k/2)$ the reference propagation constant. And we have the dispersion relation as

$$\omega^2 = k_{r,n}^2 + k_{z,n}^2. \tag{5}$$

Substituting the series (4) into the boundary condition (2) and expanding the results in a Fourier integral, we can obtain a system of linear algebraic equations for coefficients A_n:

$$\sum_{n=-\infty}^{\infty} A_n C_{mn}(k_{r,n}, \beta; \varepsilon, k) = 0, \tag{6}$$

with the coefficient C_{mn} given by

$$\begin{aligned} C_{mn} = -\int \mathrm{d}z \cdot \mathrm{e}^{-\mathrm{i}(m-n)kz} [k_{r,n} \mathrm{J}_1(k_{r,n} r_w) \\ + \mathrm{i}\varepsilon k_{z,n} W'(z) \mathrm{J}_0(k_{r,n} r_w)], \end{aligned} \tag{7}$$

where $\mathrm{J}_1(\cdot)$ is the first-order Bessel function.

To guarantee the existence of a nontrivial solution to the coefficients, the determinant of the matrix $\mathbf{C} = \{C_{mn}\}$ must vanish, such that:

$$\det \|\mathbf{C}(k_{r,n}, \beta; \varepsilon, k)\| = 0. \tag{8}$$

In the case of no corrugations, ie. $\varepsilon = 0$, Eq. (8) is reduced to $\mathrm{J}_1(k_{r,n}) = 0$, leading to the solution of radial wavenumber $k_r^{(p)}$, which is $k_r^{(p)} = 0, 3.8317, 7.0156, 10.1735\ldots$ for $p = 0, 1, 2, 3\ldots$. And the different values of p stand for different transverse modes, e.g., $k_r^{(0)}$ is the fundamental mode, $k_r^{(1)}$ is the first mode, and so on. There will appear a resonance between the pth mode of the nth harmonic and the qth mode of the n'th harmonic when

$$\beta_{pq} = -\frac{(n+n')k}{2}(1 + \eta_{pq}), \tag{9}$$

where

$$\eta_{pq} = \frac{k_r^{(q)^2} - k_r^{(p)^2}}{(n^2 - n'^2)k^2}. \tag{10}$$

If $p = q$, both the nth and the n'th harmonics have the same transverse mode and radial distribution, the Bragg resonance will arise. On the other hand, if $p \neq q$, they are different modes of non-propagated transverse standing waves, and the non-Bragg resonance will take place.[26] Here, we focus on the Bragg resonance, where $p = q$, such that

$$\beta = -\frac{(n+n')k}{2}. \tag{11}$$

According to the dispersion condition (5), the frequency of the pth mode of the nth harmonic can be formulated as

$$\omega = \sqrt{k_r^{(p)^2} + (nk+\beta)^2}, \tag{12}$$

which are called mode or reference lines and can be used to indicate mode interactions in the first Brillouin zone. The solid and dashed lines denoted by $l(p, n)$ in Fig. 2 (a) indicate the first and fundamental modes of the nth harmonic, respectively, according to the graphic method,[26] which helps demonstrate mode interactions in a waveguide. As an example, the corrugation frequency $k = \pi$ results in that the dashed lines intersect around $\omega = 1.57$ and $\omega = 3.14$ while the red solid line is away from these two intersections, which indicate the first and second Bragg stopbands of the same fundamental mode.

When we start to take the corrugation back into consideration, ie. $\varepsilon > 0$, the frequency stopbands will form at the intersection of lines $l(p, n)$, where a repulsion of levels occurs as the dashed lines in Fig. 2 (a) shows. In this type of waveguides, we have employed a sinusoidally corrugated wall with $\varepsilon = 0.1$. By using the developed numerical method,[26,27] the dispersion curves are calculated and depicted by the blue dots in Fig. 2 (a). The gray and pink areas denote the first and second Bragg stopbands, respectively. The red solid line representing the first mode is over these two stopbands and has no effect on them. To illustrate the Bragg gaps of the fundamental mode more obviously, we have built a 20-period sinusoidally walled duct with $\Lambda^* = 114$ mm and $R_0 = 57$ mm in COMSOL MultiPhysics. Setting the radiation condition to the inlet and outlet, and sound hard wall condition to the corrugated wall, we have simulated the transmission spectrum as depicted in Fig. 2 (b). We can easily find the first Bragg stopband around $\omega = 1.57$ and the second around $\omega = 3.14$, respectively, shown by the gray and pink shadows.

By deriving the different transverse mode dispersions, we have demonstrated the appearances of the first two Bragg stopbands, which are not affected by any higher modes when the wall corrugation frequency k is chosen to be π. The numerical results show that the second Bragg stopband arises distinctly because of the same fundamental mode interaction, despite the much smaller intensity in sound attenuation than the first one.

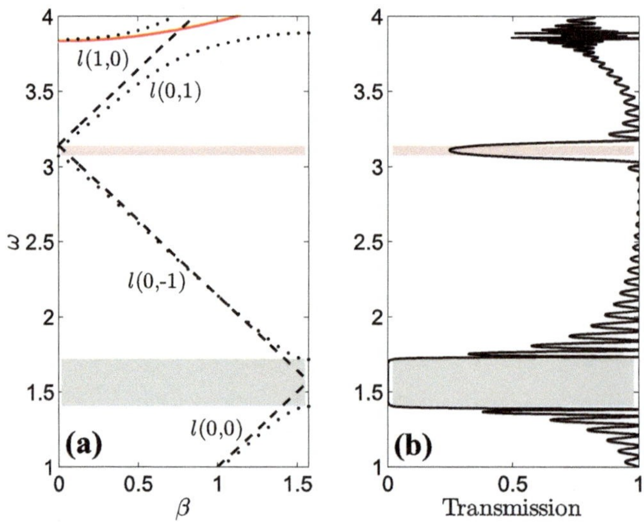

FIG. 2. The first (gray area) and second (pink area) Bragg band gaps for the sinusoidally corrugated waveguide with $k = \pi$. (a) Dispersion curves and their reference lines in the half of first Brillouin zone. The blue dots depict the dispersion curves, the blue dashed lines denote the fundamental mode, and the red solid line denotes the first mode. (b) The simulated transmission curve of the 20-period waveguide.

III. INTERACTIONS BETWEEN MODE AND BANDGAP

With $k = \pi$, we have obtained the first and second Bragg gaps without the higher mode interference. When the geometry parameters of the waveguide change, the effect of higher order modes have to be considered. Enlarging the average radius will lead to the first mode moving into the lower mode stopbands. The first mode in the first Bragg gap has been investigated and known as the indirect mode-stopband interaction[24] because the first mode has not directly had a hand in the creation of special spectra but just been transmitted in the first Bragg gap. This time, we focus on the interactions between the first mode and the second Bragg gap, which are quite different from the indirect interactions because of multimode line intersections.

Next, we will look into multimode interactions which involve more than two transverse modes. According to the graphic method,[26] multimode interactions will arise when three or more mode lines intersect at a single point or several intersections are very close together in the first Brillouin zone of periodic waveguides. We have altered the parameters of the waveguide with $k = 3.8317$, such that three lines $l(0, 1)$, $l(1, 0)$, and $l(0, -1)$ representing the three modes intersect around $\omega = 3.8317$ at the center of the first Brillouin zone when $\beta = 0$, as shown in Fig. 3 (a). In this case, the sinusoidally corrugated waveguide has geometry parameters of $\varepsilon = 0.1$, $\Lambda^* = 114$ mm, and $R_0 = 70$ mm, whose dispersion curves are depicted by the blue dots. It is found that the first mode has been moved into the second Bragg gap and that the interactions are more complex than ever. The two-mode resonance between $l(1, 0)$ and $l(1, 1)$ leads to a splitting of the first mode around $\omega = 4.28$, while the three-mode resonance among $l(0, 1)$,$l(1, 0)$, and $l(0, -1)$ leads to a three mode interaction spectrum around $\omega = 3.8317$, which is very special as there is an extraordinary passband (blue area) arising in the former second Bragg gap (pink area). The dispersion curve (blue dots) in the blue area is so close to the first mode (red solid line) that we can expect the single first mode transmission in the second Bragg gap based on direct mode-stopband interactions.

In order to better understand interactions between modes and stopbands, we need to calculate the values of five critical points marked by bold red dots in Fig. 3 (a), which are T, T^+, T^-, B^+, and B^-. As we have analyzed, these points are involved in two different resonances, one is the Bragg resonance of the first mode and the other is the three mode resonance between the fundamental and first transverse modes. Calculating the Fourier integral (7) and reserving the lowest terms yields

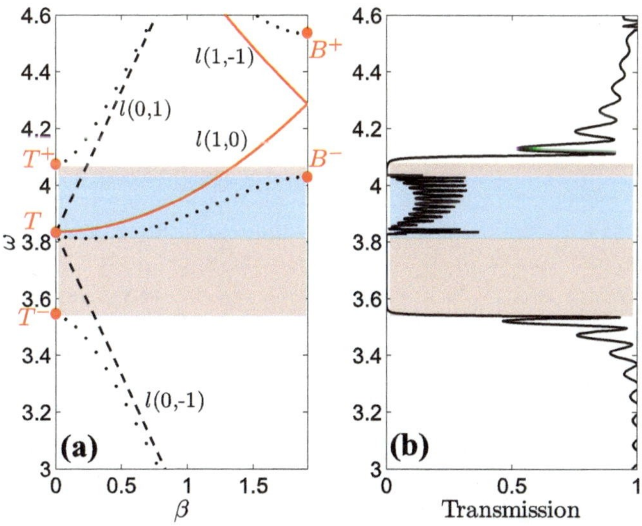

FIG. 3. Interactions between the first mode and second Bragg gap in the sinusoidally corrugated waveguide with $k = 3.8317$. (a) Dispersion curves and their reference lines in the half of first Brillouin zone. The blue dots show the dispersion curves, the blue dashed lines denote the fundamental mode, and the red solid lines denote the first mode. (b) The simulated transmission curve of the 20-period waveguide. The pink areas illustrates the band gaps whereas the blue area is the extraordinary passband arising in the former second Bragg gap.

$$C_{mn} = -\Lambda\left[k_{r,n}\mathrm{J}_1(k_{r,n})\delta_{m,n} - \varepsilon W_{m-n}J_{m,n}\right], \tag{13}$$

where

$$J_{m,n} = k_{r,n}^2\mathrm{J}_0^{(2)}(k_{r,n}) + (m-n)kk_{z,n}\mathrm{J}_0(k_{r,n}), \tag{14}$$

and

$$W_j = \int_{-0.5}^{0.5} \mathrm{d}z' W(\Lambda z')\mathrm{e}^{-\mathrm{i}2\pi jz'}. \tag{15}$$

For Bragg resonance between modes $l(1, 0)$ and $l(1, 1)$, only two spatial harmonics ($n = 0, 1$) are considered. By taking only the major Fourier components of $W_{\pm1}$ into consideration, while ignoring the higher-order terms, we can simplify Eq. (8) as

$$k_{r,0}\mathrm{J}_1(k_{r,0})k_{r,1}\mathrm{J}_1(k_{r,1}) = \varepsilon^2 W_1 W_{-1} J_{0,1} J_{1,0}. \tag{16}$$

Based on the perturbation theories, we have

$$k_{r,0} = k_r^{(1)} + \varepsilon\delta_B, \tag{17}$$

and with

$$k_{r,1} = \sqrt{k_{r,0}^2 + \beta^2 - (\beta + k)^2}, \tag{18}$$

and Eq. (16), one gets

$$\delta_B^2 = \frac{9}{4} k_r^{(1)^2} W_1 W_{-1}. \tag{19}$$

Then, we obtain

$$\delta_B = \pm\frac{3k_r^{(1)}}{2}|W_{\pm1}|, \tag{20}$$

which results in the separation of the two points $B^{\pm}$. The location of point B^{-} indicates the upper edge of the extraordinary passband.

Similarly, for the three-mode resonance among modes $l(0, 1)$, $l(1, 0)$, and $l(0, -1)$, we need to consider the three spatial harmonics of $n = 0, \pm1$. And we set

$$k_{r,0} = k_r^{(1)} + \varepsilon\delta_T, \tag{21}$$

and substitute it into the truncated Eq. (8), which is too complicated to be included here. By the aid of computers, the solution of the perturbed term can be obtained as

$$\delta_T = 0, \quad \pm k_r^{(1)}\sqrt{2}|W_{\pm1}|, \tag{22}$$

indicating the stationary of point T and the splitting of points $T^{\pm}$, which estimate the lower edge of the extraordinary passband and the edges of the second Bragg stopband, respectively. These properties of critical points have been approved by the blue dots in dispersion curves.

As we have seen, there are two solutions of the frequency splitting to the two-mode resonance, whereas three solutions including a zero value of the frequency splitting to the three-mode resonance. Therefore, beyond the traditional two-mode resonance, the three-mode interaction can lead to the more complicated frequency spectra. Nevertheless, because of the zero value of δ_T, the first mode will still appear at the original frequency and its dispersion curve will overlap with the original solid line. It suggests that the transverse mode represented by $l(1, 0)$ would not be influenced by the three-mode resonance while the other two modes represented by $l(0, 1)$ and $l(0, -1)$ interact with each other intensively. This leads to an additional passband in the frequency bandgap, as well as a expected pure first mode. The three-mode interaction can also be interpreted by the high-order transverse mode penetrating the lower-order mode forbidden band. The repulsion at intersections is related to the longitudinal wavenumber of the three interacting modes.

The transmission of the waveguide shown in Fig. 3 (b) confirms the extraordinary transmission resulting from the direct mode-stopband interaction, when we simulate the sound propagation in a rigid-walled corrugated duct with finite element method (FEM) in COMSOL MultiPhysics. The blue and pink areas denote the extraordinary passband and the former second Bragg gap, respectively. This clearly demonstrates how the mode and stopband interact directly and what the effect on the spectra is. When the first mode interact with the second Bragg stopband, not only the more intense resonance leads to the wider band gap, but also a deemed single mode transmission arises.

IV. EFFECTS OF WALL PROFILES

We have been employing the model of sinusoidally corrugated waveguide so far, but according to the perturbation results, the splitting of frequency spectrum is proportional to $W_{\pm 1}$, which is the major Fourier component of the wall function. The different wall profiles can definitely affect on the mode-stopband interactions. Then, we further investigate the influence of different wall profiles by defining the shape factor[19] as

$$S = \frac{1}{2}|W_{\pm 1}| = S_o, \tag{23}$$

where

$$S_o = \int_0^{0.5} dz' W(\Lambda z') \sin(2\pi z'), \tag{24}$$

for only the odd wall function considered here. For the even wall functions, the analysis is too similar to be included here.

Thus, we can calculate the shape factors for different wall profiles. For the sinusoidally corrugated waveguide we have considered, its shape factor $S = 0.25$. We can also investigate rectangularly corrugated waveguide and triangularly corrugated waveguide, whose shape factors are 0.3183 and 0.2026, respectively. Then, the shape factor increases in the order triangular<sinusoidal<rectangular wall profile. We retain the same parameters of $\varepsilon = 0.1$, $\Lambda^* = 114$ mm and $R_0 = 70$ mm for the three waveguides, so that three-mode interactions ensue.

On the other hand, the dimensionless frequency of five critical points introduced in Fig. 3 can be calculated as

$$\omega = k_r^{(1)}, k_r^{(1)}(1 \pm \varepsilon 2\sqrt{2}S) \tag{25}$$

and

$$\omega = \omega_B \pm \varepsilon \frac{3k_r^{(1)2} S}{\omega_B} \tag{26}$$

for the three and two mode interactions, respectively, where the Bragg resonant frequency of the first mode

$$\omega_B = \sqrt{k_r^{(1)2} + \left(\frac{k}{2}\right)^2}. \tag{27}$$

Then, the frequencies of the five points can be calculated by means of the shape factor and the results are presented in Fig. 4 by the blue circles, red squares, and green triangles for the sinusoidal, rectangular, and triangular wall profiles, respectively. It is found that the band edges are all proportional to the shape factor. The larger shape factor results in the wider stopband but the narrower passband. According to the properties of five critical points, we can estimate that the width of the second Bragg gap increases in the order triangular<sinusoidal<rectangular wall profile while the extraordinary passband width decreases in the same order. Additionally, the lower edge of the passband should be at the same frequency due to the stationary point T. Therefore, the rectangular wall profile can lead to the widest stopband and the narrowest passband.

With the FEM, we show the simulated transmissions of periodic waveguides in Fig. 5 for the sinusoidal, rectangular, and triangular wall profiles by the blue solid, red dashed, and green dash-dot lines, respectively. From these transmissions, we can conclude that as the shape factor of the wall profile increase, the width of the stopband increases and the that of the extraordinary passband decreases. Due to the larger shape factor of the rectangular wall, the passband is compressed and the enlarged stopband provide the smooth zero transmission instead of the fluctuations for the waveguides with sinusoidal and triangular walls. Physically, the rectangular wall can provide more intense interaction, which has been quantified by the shape factor (23). The estimations on the sinusoidal and triangular wall profile ducts are in accordance with the numerical results while that on the rectangular profile has

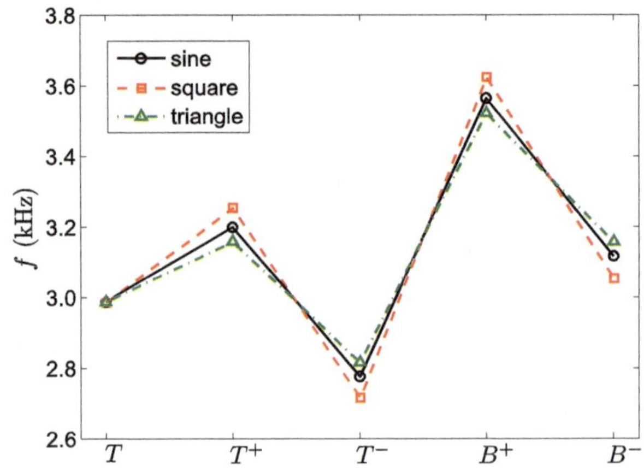

FIG. 4. Frequency locations of point T, T^+, T^-, B^+, and B^- for three different waveguides with sinusoidal, rectangular, and triangular wall profiles, denoted by the blue circles, red squares, and green triangles, respectively.

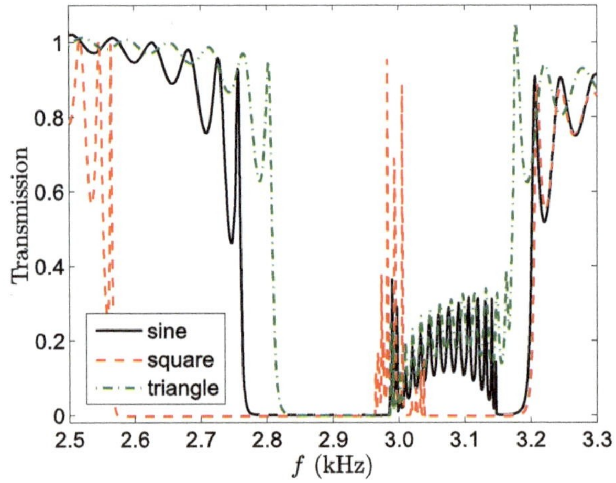

FIG. 5. Transmissions of three different corrugated waveguides with sinusoidal, rectangular, and triangular wall profiles, denoted by blue solid, red dashed, and green dash-dot lines respectively.

a few deviations. The reason is that we have only considered the major Fourier component despite the other comparable large components of the rectangular function. Even so, the proposed method based on the introduced shape factor is accurate enough to be applicable in spectrum structure estimation when the direct mode-stopband interaction occurs.

V. SINGLE FIRST MODE

We have demonstrated the mode-stopband interaction and its induced extraordinary transmission in the second Bragg gap in the three waveguides with sinusoidal, rectangular, and triangular wall profiles. According to the dispersion curves and reference lines in Fig. 2 (a), we have expected the transmitted single first mode penetrating the second Bragg gap. To verify the single first mode penetration, we have calculated the transverse distributions of the sound pressure for the three waveguides. By choosing the frequency of the highest peak in the passband and the location of the strongest longitudinal sound pressure to avoid computing errors, we have simulated the sound pressure along the radius as presented in Fig. 6. The sound pressure and radius are normalized by their own maximum. The circles, squares, and the triangles are the normalized pressures of the sinusoidal, rectangular, and triangular wall profile ducts, respectively. As the first transverse mode is a zero-order Bessel function, all the sound pressures get their maximum at $r = 0$ and oscillate when r increases.

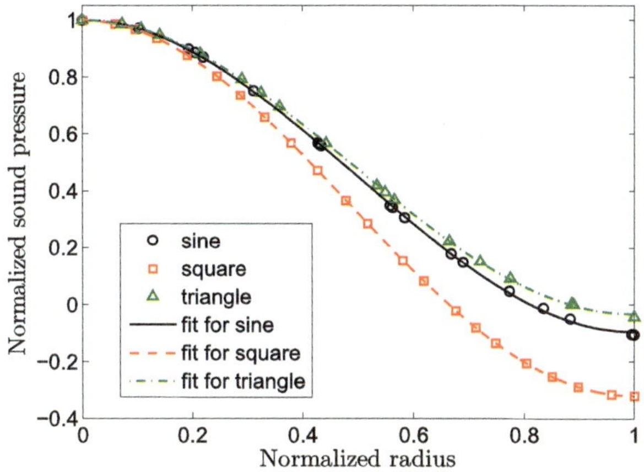

FIG. 6. Radial distributions of sound pressure for the three waveguides with sinusoidal, rectangular, and triangular wall profiles: the blue circles, red squares, and green triangles are normalized sound pressures, and the blue solid, red dashed, and green dash-dot lines are their fitting curves, respectively.

TABLE I. Coefficients of the best fits for the three waveguides.

	a_0	a_1	a_2	a_3	RMSE
Sine	0.2293	0.7802	-0.0223	0.0161	0.0049
Square	0.0648	0.9393	-0.0075	0.0037	0.0003
Triangle	0.2708	0.7364	-0.0145	0.0084	0.0029

To quantitatively obtain the mode components of sound pressures, we have performed the least-square fitting of the three sets of data with the model

$$P(r) = \sum_{p=0}^{3} a_p \mathrm{J}_0(k_r^{(p)} r), \tag{28}$$

and depict the best fit by the solid, dashed, and dash-dot lines for the three waveguides with sinusoidal, rectangular, and triangular wall profiles in Fig. 6, respectively. The optimal coefficients for the three fits are given in Table I. It is clear that the value of coefficient a_1 is much larger than the other three coefficients, verifying the propagation of the single first mode and the effective suppression of the fundamental mode in the waveguides. The extremely small a_2 and a_3 indicate that the mode-stopband interaction we proposed here highly relies on the first and fundamental modes and no higher order modes are involved, which also confirm the truncation in the fitting model (28) where the transverse modes higher than the third order are omitted.

In addition, the optimal coefficients not only confirm the existences of the single first modes in the proposed waveguides, but also exhibit their purity. The first mode possesses over 70 percent in the sinusoidal and triangular wall profile waveguides, but over 90 percent in the rectangular wall profile waveguides. The rectangularly corrugated duct is more efficient in suppression of the fundamental mode and generation of the first one. The root mean squared error (RMSE) is also included in the table for each fit to show its conformity with the original data.

VI. DISCUSSION AND CONCLUSION

We have demonstrated a type of direct mode-stopband interactions in the periodically corrugated waveguides with arbitrary wall profiles. The traditional Bragg bandgaps can always arise in a waveguide with wall undulations. When we carefully select the geometry parameters of a waveguide, the high-order transverse modes will be involved in the lower order mode stopbands. Beyond the indirect mode-stopband interactions,[24] the direct interaction leads to the multimode resonances due to the intersection of multiple mode lines or the multiple intersection approaching in the first Brillouin zone. So, the more complicated characteristic equations have to be solved. The marked critical points have been solved to indicate the creation of the single first mode penetration in the second Bragg gap.

Based on the analytical and numerical methods, we have investigated the complex spectrum caused by the direct mode-stopband interactions. There arises an extraordinary transmission in the second Bragg gap, which consists of the single first mode. In the passband, the fundamental mode is efficiently suppressed. The shape factor of different profiles provides a sufficient method to estimate the creation of spectra. The pass and stop band edges are always proportional to the shape factor. Generally, a larger shape factor can result in a narrower passband and a purer first mode. The mode fitting method has quantitatively confirmed the purity of the penetrating first mode. Because of the ubiquity of wave phenomena, the proposed interactions could be found not only in acoustics but also in optics, terahertz waves, and microwaves. The results enrich our findings in wave control physics and engineering, which will definitely benefit the applications in various fields.

ACKNOWLEDGMENTS

This work was supported by the National Natural Science Foundation of China under Grant No. 11374071, by the Fundamental Research Funds for the Central Universities of China, by the 111 project through a grant (B13015) to Harbin Engineering University, and by the Opened Fund of the Key Lab of In-fiber Integrated Optics, Ministry Education of China.

[1] J. Wen, G. Wang, D. Yu, H. Zhao, and Y. Liu, J. Appl. Phys. **97**, 114907 (2005).
[2] V. Romero-Garcia, J. V. Sanchez-Perez, and L. M. Garcia-Raffi, J. Appl. Phys. **110**, 014904 (2011).
[3] Lord Rayleigh, Philos. Mag. XXIV145–159 (1887).
[4] C. M. Soukoulis, Phys. Scripta **T66**, 146–150 (1996).
[5] M. Farhat, S. Enoch, S. Guenneau, and A. B. Movchan, Phys. Rev. Lett. **101**, 134501 (2008).
[6] L. Feng, X. P. Liu, M. H. Lu, Y. B. Chen, Y. F. Chen, Y. W. Mao, J. Zi, Y. Y. Zhu, S. N. Zhu, and N. B. Ming, Phys. Rev. Lett. **96**, 014301 (2006).
[7] X. Zhang and Z. Liu, Appl. Phys. Lett. **85**, 341 (2004).
[8] T. Gorishnyy, C. K. Ullal, M. Maldovan, G. Fytas, and E. L. Thomas, Phys. Rev. Lett. **94**, 115501 (2005).
[9] A. H. Nayfeh, J. Acoust. Soc. Am. **56**(3), 768–770 (1974).
[10] C. E. Bradley, J. Acoust. Soc. Am. **96**(3), 1844–1853 (1994).
[11] S. Giudice and G. Bernasconi, Math. Probl. Eng. 267291 (2013).
[12] M. A. Hawwa, Math. Probl. Eng. 516982 (2015).
[13] V. A. Pogrebnyak, Phys. Rev. E **58**, R5261 (1998).
[14] J. N. Munday, C. B. Bennett, and W. M. Robertson, J. Acoust. Soc. Am. **112**(4), 1353–1358 (2002).
[15] V. A. Pogrebnyak, Opt. Commun. **232**, 201–207 (2004).
[16] V. A. Pogrebnyak, E. Akray, and A. N. Kucukaltun, Appl. Phys. Lett. **86**, 151115 (2005).
[17] Z. Y. Tao, W. Y. He, Y. M. Xiao, and X. L. Wang, Appl. Phys. Lett. **92**, 121920 (2008).
[18] Y. M. Xiao, Z. Y. Tao, W. Y. He, and X. L. Wang, Phys. Rev. E **78**, 016311 (2008).
[19] J. Zhao and Z. Y. Tao, J. Sound. Vib. **332**, 6541–6551 (2013).
[20] S. V. Sorokin, J. Sound. Vib. **349**, 348–360 (2015).
[21] Y. Huang, H. M. Wang, and W. Q. Chen, J. Appl. Phys. **115**, 133501 (2014).
[22] R. B. Nielsen and S. V. Sorokin, J. Sound. Vib. **353**, 135–149 (2015).
[23] S. V. Sorokin, J. Acoust. Soc. Am. **139**(4), 1898–1908 (2016).
[24] Z. Y. Tao, W. Y. He, Y. M. Xiao, W. H. Zhang, and X. L. Wang, J. Appl. Phys. **105**, 123515 (2009).
[25] L. E. Kinsler, A. R. Frey, A. B. Coppens, and J. V. Sanders, *Fundamentals of Acoustics*, 4th ed. (Wiley, New York, 2000), p. 383.
[26] Z. Y. Tao, W. Y. He, and X. Wang, J. Sound. Vib. **313**, 830–840 (2008).
[27] J. J. Barroso, J. P. Leite Neto, and K. G. Kostov, IEEE Trans. Plasma Sci. **31**, 752 (2003).

Permissions

All chapters in this book were first published in AIP Advances, by AIP Publishing LLC.; hereby published with permission under the Creative Commons Attribution License or equivalent. Every chapter published in this book has been scrutinized by our experts. Their significance has been extensively debated. The topics covered herein carry significant findings which will fuel the growth of the discipline. They may even be implemented as practical applications or may be referred to as a beginning point for another development.

The contributors of this book come from diverse backgrounds, making this book a truly international effort. This book will bring forth new frontiers with its revolutionizing research information and detailed analysis of the nascent developments around the world.

We would like to thank all the contributing authors for lending their expertise to make the book truly unique. They have played a crucial role in the development of this book. Without their invaluable contributions this book wouldn't have been possible. They have made vital efforts to compile up to date information on the varied aspects of this subject to make this book a valuable addition to the collection of many professionals and students.

This book was conceptualized with the vision of imparting up-to-date information and advanced data in this field. To ensure the same, a matchless editorial board was set up. Every individual on the board went through rigorous rounds of assessment to prove their worth. After which they invested a large part of their time researching and compiling the most relevant data for our readers.

The editorial board has been involved in producing this book since its inception. They have spent rigorous hours researching and exploring the diverse topics which have resulted in the successful publishing of this book. They have passed on their knowledge of decades through this book. To expedite this challenging task, the publisher supported the team at every step. A small team of assistant editors was also appointed to further simplify the editing procedure and attain best results for the readers.

Apart from the editorial board, the designing team has also invested a significant amount of their time in understanding the subject and creating the most relevant covers. They scrutinized every image to scout for the most suitable representation of the subject and create an appropriate cover for the book.

The publishing team has been an ardent support to the editorial, designing and production team. Their endless efforts to recruit the best for this project, has resulted in the accomplishment of this book. They are a veteran in the field of academics and their pool of knowledge is as vast as their experience in printing. Their expertise and guidance has proved useful at every step. Their uncompromising quality standards have made this book an exceptional effort. Their encouragement from time to time has been an inspiration for everyone.

The publisher and the editorial board hope that this book will prove to be a valuable piece of knowledge for researchers, students, practitioners and scholars across the globe.

List of Contributors

R. Alipour, M. Ghoranneviss and A. Salar Elahi
Plasma Physics Research Center, Science and Research Branch, Islamic Azad University,Tehran, Iran

A. Tariq and S. Nazira
Department of Physics, University of Sargodha, 40100 Sargodha, Pakistan

H. Noori and A. H. Ranjbar
Faculty of Physics, Shahid Bahonar University of Kerman, Kerman 7616914111, Iran

M. Mahjour-Shafiei
Department of Physics, University of Tehran, Tehran 1439955961, Iran

Li Zhang, Haoze Chen, Shan Feng, Linxian Li and Jianye Chen
Department of physics, China Jiliang University, Hangzhou 310018, China

Yuke Li
Department of physics, Hangzhou Normal University, Hangzhou 310018, China

Yu Xu and Fu-Tian Liang
Department of Modern Physics and Hefei National Laboratory for Physical Sciences at Microscale, University of Science and Technology of China, Hefei 230026, China
Chinese Academy of Sciences (CAS) Center for Excellence and Synergetic Innovation Center in Quantum Information and Quantum Physics, University of Science and Technology of China, Shanghai 201315, China
State Key Laboratory of Particle Detection and Electronics, University of Science

Yu-Huai Li, Sheng-Kai Liao, Qi Shen and Cheng-Zhi Peng
Department of Modern Physics and Hefei National Laboratory for Physical Sciences at Microscale, University of Science and Technology of China, Hefei 230026, China
Chinese Academy of Sciences (CAS) Center for Excellence and Synergetic Innovation Center in Quantum Information and Quantum Physics, University of Science and Technology of China, Shanghai 201315, China

Wei Chen and Hao Liang
Department of Modern Physics and Hefei National Laboratory for Physical Sciences at Microscale, University of Science and Technology of China, Hefei 230026, China
State Key Laboratory of Particle Detection and Electronics, University of Science and Technology of China, Hefei 230026, China

R. M. Hedlof and C. A. Ordoneza
Department of Physics, University of North Texas, Denton, Texas 76203, USA

Pengcheng Yu, Jinxiang Cao, Zhongkai Zhang and Xiao Zhang
Department of Modern Physics, University of Science and Technology of China, Hefei, Anhui 230026, China

Yu Liu and Jiuhou Lei
CAS Key Laboratory of Geospace Environment, School of Earth and Space Science, University of Science and Technology of China, Hefei, Anhui 230026, China

A. Abdelmalek and Z. Bedrane
Physics Department, Theoretical Physics Laboratory, Tlemcen University, 13000 Tlemcen, Algeria

B. Sotillo, V. Bharadwaj, S. Pietralunga, R. Ramponi and S. M. Eaton
IFN-CNR and Dipartimento di Fisica, Politecnico di Milano, 20133 Milano, Italy

E.-H. Amara
Laser Material Processing Team, Centre for Advanced Technologies Development, CDTA, PO. Box 17, Baba-Hassen 16303, Algiers, Algeria

Sara Moradi
Laboratory for Plasma Physics - LPP-ERM/KMS, Royal Military Academy, 1000 Brussels, Belgium

Bogdan Teaca
Applied Mathematics Research Centre, Coventry University, Coventry CV1 5FB, United Kingdom

Johan Anderson
Department of Space, Earth and Enivronment Sciences, Chalmers University of Technology, SE-412 96 Göteborg, Sweden

G. Hassnain Jaffaria and Wardah Mahmood
Department of physics, Quaid-i-Azam University, Islamabad 45320, Pakistan

Jun Cui, Jiehui Liu, Yiwei Mao and Xiaozhou Liu
Key Laboratory of Modern Acoustics, Institute of Acoustics and School of Physics, Collaborative Innovation Center of Advanced Microstructures, Nanjing University, Nanjing 210093, P. R. China

Yifeng Li
College of Computer Science and Technology, Nanjing Tech University, Nanjing 211800 P. R. China

Taoreed O. Owolabi
Physics Department, King Fahd University of Petroleum and Minerals, Dhahran Postal code 34464, Kingdom of Saudi Arabia
Physics and Electronics Department, Adekunle Ajasin University, Akungba Akoko Postal code 342111, Ondo State Nigeria

Kabiru O. Akande
Institute for Digital Communications, School of Engineering, University of Edinburgh, Edinburgh Postal code EH8 9AB, United Kingdom

Sunday O. Olatunji
Computer Science Department, College of Computer Science and Information Technology, Imam Abdulrahman Bin Faisal University, Dammam Postal code 31433, Saudi Arabia

Nahier Aldhafferi and Abdullah Alqahtani
Computer Information System Department, College of Computer Science and Information Technology, Imam Abdulrahman Bin Faisal University, Dammam Postal code 31433, Saudi Arabia

V. A. Andrianov and A. L. Erzinkian
Skobeltsyn Institute of Nuclear Physics, Lomonosov Moscow State University, Leninskie gory 1(2), GSP-1, Moscow 119991, Russian Federation

L. I. Ivleva and P. A. Lykov
Prokhorov General Physics Institute RAS,Vavilov Str., 38, Moscow 119991, RussianFederation

Ziqiang Yang, Zongjun Shi, Lin Meng, Dagang Liu and Wenjie Fu
School of Physical Electronics, University of Electronic Science and Technology of China, 610054 Chengdu, Sichuan, P. R. China

Feng Lan
School of Physical Electronics, University of Electronic Science and Technology of China, 610054 Chengdu, Sichuan, P. R. China
Department of Electrical Engineering and Computer Science, University of Michigan, Ann Arbor, MI 48109-2121, USA

Pinaki Mazumder
Department of Electrical Engineering and Computer Science, University of Michigan, Ann Arbor, MI 48109-2121, USA

Xuan-Ting Ji
School of Electronic, Electrical and Communication Engineering, University of Chinese Academy of Sciences, Beijing 100049, China

Hai-Zhou Lu
Institute for Quantum Science and Engineering and Department of Physics, South University of Science and Technology of China, Shenzhen 518055 China

Zhen-Gang Zhu
School of Electronic, Electrical and Communication Engineering, University of Chinese Academy of Sciences, Beijing 100049, China
Theoretical Condensed Matter Physics and Computational Materials Physics Laboratory, School of Physical Sciences, University of Chinese Academy of Sciences, Beijing 100049, China

Gang Su
Theoretical Condensed Matter Physics and Computational Materials Physics Laboratory, School of Physical Sciences, University of Chinese Academy of Sciences, Beijing 100049, China
Kavli Institute of Theoretical Sciences, University of Chinese Academy of Sciences, Beijing 100190, China

Liyong Cui, Guang Lu, Shan Zhang, Fen Liu, Yanqing Xin, Kunlun Wang and Tianlin Yang
School of Space Science and Physics, Shandong University at Weihai, Weihai 264209, P. R. China

Guiqiang Du
School of Space Science and Physics, Shandong University at Weihai, Weihai 264209, P. R. China
State Key Laboratory of Surface Physics and Department of Physics, Key Laboratory of Micro- and Nano-Photonic Structures (Ministry of Education), Fudan University, Shanghai 200433, P. R. China

Li Wang
Beijing Institute of Space Mechanics & Electricity, Beijing 100094, P. R. China

Luo Min
Department of Physics, Shanghai Second Polytechnic University, Shanghai 201209, China

Yu E. Xu
Department of Electronic Engineering, Shang Hai Jian Qiao University, Shanghai 201306, China

Xi Yu Song
Key Laboratory of Polar Materials and Devices, East China Normal University, Shanghai 200241, China

A. S. Kurdyubov, A. V. Trifonov, I. Ya. Gerlovin and I. V. Ignatiev
Spin Optics Laboratory, St. Petersburg State University, 1 Ul'anovskaya, Peterhof, St. Petersburg 198504, Russia

A. V. Kavokin
Spin Optics Laboratory, St. Petersburg State University, 1 Ul'anovskaya, Peterhof, St. Petersburg 198504, Russia
CNR-SPIN, Viale del Politecnico 1, I-00133 Rome, Italy
School of Physics and Astronomy, University of Southampton, SO17 1NJ Southampton, United Kingdom

Wenhui Maa and Fei Wang
Department of Physics, Shantou University, Shantou, Guangdong 515063, People's Republic of China

Jinhua Han and Gang Guo
Department of Nuclear Physics, China Institute of Atomic Energy, Beijing 102413, China
National Innovation Center of Radiation Application, Beijing 102413, China

Kishan Kumara and M. K. Mishrab
Department of Physics, University of Rajasthan, Jaipur 302004, India

Vivek Dixit and Seong-Gon Kim
Department of Physics and Astronomy, Mississippi State University, Mississippi State, MS 39762, USA and Center for Computational Sciences, Mississippi State University, Mississippi State, MS 39762, USA

Jihoon Park
Korea Institute of Materials Science, Changwon, Gyeongnam 51508, Republic of Korea

Yang-Ki Hong
Department of Electrical and Computer Engineering and MINT Center, The University of Alabama, Tuscaloosa, AL 35487, USA

Liwen He, Jia Lou, Aibing Zhang, Jianke Du and Ji Wang
Piezoelectric Device Laboratory, Department of Mechanics and Engineering Science, Ningbo University, Ningbo, Zhejiang 315211, China

Huaping Wu
Key Laboratory of E&M (Zhejiang University of Technology), Ministry of Education & Zhejiang Province, Hangzhou 310014, China

Ya-Xian Fan
Key Lab of In-fiber Integrated Optics, Ministry Education of China, Harbin Engineering University, Harbin 150001, People's Republic of China

Jiu-Ling Xue and Zhi-Yong Tao
Key Lab of In-fiber Integrated Optics, Ministry Education of China, Harbin Engineering University, Harbin 150001, People's Republic of China
Physics Research Center, College of Science, Harbin Engineering University, Harbin 150001, People's Republic of China

Huan Liu, Ting Liu and Fu-Hou Zheng
Physics Research Center, College of Science, Harbin Engineering University, Harbin 150001, People's Republic of China

Index

P

R

S

T

V

W